SYSTEMS AND TELECOMMUNICATIONS

SECURITY, ACCURACY, AND PRIVACY IN COMPUTER SYSTEMS

About the prevention of unauthorized access to computers and data banks, embezzlement, crime, sabotage, invasion of privacy, and the maintenance of accuracy.

DESIGN OF MAN-COMPUTER DIALOGUES

A guide to the design of man-machine dialogues; detailed examination of the many types of real-time man-computer interface, especially for commercial and management-information systems.

SYSTEMS ANALYSIS FOR DATA TRANSMISSION

A detailed guide to the design of data transmission systems. Terminal, network, user, software, and system considerations. Examples of the design calculations needed. Lists of formulae and tables for design.

THE COMPUTERIZED SOCIETY

Euphoria; Alarm; Protection Action. An appraisal of the impact of computers on society over the next fifteen years, and the steps that can be taken to direct it into the most beneficial channels. (with Adrian Norman)

FUTURE DEVELOPMENTS IN TELECOMMUNICATIONS
Second Edition

An exploration of the foreseeable future in a technology that has reached a period of very rapid change.

PROGRAMMING REAL-TIME COMPUTER SYSTEMS

Programming mechanisms, program testing tools and techniques, problems encountered, implementation considerations, project management.

**TELEPROCESSING
NETWORK
ORGANIZATION**

Prentice-Hall
Series in Automatic Computation
George Forsythe, editor

ARBIB, *Theories of Abstract Automata*
BATES AND DOUGLAS, *Programming Language/One*
BAUMANN, FELICIANO, BAUER, AND SAMELSON, *Introduction to ALGOL*
BLUMENTHAL, *Management Information Systems*
BOBROW AND SCHWARTZ, EDS., *Computers and the Policy-Making Community: Applications to International Relations*
BOWLES, ED., *Computers in Humanistic Research*
CESCHINO AND KUNTZMANN, *Numerical Solution of Initial Value Problems*
CRESS, DIRKSEN, AND GRAHAM, *FORTRAN IV with WATFOR*
DESMONDE, *Computers and Their Uses*
DESMONDE, *A Conversational Graphic Data Processing System: The IBM 1130/2250*
DESMONDE, *Real-Time Data Processing Systems: Introductory Concepts*
EVANS, WALLACE, AND SUTHERLAND, *Simulation Using Digital Computers*
FIKE, *Computer Evaluation of Mathematical Functions*
FIKE, *PL/I for Scientific Programmers*
FORSYTHE AND MOLER, *Computer Solution of Linear Algebraic Systems*
GAUTHIER AND PONTO, *Designing Systems Programs*
GOLDEN *FORTRAN IV: Programming and Computing*
GOLDEN AND LEICHUS, *IBM 360: Programming and Computing*
GORDON, *System Simulation*
GREENSPAN, *Lectures on the Numerical Solution of Linear, Singular, and Nonlinear Differential Equations*
GRISWOLD, POAGE, AND POLONSKY, *The SNOBOL 4 Programming Language*
GRUENBERGER, ED., *Computers and Communications—Toward a Computer Utility*
GRUENBERGER, ED., *Critical Factors in Data Management*
HARTMANIS AND STEARNS, *Algebraic Structure Theory of Sequential Machines*
HULL, *Introduction to Computing*
HUSSON, *Microprogramming: Principles and Practices*
JOHNSON, *Structure in Digital Computer Systems: An Introduction*
KIVIAT, VILLANUEVA, AND MARKOWITZ, *The SIMSCRIPT II Programming Language*
LOUDEN, *Programming the IBM 1130 and 1800*
MARTIN, *Design of Real-Time Computer Systems*
MARTIN, *Programming Real-Time Computer Systems*
MARTIN, *Telecommunications and the Computer*
MARTIN, *Teleprocessing Network Organization*
MARTIN AND NORMAN, *The Computerized Society*
MCKEEMAN ET AL., *A Compiler Generator*
MINSKY, *Computation: Finite and Infinite Machines*
MOORE, *Interval Analysis*
PYLYSHYN, *Perspectives on the Computer Revolution*
PRITSKER AND KIVIAT, *Simulation with GASP II: A FORTRAN Based Simulation Language*
SAMMET, *Programming Languages: History and Fundamentals*
SNYDER, *Chebyshev Methods in Numerical Approximation*
STERLING AND POLLACK, *Introduction to Statistical Data Processing*
STROUD AND SECREST, *Gaussian Quadrature Formulas*
TAVISS, *The Computer Impact*
TRAUB, *Iterative Methods for the Solutions of Equations*
VARGA, *Matrix Iterative Analysis*
VAZSONYI, *Problem Solving by Digital Computers with PL/I Programming*
WILKINSON, *Rounding Errors in Algebraic Processes*
ZIEGLER, *Time-Sharing Data Processing Systems*

TELEPROCESSING NETWORK ORGANIZATION

James Martin

STAFF MEMBER
IBM SYSTEMS RESEARCH INSTITUTE

Prentice-Hall, Inc., Englewood Cliffs, New Jersey

© 1970 by
PRENTICE-HALL, INC.
Englewood Cliffs, N.J.

All rights reserved. No part of this
book may be reproduced in any form or
by any means without permission in
writing from the publisher.

Current printing (last digit):

15 14 13 12

13-902452-2
Library of Congress Catalog Card Number: 77-102101
Printed in the United States of America

PRENTICE-HALL INTERNATIONAL, INC., *London*
PRENTICE-HALL OF AUSTRALIA, PTY. LTD., *Sydney*
PRENTICE-HALL OF CANADA, LTD., *Toronto*
PRENTICE-HALL OF INDIA PRIVATE LIMITED, *New Delhi*
PRENTICE-HALL OF JAPAN, INC., *Tokyo*
PRENTICE-HALL OF SOUTHEAST ASIA (PTE.) LTD., *Singapore*

ACKNOWLEDGMENTS

Any book about the state of the art in a complex technology draws material from a vast number of sources. While many of these are referenced in the text, it is impossible to include all of the pioneering projects that have contributed to the new uses of telecommunications. To the many systems engineers who contributed to this body of knowledge, the author is indebted.

The author is very grateful for the time spent reviewing and criticizing the manuscript by Mr. J. W. Greenwood in New York and Dr. D. R. Doll at the University of Michigan, Ann Arbor. The concepts in it were discussed in seminars organized by NOMA in Japan, whose hospitality I shall always remember. The long-suffering students of IBM's Systems Research Institute in New York also used the manuscript and suggested many changes. Miss Charity Anders (Fig. 1.10) prepared the index. To all these the author is indebted. Mr. R. B. Edwards' staff helped enormously in typing and reproducing the manuscript. The author is particularly grateful to Miss Cora Tangney for her help in this.

Last, and perhaps most important, the author is again indebted to Dr. E. S. Kopley, Director of the IBM Systems Research Institute for his constant encouragement. Without the environment that he created, this work would not have been completed.

JAMES MARTIN

New York

STATEMENT OF INTENT

In years ahead an understanding of data transmission will be of great importance to the systems analyst. Our systems and to some extent our society will be dominated by the ability of separate machines to transmit information to one another.

This book is an attempt to set into perspective the principles behind the different types of data transmission hardware and organization. It discusses control methods and mechanisms, with illustrations of different types of networks. It explains the techniques possible for organizing the flow of data on today's telecommunication lines.

We may state the main problem of teleprocessing network organization simply: If we have a data transmission network, possibly with long lines and possibly with many terminals, how can we minimize the cost of that network and still have it work effectively?

Many time-sharing systems and commercial real-time systems today have had a small number of terminals. They have given a valuable service to a small community of users. The industry now needs to extend that service to a *large* community. A corporation would like all of its locations to have on-line access to a scientific computing system. It would like to deploy many hundreds of terminals on its factory shop floors for production control. We would like doctors in certain hospitals throughout the country to have access to a medical data base. Airlines must connect offices around the world to one reservation system. The police need a vast network of information terminals. All of a firm's sales offices should be connected to a central computer. . . .

How do we organize the data transmission links so that they are reliable, give fast service to the user, and are not too costly? How do we achieve a fast response time on a network with many users? Furthermore, how do we organize these links and keep the cost of the terminal low?

Can we lessen today's telecommunication bill? Many firms in the U.S.A. are paying the common carriers a million dollars a year for data transmission, and this will increase considerably in the next decade. But small firms also need to organize their transmission more efficiently.

To answer these problems a multitude of devices and techniques have been created. Many more are on their way. Some systems have more control characters flowing on their lines than data. The choices facing the systems analyst in the area of network design are becoming bewilderingly complex.

The material in this book is the basis of a course given at IBM's Systems Research Institute and it is thought important that such a course be given in universities and industry elsewhere. The book may be used in conjunction with the author's other books, and is intended to be complementary to the author's *Telecommunications and the Computer*, Prentice-Hall, 1969. Design calculations and algorithms are not discussed in this book, but are included in a more comprehensive work by the author entitled *Systems Analysis for Data Transmission*, Prentice-Hall, in press.

It is intended that some of the more detailed sections be used by computer personnel for repeated reference, and that the book will be of value to them in forwarding their careers. It is hoped, therefore, that this book will be owned by such people rather than meeting the sad fate of only being borrowed from a library and read once.

J.M.

CONTENTS

1 TYPES OF COMMUNICATION LINKS   1

Telephone Channels   2
Leased Versus Switched Lines   3
Bandwidth of a Voice Channel   4
Signaling   5
Advantages of Private Lines   6
Simplex, Duplex, and Half-Duplex Lines   8
Voice and Subvoice Line Types   9
Line Conditioning   12
Wideband and Telpak Tariffs   12
Telex   14
Teletypewriter Exchange Service   15
Modulation   16
Modems and Data Sets   17
Three Types of Modulation   20
Transmission Without Modems   22
Acoustical Coupling   22
Tariffs Including Data Sets   23

2 INFORMATION CODING   25

Escape Characters   25
Parity Bits   28
Control Bits and Characters   28
Binary Coded Decimal   31
N-Out-of-M Codes   33

ix

American Standards Association Code 33
Pseudo-Baudot Characters 36
Transparent Codes 37

3 MODES OF TRANSMISSION 38

Full Duplex Vs. Half Duplex 38
Parallel Vs. Serial Transmission 40
Multitone Transmission 40
Synchronous Vs. Asynchronous Transmission 43
Asynchronous (START–STOP) Transmission 43
Synchronous Transmission 50
Block Structure 54
Synchronization Requirements 57
High-Speed Pulse Stream 58
Relative Advantages and Disadvantages 59

4 ERRORS AND THEIR TREATMENT 62

Numbers of Line Errors 62
Possible Approaches 63
Calculations of the Effect of Errors 64
Detection of Errors 65
Dealing with Errors 66
Error-Correcting Codes 66
Loop Check 67
Retransmission of Data in Error 67
Error Control on Radio Circuits 67
How Much is Transmitted? 68
Transmission Error-Control Characters 72
Odd–Even Record Count 73
An Example of a Typical Error-Control System 74

5 ERROR-DETECTING CODES 76

Criteria for Choice of Code 76
Error-Correcting Codes 77
Parity Checks 79
The U.S. ASCII Recommendation 80
M-Out-of-N Codes 81

Polynomial Codes 83
Error-Detection Probabilities 86
Results Obtained in Practice 91
Encoding and Decoding Circuits 91
Polynomial Checking on Variable-Length Messages 94
To Achieve a Very High Measure of Protection 94
Bibliography 95

6 POINT-TO-POINT LINE CONTROL 96

Getting Started 97
Establishing Synchronization 99
Control Characters 99
Idling, Waiting for a Signal 101
Requesting Permission to Transmit 103
Confirmation of Correct Receipt 104
Error Signals 105
Conversational Transmission 105
Modem Turn-Around Time 108
Transmission of Data in Both Directions at Once 111
Housekeeping Core Planes 112
Code Translation 113

7 MEANS OF LOWERING COMMUNICATION NETWORK COSTS 118

1. Systems with Very Short Private Lines 119
Terminal Control Units 121
2. Short Public Lines 123
3. Long Lines 127
Techniques for Minimizing Network Cost 129

8 PRIVATE EXCHANGES 139

9 POLLING AND MULTIPOINT LINE CONTROL 153

Buffered Terminals 155
Response Time 156
Modems 157

Selective Calling   157
Contention and Polling   158
Line Control   159
Time-Out   163
Two Types of Polling   165
An Illustration of Hub Polling   167
Advantages and Disadvantages of Hub Polling   169

10 TERMINAL CONTROL UNITS   170

Polling   175
General Poll   181

11 REMOTE MULTIPLEXORS   184

Two Methods of Multiplexing   186
Frequency-Division Multiplexing   186
Time-Division Multiplexing   188
Multiplexing with Multidrop Lines   191
Advantages   193

12 HOLD-AND-FORWARD CONCENTRATORS   195

Basic Functions of a Concentrator   196
Example   201
Simple Line Control   206
Response Time Requirements   206
Error Control   206
Multidrop Low-Speed Lines   207
Complex Line Control   209
Number of Control Characters   210

13 REMOTE LINE COMPUTERS   218

Canned Responses   220
Commercial Systems   225
Systems with Complex Conversations   229

14 MESSAGE SWITCHING — 233

Torn-Tape Switching Centers 234
Semiautomatic Switching Centers 237
Computers for Message Switching 238
Functions of the System 239
Very Large Networks 243

15 DISTRIBUTED INTELLIGENCE — 246

Locations for Subsystems 246
Duplication of Hardware 249
Functions to be Performed 250
Availability 255

GLOSSARY — 259

INDEX — 283

1 TYPES OF COMMUNICATION LINKS[1]

The facts about the transmission links which most concern a systems analyst are the cost of the links and the rate at which data may be sent over them. This can have a major effect on the feasibility of different systems approaches. All the common carriers issue a price list describing their facilities. Let us begin our description of communication networks with a summary of the links that are available.

The illustrations in the following summary are taken from the offering of the carriers in the United States and Britain. Those of other countries are broadly similar. Countries with less well-developed data transmission have less to offer in the way of wideband (high-capacity) facilities.

The services that a common carrier offers to the public are described in tariffs. A tariff is a document which, in the United States, is required by the regulating bodies that control the carriers. The United States Federal Communications Commission must eventually approve all interstate facilities, and similar state commissions control those within state boundaries. By law, all tariffs must be registered with these bodies. In most other countries, the telecommunication facilities are set up by government bodies and thus are directly under their control.

In the United States the subject of communication rates has become complex. The amount and structure of charges differ from one state to another. In other countries, such as England and Germany, the rates for more conventional channels remain relatively straightforward; however, as the carriers there are government organizations, they are not obliged to

---

[1] This opening chapter summarizes material discussed in more detail in the companion book *Telecommunications and the Computer* by James Martin, Prentice-Hall, Inc. 1969. It is recommended that the reader should read that book in conjunction with this.

publish tariffs for all of their facilities. The price for less common channels, such as broadband, may have to be obtained by a special request to the carrier. In general it is desirable, when one is designing a system, that the carrier in question be called in to quote a price for the facilities needed.

TELEPHONE CHANNELS
When from your home telephone you dial a number in the same locality, your call goes to a nearby public exchange. (Referred to as a *central office* in American parlance.) The equipment in this building connects the wires from your telephone to those of the party you call. These wires are permanently connected from your local central office to your telephone and are referred to as a *loop*. No other subscriber uses them, unless you have a party line. Thousands of such wire connections lie under the streets of a city in cables like that in Fig. 7.6.

Loops to a company building normally terminate at a switchboard, which has several extension lines to telephones in the building. The switchboard may be manual, with interconnections made by an operator using cords, or it may be automatic, in which case the user can dial his own connections. Switchboards at such premises are described as private branch exchanges (PBX), private automatic exchanges (PAX) or sometimes private automatic branch exchanges (PABX). Use of private exchanges in data-processing systems is discussed in Chapter 8.

A circuit between two switching equipments is referred to as a *trunk*. When you dial a person whose telephone is not connected to the same central office as your own, your call is routed over an *interoffice trunk*. The switched *public network* consists of a complicated system of switching offices and interconnecting trunks, which will be described in a later chapter. Trunks between switching offices are normally designed to carry many telephone conversations, not just one as on the local loop.

A *tie line*, or tie trunk, is a private, leased communication line between two or more private branch exchanges. Many companies have a leased system of telecommunication lines with switching facilities. To telephone a person in a distant company location, an employee must first obtain the appropriate tie line to that person's private branch exchange. On an automatic system this is done by dialing a tie-line code before the extension number.

The tie line (or lines) to that location may be busy, in which case a "busy tone" ("engaged signal," in British parlance) can be heard before the extension number is dialed.

As we will see in subsequent chapters, tie lines often form an important feature of data processing networks. All of the lines above can be used for data transmission as well as voice transmission. Where one voice line is used the transmission speed is limited to a few thousand bits per second, depending, as will be discussed later, on the equipment used at each end of the line. This is the data capacity of a telephone line. Often, however, the tie-line system has more than one voice line connecting two locations. At the time of writing, the common carrier tariffs in the United States make a channel giving 60 voice lines (or their equivalent) considerably less expensive than leasing 60 individual voice channels. If such a channel is used, a group of voice lines may be taken over *together* for data transmission at a speed many times higher than that possible over a voice channel.

One typical large corporation uses its tie-line groups in this manner as a broadband data network during its second and third shifts. The lines are transferred to data automatically, as a group, and so telephone calls in progress on those lines may possibly be interrupted. The corporation tie-line directory contains the following note:

> "When a transfer is about to occur, callers will hear a special interrupt tone that is introduced during a conversation. Callers hearing this tone should hang up and redial the call after a 30-second interval. Normally, these transfers will only occur after 6:00 p.m. Eastern Time."

LEASED VERSUS SWITCHED LINES   Voice lines and telegraph lines can be either switched through public exchanges (central offices), or permanently connected. Facilities for switching broadband channels are also coming into operation in some countries, although these channels, today, are usually permanent connections.

When you dial a friend and talk to him on the telephone, you speak over a line connected by means of the public exchanges. This line, referred to as a "public" or "switched" line, could be used for the transmission of data. Alternatively, a "private" or "leased" line could be connected permanently or semipermanently between the transmitting machines. The private line may be connected via the local switching office, but it would not be connected to the switch-gear and signaling devices of that office. An inter-office private connection would use the same physical links as the switched circuits. It would not, however, have to carry the signaling that is needed on a switched line. This is one reason why it is possible to achieve a higher rate of data transmission over a private line. Another reason is that private lines can be carefully balanced to provide the high quality that makes higher-speed data transmission possible.

There are also in operation several large *private* line systems that are also *switched* (Chapter 8). It is possible to engineer these to the same quality as private lines, and hence provide a switching system of better quality than the public network.

Some private line systems are wholly owned by their users rather than leased. Typical of such systems are communication links within a factory, private microwave connections and other radio links, lines along railroad tracks, and, today, in some laboratories, private wiring of terminals to a time-sharing computer.

As you can either dial a telephone connection or else have it permanently wired, so it is with other types of lines. Telegraph lines, for example, which have a much lower speed of transmission than is possible over voice lines, may be permanently connected, or may be dialed like a telephone line using a switched public network. Telex is such a network; it exists throughout most of the world, permitting transmission at 50 bits per second. Telex users can set up international connections to other countries. Some countries have a switched public network, operating at a somewhat higher speed than Telex but at less speed than telephone lines. In the United States, the TWX network gives speeds up to 150 bits per second. TWX lines can be connected to Telex lines for overseas calls. Also, certain countries are building up a switched network for very high-speed (wideband) connections. In the United States, Western Union has installed the first sections of a system in which a user can indicate *in his dialing* what capacity link he needs.

BANDWIDTH OF A VOICE CHANNEL   The signal-carrying capacity of communication links can be described in terms of the frequencies that they will carry. A certain physical link might, for example, transmit energy at frequencies ranging from 300 hertz to 150,000 hertz. (The word *hertz* has replaced "cycles per second" in describing frequency and bandwidth. Their meaning is identical.) Above 150,000 and below 300 hertz, the signal is too much attenuated to be useful. The range of frequencies is described as the *bandwidth* of the channel. The bandwidth is 149,700 ($=150,000-300$) hertz (or cycles per second). In fact, the upper cutoff point is not as sharp as is suggested by this, and we would probably say a bandwidth of 150 kilohertz.

Figure 1.1 shows the attenuation of different frequencies on a typical voice channel. It will be seen that between about 300 and 3400 hertz (cycles per second) different frequencies are attenuated roughly equally. Frequencies outside these limits are not usable, and hence we would say that this channel had a bandwidth of 3100 hertz.

The quantity of data that can be transmitted over a channel is approximately proportional to the bandwidth.[2]

The frequencies transmitted in Fig. 1.1 are not sufficient to reproduce the human voice *exactly*. They are, however, enough to make it intelligible and to make the speaker recognizable. This is all that is demanded of the telephone system. Hi-fi enthusiasts strive to make their machines reproduce frequencies from 30 to 20,000 hertz. If the telephone system could

---

[2] *Telecommunications and the Computer*, Chapters 10 and 11.

transmit this range, then we could send high fidelity music over it. Sending music over the channel in Fig. 1.1 would clip it of its lower and higher frequencies, and it would sound less true to life than over a small transistor radio.

The physical media that are used for telecommunications all have a bandwidth much larger than that needed for one telephone conversation, so between towns one link is made to carry as many voice channels as possible. The bandwidth of one physical channel is electronically cut up into slices of 4000 hertz and each of these slices becomes one voice channel.[3] The result is shown in Fig. 1.1. The frequencies of Fig. 1.1 fit easily into the 4000 hertz slice.

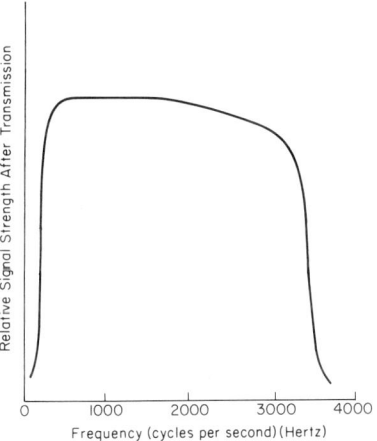

Fig. 1.1. Variation in signal strength with frequency after transmission over a typical voice line.

To transmit data over the telephone line, then, we must manipulate it electronically so that it fits into the frequencies of Fig. 1.1. This is done by a device called a *modem*, which will be discussed shortly.

Nonvoice channels have bandwidths different from that in Fig. 1.1—subvoice grade channels are lower in bandwidth, and broadband channels are higher. If desirable, channels of extremely high bandwidth can be obtained.

SIGNALING

We must add one complication to Fig. 1.1. On *public* voice lines we are not completely free to use all of the bandwidth shown. Certain frequencies are used by the telephone company for their own signaling, and we must not interfere with this.[4]

Figure 1.2 shows the signaling frequencies used on Bell System and British telephone lines. Unfortunately the frequencies are not the same, and in some other countries they are different from either of these. Signals are sent at these frequencies to indicate, for example, when a subscriber puts down his telephone so that the line may be disconnected.

---

[3] *Telecommunications and the Computer*, Chapters 10 and 15.
[4] *Telecommunications and the Computer*, Chapter 17.

6 TYPES OF COMMUNICATION LINKS

If a data-processing machine happens to transmit at the signaling frequency while sending little or no current at other frequencies, then it could interfere with the operation of the line. Many data-processing machines, therefore, are designed to avoid the signaling frequencies completely. This limits the bandwidth available for transmission and hence the speed at which data can be sent. It is possible to design a modem that randomizes the signal, smearing it across the available bandwidth like the human voice so that it can never be mistaken for line signaling. However, this increases the modem cost.

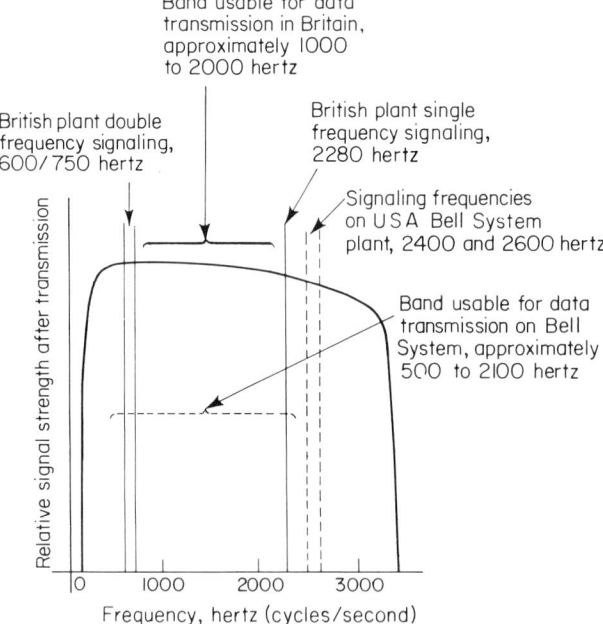

Fig. 1.2. On the public network, the telephone bandwidth shown in Fig. 1.1 has signaling frequencies inside it. "In-band" signaling frequencies on British telephone plant are shown as solid lines. Bell System signaling frequencies are shown as dashed lines. Any means used for sending data over these channels must avoid interfering with the signaling.

ADVANTAGES OF PRIVATE LINES

The private line has thus certain advantages for data transmission over the switched connections:

1. If it is to be used for more than a given number of hours per day, it is less expensive than the switched line. If it is used for only an

hour or so per day, then it is more expensive. The break-even point depends upon the actual charges, which in turn depend upon the mileage of the circuit, but it is likely to be of the order of several hours per day. This is clearly an important consideration in designing a data transmission network.
2. Private lines can be specially treated or "conditioned" to compensate for the distortion that is encountered on them. In this way the number of data errors can be reduced or, alternatively, a higher transmission rate can be made possible. The switched connection cannot be conditioned beforehand in the same way, because it is not known what path the circuit will take. Dialing one time is likely to set up a quite different physical path from that obtained by dialing at another time, and there are a large number of possible paths. Line termination equipment has been designed now that conditions dynamically and adjusts to whatever connection it is used on. When this comes into more common use, it will enable higher speeds to be obtained over switched circuits. Many carriers either have separate tariffs for conditioned telephone lines and unconditioned lines, or charge extra for conditioning.
3. Switched voice lines usually carry signaling within the bandwidth that would be used for data (at frequencies such as those shown in Fig. 1.2). Data transmission machines must be designed so that the form in which the data are sent cannot interfere with the common carrier's signaling. With some machines this also makes the capacity available for data transmission somewhat less than that over a private voice line. A common rate over a switched voice line in the 1960's has been 1200 bits per second, whereas 2400 bits per second has been common over a specially conditioned, leased voice line. Machines equipped with suitable modems transmit satisfactorily at 4800 over the leased line. Because of improved modem designs, however, it is likely that 3600 bits per second over switched voice lines and 9600 bits per second over conditioned, leased voice lines will become common in the 1970's. Already some modems transmit at higher speeds than 3600 over public voice lines.
4. The leased line may be less perturbed by noise and distortion than the switched line. The switching gear can cause impulse noise which causes errors in data. This is a third factor that contributes to a lower error rate for a given transmission speed on private lines.

The cost advantage of switched lines will dominate if the terminal has only a low usage. Also the ability to dial a distant machine gives great flexibility. Different machines can be dialed with the same terminal, perhaps offering quite different facilities. A typewriter terminal used at

8 TYPES OF COMMUNICATION LINKS

one time by a secretary for computer-assisted text editing may at another time be connected to a scientific time-sharing system and at another time may dial a computer-assisted teaching program. Machine availability is another consideration. If one system is overloaded or under repair, the terminal user might dial an alternative system. Often this dialing is done over the firm's own leased tie lines.

SIMPLEX, DUPLEX, AND HALF-DUPLEX LINES In designing a data-processing system it is necessary to decide whether the line must transmit in one direction only or in both directions. If the latter, will the machines transmit in both directions at the same time?

Transmission lines are classed as simplex, half duplex, and full duplex. In North America, these terms have the following meanings:

Simplex lines transmit in one direction only.

Half-duplex lines can transmit in either direction, but only in one direction at once.

Full-duplex lines transmit in both directions at the same time. One full-duplex line is thus equivalent to two simplex or half-duplex lines used in opposite directions.

The above are the meanings in current usage throughout most of the world's computer industry. Unfortunately, however, the International Telecommunications Union, an organization that has determined most of the world's standards in telecommunications, defines the first two terms differently—as follows:

Simplex (circuit). A circuit permitting the transmission of signals in either direction but not simultaneously.

Half duplex (circuit). A circuit designed for duplex operation but which on account of the nature of the terminal equipment can be operated alternately only.

"Simplex" and "half duplex" are thus used differently by European telecommunications engineers and computer manufacturers (especially American ones) using such facilities.

Throughout this book, the words will be used with the former meanings exclusively.

Simplex lines (American meaning) are not generally used in data transmission because, even if the data are being sent in only one direction, control signals are normally sent back to the transmitting machine to tell it that the receiving machine is ready, or is receiving the data correctly.

Commonly, error signals (positive or negative acknowledgment) are sent back so that there can be retransmission of messages damaged by communication line errors. Many data transmission links use half-duplex lines. This allows control signals to be sent and two-way "conversational" transmission to occur. On some systems full-duplex lines can give more efficient use of the lines at little extra line cost. A full-duplex line often costs little more than a half-duplex line. Data transmission machines that can take full advantage of full-duplex lines are more expensive, however, than those that use half-duplex lines. Half-duplex transmission is, therefore, more common at present, although this situation might well change.

VOICE AND SUBVOICE LINE TYPES

The main categorization of line types relates to the speed of transmission possible. The line may be specifically adapted by the common carrier for data transmission, in which case a fixed transmission speed may be quoted in the description of the line. On the other hand, it may be a normal voice, or other, line with no special devices for data. In this case the speed obtained over the line will depend upon the modems (discussed shortly) with which the user sends data over the line.

Table 1.1 lists the main types of communication links in order of increasing speed. The speeds have been listed in terms of the number of data bits per second that may be sent over the line. Communication lines fall into one of three categories of speed:

1. *Subvoice grade.* Lines designed for telegraph and similar machines transmitting at speeds ranging, in the United States, from 45 to 150 bits per second. Some countries have lines of higher speed than their telegraph facilities, but still much slower than the capacity of voice lines. Britain, for example, has a Datel 200 service operating at 200 bits per second. Japan and other countries have a similar 200 bps service. All of these lines are today commonly obtained by subdividing telephone channels.

2. *Voice-band.* Telephone channels normally transmit today at speeds from 600 to 4800 bits per second. Speeds of 9600 and possibly higher may become common in the near future. Dial-up telephone lines are commonly used for speeds of 1200 or 2400 bits per second, today. Speeds up to 3600, however, may soon become common on the public network. 4800 is already possible with a reasonable error rate. Telephone organizations in some other countries have not yet permitted the use of such high speeds over their telephone lines. In many countries 600 or 1200 bits per second is still the maximum.

10 TYPES OF COMMUNICATION LINKS

Table 1.1. THE MAIN TYPES OF COMMUNICATIONS LINKS IN USE TODAY*

| | SPEED (bits per second) | UNITED STATES ||| UNITED KINGDOM | HALF DUPLEX OR FULL DUPLEX |
		AT&T (New terminology)	AT&T (Old terminology)	WESTERN UNION		
Subvoice grade	45	Type 1002	Schedule 1	Class A		FDX/HDX
	50				Telegraph, Tariff H	FDX
	55	Type 1002	Schedule 2	Class B		FDX/HDX
	75	Type 1005	Schedule 3	Class C		FDX/HDX
	100				Datel 100, Tariff J	FDX
	150	Type 1006	Schedule 3A			FDX
	180			Class D		FDX/HDX
	200				Datel 200	FDX
Voice grade	600			Broadband exchange Schedule 1		FDX
	1200			Broadband exchange Schedule 2		FDX
	1200	Type 3002	Schedule 4	Class G	Datel 600 Tariff S	FDX/HDX
	1400	Type 3002 plus C1 conditioning	Type 3003	Class E	Tariff S with line improvements	FDX/HDX
	2400	Type 3002 plus C2 conditioning	Type 3004	Class F	Ditto	FDX/HDX
	4800	Type 3002 plus C4 conditioning	Type 3005	Class H		FDX/HDX
Wideband	19,200 fixed	Type 8803				FDX
	50,000	Type 8801		Wideband channel	Special quotation	FDX
	105,000	Type 5700 or 5800	Telpak C	Telpak C	Special quotation	FDX
	240,000					
	500,000	Type 5800	Telpak D	Telpak D	Special quotation	FDX

* The terminology for these line types has been subject to many changes recently, and may well change again. Also the data speeds, in bits per second, on some line types are likely to improve with better modem design. For up-to-the-minute information, the reader should refer to the Common Carrier or FCC tariffs.

3. *Wideband.* Lines giving speeds much higher than voice channels, using facilities that carry many simultaneous telephone calls. Speeds up to about 500,000 bits per second are in use, and higher bit rates are possible if required.

All of these line types may be channeled over a variety of different physical facilities. This chapter, and indeed the tariffs themselves normally, say nothing about the medium used for transmission. It could equally well be wire, coaxial cable, microwave radio, or even satellite. The transmission over different media is organized in such a way that the channels obtained have largely the same properties—same capacity, same noise level, and same error rate. The user usually cannot tell whether he is using a microwave link, coaxial cable, or pairs of open wires stretched between telephone poles. Only satellite transmission requires different data-handling equipment, and here only because a delay of $\frac{1}{3}$ of a second or so is encountered because of the great distance.

In the terminology of the United States telephone companies, Schedule 4 used to refer to voice lines. This was replaced by the term Series 3000 lines. A Type 3002 line refers to a voice line used for data transmission. Western Union refers to it as a Class G line. Type 3002 lines with conditioning were referred to by AT&T as Type 3003, 3004, and 3005 lines; however, these terms have now been dropped, and one refers to a "Type 3002 line with Type C1, C2, or C4 conditioning." Western Union refers to lines with those three levels of conditioning as Class E, Class F, and Class H lines. The speed at which one can transmit over such lines varies with the type of modem that is used.

In AT&T terminology, Schedules 1, 2, 3, and 3A used to refer to lines of much lower capacity which were obtained by combining the transmission of several separate channels on one voice link. These are now called Series 1000 lines. A Type 1002 line refers to 45 and 55 bits-per-second teletypewriter lines, which used to be called Schedule 1 and 2 lines. A Type 1005 line is the old Schedule 3 line giving teletypewriter service at 100 bits per second. A Type 1006 line permits data transmission at 150 bits per second; this used to be called Schedule 3A.

In Western Union tariffs, teletype and other low-speed lines are listed as Class A, B, C, and D. Classes A, B, and C are equivalent to AT&T's Schedules 1, 2, and 3 (now Types 1002 and 1005). This is all summarized in Table 1.1.

In the United Kingdom, today's teletype circuits can be used to transmit at 50 or 100 bits per second (full duplex). These are referred to as Tariff H and Tariff J lines, respectively. The 100 bits-per-second links are new and are available only between selected locations. This is referred to as the *Datel 100* service. The United Kingdom voice lines are used, split into

12 TYPES OF COMMUNICATION LINKS

separate channels, to give a *Datel 200* service at 200 bits per second, a *Datel 300* service in which multifrequency signaling permits the transmission of up to 20 characters per second (only), and a *Datel 600* service with a speed range of 600 to 1200 bits per second. The latter is referred to as Tariff S. A conditioned voice line giving speeds of 2000 bits per second is referred to as *Datel 2000*. The voice line can sometimes be leased for part of the day only, and this is referred to as Tariff E. If it is leased for two hours per day (the minimum), its cost will be approximately a quarter of Tariff S.

LINE CONDITIONING

As has been mentioned, private, leased, voice lines can be *conditioned* so that they have better properties for data transmission. Tariffs specify maximum levels for certain types of distortion. *An additional charge is made by most carriers for lines that are conditioned.*

The American Telephone and Telegraph Company, for example, has three types of conditioned voice lines, the conditioning being referred to as Types C1, C2, and C4 (formally Types 4A, 4B, and 4C). A line ideal for data transmission would have an equal drop in signal voltage for all frequencies transmitted. Also, all frequencies would have the same propagation time. This is not so in practice. Different frequencies suffer different attenuation and different signal delay. Conditioning attempts to equalize the attenuation and delay at different frequencies. Standards are laid down in the tariffs for the measure of equalization that must be achieved. The signal attenuation and delay at different frequencies must lie within certain limits for the Type C1, C2, and C4 conditioning.[5] The result of the conditioning is that a higher data speed can be obtained over that line, given suitable line termination equipment (modems).

WIDEBAND AND TELPAK TARIFFS

TELPAK is a private-line, "bulk" communications service offered in the United States by the telephone companies and Western Union. It transmits high-volume, point-to-point communications in various forms—voice, telephotograph, teletypewriter, control, signaling, facsimile, and data.

The service, introduced in 1961, is designed for businesses and government agencies with large private-line communications requirements. TELPAK has a flexible capacity and can be tailored to the customer's needs; it can provide, if required, wideband communications, or it can be subdivided for use with teletypewriter equipment. A base capacity channel could be employed for high-speed transmission of such data as magnetic tape, computer memory, and facsimile.

---

[5] *Telecommunications and the Computer*, Chapter 12.

The TELPAK customer pays a monthly charge based on the capacity of the communications channel he selects, the number of airline miles between locations, and the type and quantity of channel terminals. He has use of this channel on a full-time basis.

There were originally four sizes of TELPAK channels: TELPAK A, B, C, and D. However, in 1964 the Federal Communications Commission ruled that rates for TELPAK A (12 voice circuits) and TELPAK B (24 voice circuits) were discriminatory in that a large user could obtain a group of channels at lower cost per channel than a small user, who could not take advantage of the bulk rates. In 1967 the TELPAK A and B offerings were eliminated.

TELPAK A *had* a base capacity of 12 voice channels (full duplex).
TELPAK B *had* a base capacity of 24 voice channels (full duplex).
TELPAK C has a base capacity of 60 voice channels (full duplex).
TELPAK D has a base capacity of 240 voice channels (full duplex).

Each TELPAK voice channel can itself be subdivided into one of the following:

1. Twelve teletype channels, half or full duplex (75 bits per second).
2. Six class D channels, half or full duplex (180 bits per second).
3. Four AT&T Type 1006 channels, half or full duplex (150 bits per second).

There cannot be mixtures of these channel types in a voice channel. TELPAK C can transmit data at speeds up to 105,000 bits per second; TELPAK D has a potential transmission rate of 500,000 bits per second. Line termination equipment is provided with these links, and each link has a separate voice channel for coordination purposes.

The TELPAK channels thus serve two purposes. First, they provide a wideband channel over which data can be sent at a much higher rate than over a voice channel. Second, they provide a means of offering groups of voice or subvoice lines at reduced rates—a kind of discount for bulk buying.

Suppose that a company requires a 50,000-bit-per-second link between two cities, and also 23 voice channels and 14 teletypewriter channels; or perhaps 30 voice channels and no teletypewriter links. Then it would be likely to use the TELPAK C tariff. In leasing these facilities it would have some unused capacity. If it wishes it can make use of this at no extra charge for mileage, though there would be a terminal charge.

Government agencies and certain firms in the same business whose rates and charges are regulated by the government (e.g., airlines and railroads) may share TELPAK services. Airlines, for example, pool their needs for voice and teletypewriter channels. An intercompany organization purchases the TELPAK services and then apportions the channels to individual airlines. Most of the lines channeling passenger reservations to a distant office where bookings can be made are TELPAK lines, and so also are the lines carrying data between terminals in those offices and a distant reservations computer. There has been some demand to extend these shared TELPAK facilities to other types of organizations that could benefit from them by sharing, but in the late nineteen-sixties this was not permissible.

TELPAK originally was proposed as an interstate service, but since then it has become generally available intrastate as well.

Although not a TELPAK offering, Series 8000 is another "bulk" communications service in the United States that offers wideband transmission of high-speed data, or facsimile, at rates up to 50,000 bits a second; the customer has the alternative of using the channel for voice communication up to a maximum of 12 circuits. A Type 8801 link, part of this series, provides a data link at speeds up to 50,000 bits per second with appropriate terminating data sets and a voice channel for coordination. A Type 8803 link provides a data link with a fixed speed of 19,200 bits per second, and leaves a remaining capacity that can be used either for a second simultaneous 19,200-bits-per-second channel or for up to five voice channels. These links must connect only two cities. The separate channels cannot terminate at intermediate locations.

Most countries outside North America also offer tariffs similar to the Series 8000, and in most locations quotations for higher speeds can be obtained on request. Obtaining a wideband link in many such countries can be a slow process. This is particularly so if the termination is required in a small town or rural area rather than in a city to which such links already exist. No doubt, as the demand for such facilities increases, so the service of the common carriers in providing them will improve.

TELEX

Telex is a worldwide switched public teleprinter system. It operates at 66 words per minute (50 bits per second), and uses the Baudot code (Fig. 2.1). It is operated in the United States by Western Union. Any teleprinter on the system can dial any other teleprinter in that country, and Telex machines can be connected internationally without speed or code conversion. The United States can dial Canada and Mexico directly, but to other countries operator intervention is needed. Some countries permit the Telex facilities to be used for other forms of dial-up data transmission. Each Telex call is billed on a time and distance basis.

Each subscriber has an individual line and his own number, as with the conventional telephone service. His teleprinter is fitted with a dial, like a telephone, with which he can dial other subscribers. The teleprinter used may or may not have paper-tape equipment also. The teleprinter can be unattended. When a message is sent to an unattended teleprinter, it will switch itself on, print the message, and then switch itself off. Figure 3.6 shows a typical Telex installation.

TELETYPEWRITER EXCHANGE SERVICE The Bell System and other telephone companies offer a service in the United States and Canada that is competitive with Telex. Again, each subscriber has a dial-up teletypewriter with his own number listed in a nationwide directory. This service is called the Teletypewriter Exchange Service (TWX), and it uses the telephone circuits combined with several TWX channels so that they can be sent over one voice channel. The combining or "multiplexing" is done at the local switching office, where the DC signals are changed to equivalent bursts of appropriate frequencies. The link between the local switching office and the subscriber is often a conventional telephone line, and in this case the teletypewriter needs a data set to convert the DC signals to appropriate frequencies in the voice range.

Other manufacturers' data transmission equipment can be connected to TWX lines, and can transmit at speeds up to 150 bits per second, half or full duplex. This requires a special terminal arrangement at additional cost. There are three types of access lines to the TWX network, as follows:

1. TTY-TWX.
   This is an access line with a teletypewriter (e.g., that in Fig. 3.6) provided by the common carrier. The speeds of transmission are either 6 characters per second in Baudot code, or 10 characters per second in Data Interchange Code (DIC).
2. CPT-TWX.
   CPT stands for "customer provided terminal"; to this access line the customer can attach any device operating with one of the above two speeds and codes and adhering to normal TWX line control. The device could be a computer with an appropriate adapter on its input/output channel.
3. CE TWX (formerly called "TWX Prime").
   CE stands for "customer equipment." This can now be any device and is not restricted to a specific code or character speed. Two TWX subsystems are accessible, one which operates at speeds up to 45 bits per second, and the other up to 150. A CE-TWX terminal can communicate only with another CE-TWX terminal.

TWX directories are published listing TTY-TWX and CPT-TWX subscribers.

16   TYPES OF COMMUNICATION LINKS

MODULATION   Data entering or leaving data-processing machines are normally binary in form, and consist of rectangular pulses something like those in Fig. 1.3. The line over which we may wish to transmit this information might have properties such as those represented by Fig. 1.1.

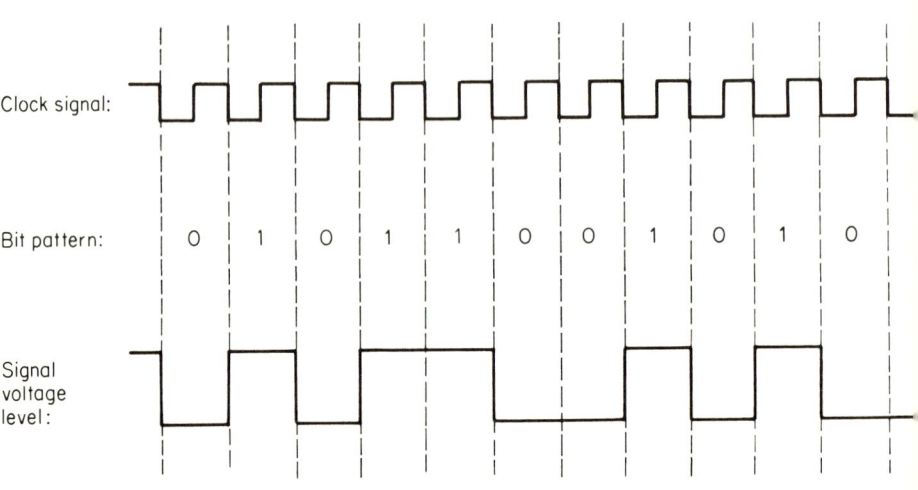

Figure 1.3

Two problems become apparent. First, the line represented by Fig. 1.1 does not transmit DC current. Frequencies below 200 hertz are severely attenuated. A data pattern in which every bit is a "1," for example, would not be transmitted. Second, high frequencies are attenuated, and this fact alone would cause our square-edged pulses to become distorted. The faster the bit rate, the greater would be the distortion.

The square-edged pulse train is, therefore, manipulated electronically to make it fit as well as possible into the transmission frequencies of Fig. 1.1. In a typical system, a "carrier" is used which is a single-frequency signal in the middle of the band available for transmission. The carrier is modified in some way by the data to be sent so that it "carries" the data. This process is referred to as "modulation."

As is partially seen in Fig. 1.1, there is a certain range of frequencies that travel without much distortion over telephone circuits. A frequency of 1500 cycles per second, for example, is near the middle of the human voice range. *Modulation* employs these voice frequencies to carry data that would otherwise suffer too much distortion. Thus a sine wave of 1500 cycles per second may be used as a *carrier* on which the data to be sent are superimposed in the manner shown in Fig. 1.4.

In addition to making it possible to send signals with a DC component over channels that will not transmit direct current, modulation achieves two ends: first, it reduces the effects of noise and distortion, and second, it increases the possible signaling speed. Using simple modulation devices, one can send computer data without undue distortion over the voice circuits and other communication lines of the world.

Modulation is also used by the telephone companies to pack many voice channels into one high-frequency signal which can be sent over cables or microwave links. It is one of the techniques of "multiplexing" many voice channels onto one physical transmission link. In this case, it is the voice signal that is used to modulate the carrier wave. Many such carrier waves, separated in frequency by about 4000 cycles per second, are transmitted together.

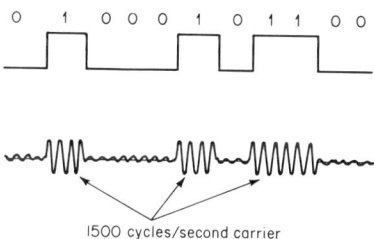

Fig. 1.4. Amplitude modulation of a frequency at the center of the voice band.

If we use modulation to send data over a voice channel on such a system, the data pass through two modulation processes. In fact, there are often more than two, and the data waveform is manipulated by a variety of electronic processes before it eventually emerges after transmission in its original form.

Figure 1.4 illustrates one type of modulation, *amplitude modulation*, in which a "1" is represented by a high amplitude sine wave at the carrier frequency and a "0" is represented by a lower amplitude wave of the same frequency. Other types of modulation are described below.

MODEMS AND DATA SETS

In order to achieve modulation, the binary output from the data-processing machine must enter a "modulator," which produces the appropriate sine wave and modifies it in accordance with the data. This produces a signal suitable for sending over voice circuits, and whatever manipulation the electronics do to the human voice, they can also do to this signal and the data will still be recoverable. At the other end of the communication line the carrier must be "demodulated" back to binary form. The circuitry for modulating and demodulating is usually combined into one unit, referred to by the abbreviated term *modem*.

The modem, a unit slightly larger than a domestic radio set, is connected to the data-processing machine, and it is then able to transmit data over normal telephone, or other, lines as in Fig. 1.5.

Modems are made both by the computer manufacturers and by the telephone companies. They are sometimes also called *data sets*. Figure 1.6 shows some typical data sets.

18 TYPES OF COMMUNICATION LINKS

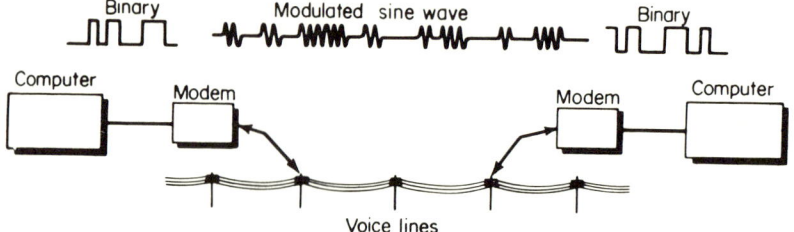

Fig. 1.5. The use of modems.

The GPO Datel Modem No. 2 for transmitting up to 200 bits per second over Datel 200 lines in the United Kingdom.

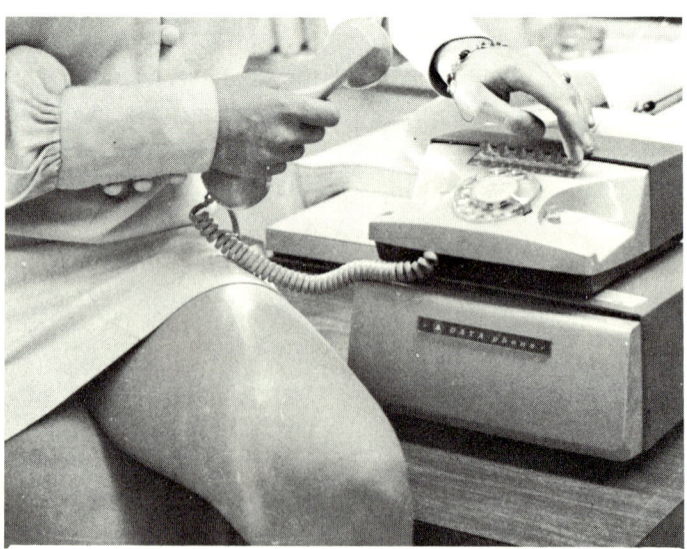

The Bell System Modem No. 103 A for transmitting at speeds up to 300 bits per second over public dial-up voice lines. A similar modem is used for TWX lines.

Fig. 1.6. Typical modems.

TYPES OF COMMUNICATION LINKS   19

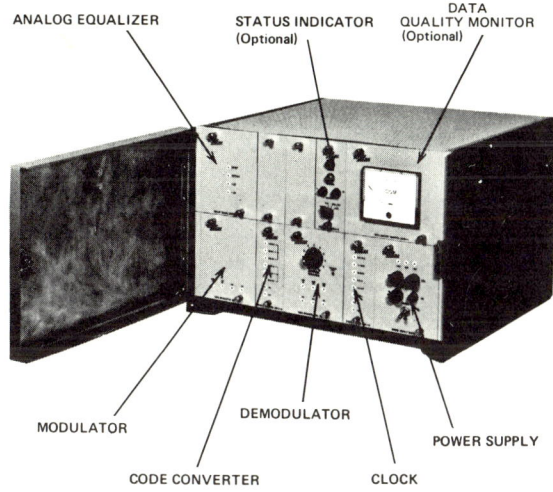

The Rixon SEBIT-48C Modem designed to transmit at 4800 bits per second over a leased conditioned voice line (e.g. Bell Type 3002 with C4 conditioning). Rixon makes other modems that transmit at speeds up to 2400 bits per second over dial-up voice lines and 9600 over leased, conditioned voice lines. They also make modems that multiplex several channels into one voice channel as illustrated in Fig. 11.2.

The Datamax QB48 Modem incorporates its own forward operating error correction. This and its instantaneous dynamic line equalization enable it to achieve high speeds over dial-up voice lines. Datamax has designed a set of similar modems which offer a choice of low speed, low error rate, or high speed, higher error rate over the public network as follows:
A. 1200 bits per second with less than one error in 10^9 bits.
B. 4800 bits per second with less than one error in 10^6 bits.
C. 9600 bits per second with less than one error in 10^3 bits.

Many computer centers now have a large number of dial-up lines coming into them This rack of several hundred Bell System data sets connects tie-lines and toll telephone lines to the computers.

Fig. 1.6. Typical modems.

20 TYPES OF COMMUNICATION LINKS

A variety of different modulation processes is possible.[6] There is much scope for ingenuity in the design of modems, and the increasing speed at which data can be sent over telephone lines is largely due to improving modem design. Often the data-processing system designer is faced with a choice of modems he may use. A further function of the modem is to protect the common carrier lines from undesirable signals that might cause interference with other users, or with the network's signaling system.

THREE TYPES OF MODULATION Two other ways of modulating a sine wave carrier are now in more common use for carrying data than amplitude modulation. These are referred to as *frequency modulation* and *phase modulation*.

1. *Frequency Modulation*

Instead of the 0 and the 1 being represented by two different amplitudes, they are represented by two different frequencies f_1 and f_2, Fig. 1.7.

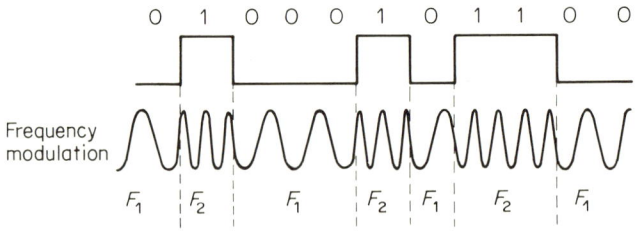

Figure 1.7

2. *Phase Modulation*

The 0 and the 1 are represented by sine wave carriers 180 degrees out of phase with one another as in Fig. 1.8.

The previous figures all relate to binary transmission. It is possible, however, to use a waveform with *more than two* states as the modulating waveform. This is shown for amplitude modulation in Fig. 1.9. "Di-bits" are used to give four possible states in the modulated waveform. This approximately doubles the transmission rate that can be achieved with the modem, but it also considerably increases its susceptibility to noise. As amplitude modulation is already susceptible to noise, di-bits are not normally used with this; however, they are frequently used with phase modulation.

The designer of a modem is seeking a workable compromise between the quantity of data that can be packed into the transmission and the ability of the modem to decode it correctly in the presence of noise and distortion.

---

[6] *Telecommunications and the Computer*, Chapter 13.

In selecting a modem for a system, the systems analyst has to strike an appropriate balance between *cost of modem, speed,* and *susceptibility to error.*

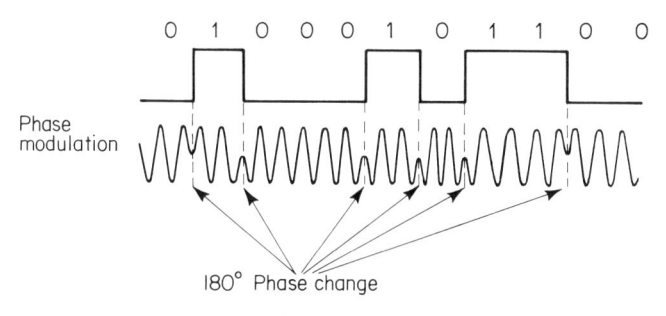

Figure 1.8

On some systems, the terminals are slow as compared with line capacity, and so a slow and inexpensive modem may be selected. On others, the maximum protection against errors is needed. Where long and therefore expensive lines are in use, it can be desirable to pack the maximum amount of data onto the lines, and so speed may be the prime consideration. It may be worthwhile to use the fastest modem even though it is more error-prone. An elaborate error-detection code may be used to compensate for the use of a modem that increases, perhaps doubles, the data speed. As will be seen in Chapter 5, it is possible to devise error-detecting codes for communication lines, that are highly effective without excessive redundancy.

Sometimes, on the other hand, relatively slow terminals are used on voice lines—often because of the convenience of using telephone dialing on the public network, or of using a company's telephone tie line. A voice line is often used to dial a time-sharing computer regardless of slow terminal speed. A variety of terminals transmitting at 14.8 characters per second—a speed geared to the technology of electric typewriters—are used, for example; yet the public phone line is capable of transmitting 120 such characters, or more with good modems. Speed, then, is certainly not a consideration in the design of a modem for using such terminals on a telephone line. Low cost and error-free operation dominate the design instead.

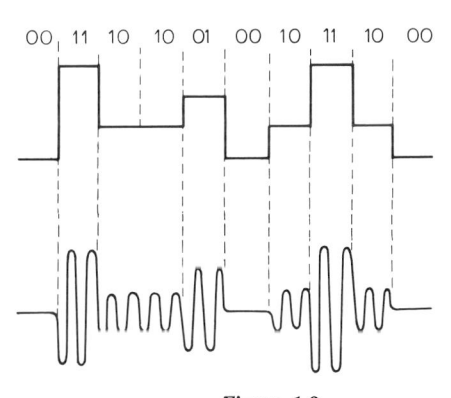

Figure 1.9

The increasing choice of modems on the market and in the laboratory will give the systems engineer more scope in selecting between these criteria to suit his particular data-processing environment.

TRANSMISSION WITHOUT MODEMS

Modems are not necessarily used on lines privately laid, though here, still, they can increase greatly the speed of transmission. Many computer users need to have data transmission lines within their own premises, and privately owned lines linking two buildings near to each other. The terms "in-plant" and "out-plant" system are used. The former means that common carrier lines are not used and usually the communication links are within one plant or localized area. "Out-plant" implies that common carrier lines are used. In-plant lines are normally a straightforward copper path, possibly coaxial cable, connecting the points in question. Private links of this type are often installed by a firm's own engineers. Sometimes they are also provided by telecommunication companies, but external to any major telecommunication network.

Devices that use these lines often operate by the simple making and breaking of relay contacts, or the sending of rectangular pulse trains such as that in Fig. 1.3. No modulation is needed. Over a wire pair a few miles in length, DC pulses can be sent at speeds up to about 300 bits per second. The distortion of the signals makes it impractical to send data in this form at speeds much higher than 300 bits per second over an ordinary pair of wires, except over short distances or unless closely-spaced repeaters reconstruct the pulse stream—a very powerful technique which will be discussed later. 300 bits per second, however, is a useful speed for many computer applications. The speed could be increased greatly by using small coaxial cables rather than wire pairs. In many systems a large number of typewriter-speed terminals within a localized area, say three miles across, could be connected to a time-sharing system or to a concentrator without modems. Although today most common carrier lines require modems, it is possible that the wire pairs that connect a central office to all locations with a telephone could be used over a limited area for DC signaling as in the earlier days of telegraphy. A low-cost private branch exchange for data signals used in this way has been developed.

A modem for low-speed transmission typically costs about $20 to $40 monthly rental. A time-sharing system with 500 low-speed terminals (only a few of which are in use at any one time) would be likely to pay, then, about $20,000 to $40,000 per month for modems. Figure 1.6 shows a rack of such modems for a typical time-sharing system. If DC signaling could be used, this high cost would be avoided.

ACOUSTICAL COUPLING

A very inexpensive form of data transmission device sends audible tones down the telephone line, such as those generated by a push-button telephone. This will be described more fully in Chapter 3. A machine using audible tones need not necessarily be physically wired to the line. It can make sounds that are picked up directly by the telephone handset. This is referred to as *acoustical coupling*. The sounds used must be at those frequencies having low attenuation in Fig. 1.1. The sounds are reproduced at the far end of the connection by a telephone earpiece. They are then converted back into data signals. Acoustical coupling is somewhat less efficient than direct coupling. At the time of writing, it is used for transmitting between relatively slow machines such as typewriter-like terminals.

In a typical acoustical-coupling device the telephone handpiece fits in a special cradle as shown in Fig. 1.10. Acoustical couplers are generally less expensive than modems. Another advantage of acoustical coupling is that the terminal can easily be made portable. A small terminal could be made to transmit to a computer from a public call box. Nondigital machines also use acoustical coupling. Documents can be copied at a distance with a Xerox machine in this way.

Although there is no electrical connection to the telephone lines, it is still possible for acoustically coupled machines to interfere with the public network's signaling, as with a directly coupled device. Also, severe cross talk can be caused on a telephone link that carries many voice channels, by the transmission of a continuous frequency, as with a repetitive data pattern. It is, therefore, desirable that the coupling device randomize the signal before it is sent, though most such devices do not do so today.

By 1970 acoustical couplers were extensively used in the United States. In many other countries, however, the telecommunication authorities would not permit their use. This prohibited the variety of applications of small portable terminals—such as from public coin-operated telephones.

TARIFFS INCLUDING DATA SETS

The modem or data set, which is essential to most transmission of data, is sometimes provided by the common carrier as part of the service. The link purchased in that case would include the data set cost in its price. Sometimes the cost of the line is quoted separately from the data set cost. If the tariff includes the data set, it will state the speed or speeds in bits per second at which the line will transmit.

An example is the Bell System Dataphone service. A data set (Fig. 1.6) is provided that connects to the data-processing machine. A call may be dialed by an operator with the data set switched to "Voice." When she has established the call she hears a characteristic whistling tone from the data set at the dialed machine and she then switches the data set to "Data." The GPO *Datel* Services are such tariffs in Britain, the word *Datel* implies that the modem is provided.

24 TYPES OF COMMUNICATION LINKS

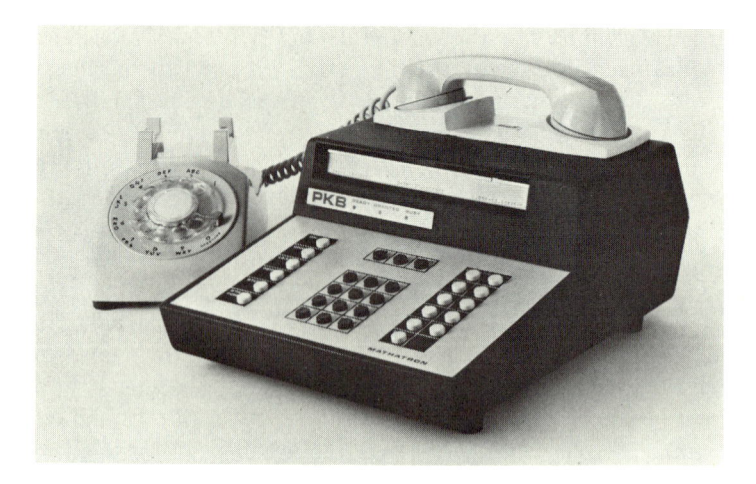

A small portable terminal with a built-in acoustical coupler, which can be connected to any telephone. This is Marathon's PKB electronic calculator in which this remote keyboard is connected to a computer.

A terminal that operates at 14.8 bits per second being connected to a distant computer via a normal office telephone and an inexpensive acoustical coupler.

Fig. 1.10. Typical examples of acoustical couplers. These permit a terminal to transmit over a telephone line without a direct electrical connection. Data is converted into sounds that are transmitted over the telephone. Acoustical couplers are inexpensive and can make a terminal portable. The telephone company does not have to install these couplers.

2 INFORMATION CODING

Ever since the earliest swinging telegraph needles and the first waving of semaphore flags it has been necessary to use codes to translate signals transmitted into digits, letters, or words. Figure 2.1 shows a conventional method of interpreting the 1 and 0 pulses of telegraphy into meaningful information. A variety of other such data transmission codes is used, and machines have been built to translate these into printed words or store them in computer memories. This chapter gives examples of some of the codes in common use.

If we transmit n binary pulses, such as the telegraph pulses in Fig. 1.3, we can, in theory, code 2^n different combinations into these pulses. The "characters" that are sent by data transmission often contain five, six, seven or eight bits. Five bits can give 32 different characters, six bits 64, seven bits 128, and eight bits 256. A seven-bit code is thus used, for example, to transmit a range of up to 128 characters. It would not always transmit 128 characters, because some of the combinations are reserved for special-purpose *control characters*, which have functions such as indicating the end of a record, making a teleprinter carriage return, and a variety of other operations.

ESCAPE CHARACTERS

Telegraphy has commonly used a five-bit code, as shown in Fig. 2.1. This would restrict the number of characters to 32 if used normally; however, *letters shift* and *figures shift* characters are used to extend the range. When a *figures shift* character is sent, the characters that follow it are upper case characters until a *letters shift* character is sent. Similarly, the characters following a *letters shift* are letters—until a *figures shift* is sent. The *letters shift* and *figures shift* characters must be recognized in either case.

• Denotes positive current

Start	1	2	3	4	5	Stop	Lower case	CCITT standard international telegraph alphabet No.2	U.S.A. teletype commercial keyboard	AT & T fractions keyboard	Weather keyboard
	•	•				•	A	—	—	—	↑
	•			•	•	•	B	?	?	$\frac{5}{8}$	⊕
		•	•	•		•	C	:	:	$\frac{1}{8}$	○
	•			•		•	D	Who are you?	$	$	↗
	•					•	E	3	3	3	3
	•		•	•		•	F	Note 1	!	$\frac{1}{4}$	→
		•		•	•	•	G	Note 1	&	&	↘
			•		•	•	H	Note 1	#		↓
		•	•			•	I	8	8	8	8
	•	•		•		•	J	Bell	Bell	'	↙
	•	•	•	•		•	K	(	(	$\frac{1}{2}$	←
		•		•	•	•	L	)	)	$\frac{3}{4}$	↖
			•	•	•	•	M	.	.	.	.
			•	•		•	N	,	,	$\frac{7}{8}$	⦶
			•	•	•	•	O	9	9	9	9
		•	•		•	•	P	0	0	0	∅
	•	•	•		•	•	Q	1	1	1	1
		•		•		•	R	4	4	4	4
	•		•			•	S	,	,	Bell	Bell
					•	•	T	5	5	5	5
	•	•	•			•	U	7	7	7	7
		•	•	•	•	•	V	=	;	$\frac{3}{8}$	⦶
	•	•			•	•	W	2	2	2	2
	•		•	•	•	•	X	/	/	/	/
	•		•		•	•	Y	6	6	6	6
	•				•	•	Z	+	"	"	+
					•	•	Blank				—
	•	•	•	•	•	•	Letters shift				↓
	•	•		•	•	•	Figures shift				↑
			•			•	Space				■
				•		•	Carriage return				<
		•				•	Line feed				≡

Note 1. Not allocated internationally; available to each country for internal use.

Fig. 2.1. Baudot five-bit telegraphy code.

Standard code	Paper tape code	Data character
8765 4321		
0100 0000	1100 0000	Master space
1100 0001	0100 0001	Upper case
1100 0010	0100 0010	Lower case
0100 0011	1100 0011	Line feed
1100 0100	0100 0100	Carriage return
0100 0101	1100 0101	Space
0100 0110	1100 0110	A
1100 0111	0100 0111	B
1100 1000	0100 1000	C
0100 1001	1100 1001	D
0100 1010	1100 1010	E
1100 1011	0100 1011	F
0100 1100	1100 1100	G
1100 1101	0100 1101	H
1100 1110	0100 1110	I
0100 1111	1100 1111	J
1101 0000	0101 0000	K
0101 0001	1101 0001	L
0101 0010	1101 0010	M
1101 0011	0101 0011	N
0101 0100	1101 0100	O
1101 0101	0101 0101	P
1101 0110	0101 0110	Q
0101 0111	1101 0111	R
0101 1000	1101 1000	S
1101 1001	0101 1001	T
1101 1010	0101 1010	U
0101 1011	1101 1011	V
1101 1100	0101 1100	W
0101 1101	1101 1101	X
0101 1110	1101 1110	Y
1101 1111	0101 1111	Z
1110 0000	1010 0000	)
0110 0001	0010 0001	—
0110 0010	0010 0010	+
1110 0011	1010 0011	<
0110 0100	0010 0100	=
1110 0101	1010 0101	>
1110 0110	1010 0110	—
0110 0111	0010 0111	S
0110 1000	0010 1000	(*)
1110 1001	1010 1001	(
1110 1010	1010 1010	"
0110 1011	0010 1011	:
1110 1100	1010 1100	?
0110 1101	0010 1101	!
0110 1110	0010 1110	.
1110 1111	1010 1111	⊕ (Stop)
0111 0000	0011 0000	0
1111 0001	1011 0001	1
1111 0010	1011 0010	2
0111 0011	0011 0011	3
1111 0100	1011 0100	4
0111 0101	0011 0101	5
0111 0110	0011 0110	6
1111 0111	1011 0111	7
1111 1000	1011 1000	8
0111 1001	0011 1001	9
0111 1010	0011 1010	'
1111 1011	1011 1011	;
0111 1100	0011 1100	/
1111 1101	1011 1101	.
1111 1110	1011 1110	□ Special
0111 1111	0011 1111	Idle
0001 1010	1001 1010	WRU
1011 1111	1111 1111	Delete

Sequence of bits in serial transmission ← (points to Standard code)

Parity bit, Control bit, Data bits (labels pointing to bit groupings)

These are control characters because the seventh bit of the standard code is 0 (The sixth and seventh bits of the tape code are the same)

Fig. 2.2. Fieldata code used in military data transmission.

28 INFORMATION CODING

This is an example of what is known as an *escape* mechanism in a code. The "letters shift" and "figures shift" are referred to as *escape* characters. By using escape characters or combinations of characters, the total number of possible characters in a code can be greatly increased. Sometimes the escape character changes the meaning of all of the other characters following it until a further escape character is received, as in the above example. In other codes the escape character changes the meaning of only the one character that follows it. This enables special characters to be inserted at the wish of the user. This is the case in the military *Fieldata* telegraphy code shown in Fig. 2.2. The character labeled "Special" here is the escape character, and the character immediately following it has a special alternate meaning.

PARITY BITS

Many data transmission codes use an extra bit, called a parity bit, in each character for checking purposes. This is added so that the total number of 1 bits in the character transmitted will be an odd (or in some codes, even) number. For example, if a character without the parity bit is 0100010, then an extra 1 bit is added to make it 01000101, with an odd number of 1's. On the other hand, if it is 0110010, then a 0 bit is added to again produce an odd number of 1's. The receiving machine detects whether there is an odd number of 1's. If not, it knows that noise or distortion on the line has lost or added a bit. It either notes this error or instigates retransmission of that data. This is sometimes called *vertical redundancy checking*.

Unfortunately, noise on the communication line sometimes changes more than one bit, and this lessens the effectiveness of parity checking. If two 0 bits are both changed to 1's, the parity check will not detect this. Further checking facilities are needed, as discussed in Chapter 4.

The Fieldata code given in Fig. 2.2 uses *odd* parity checking as described above for normal transmission. However, it uses even parity checking for its paper tape; in other words, there must be an even number of holes in the tape for each character. This then complies with the convention that a section of tape with all holes punched is valid, but contains no data. To delete incorrect data one punches all the holes of those characters. The reading machine then passes over this section, taking no action.

CONTROL BITS AND CHARACTERS

A variety of control functions is needed in data transmission, some of which were discussed above. It may be necessary, for example, to make a distant printer space or feed a line. In some transmission it is necessary to indicate end of record, end of text, and so on. Figure 2.1 lists five control characters, at the bottom of the list, and these must be recognizable in either figures shift or letters shift. In a similar manner the more extensive code shown in Fig. 2.3 must be able to control distant machines,

INFORMATION CODING   29

DATA CHARACTERS:

Normal shift

Character:	S	B	A	8	4	2	1	Parity
1	0	0	0	0	0	0	1	0
2	0	0	0	0	0	1	0	0
3	0	0	0	0	0	1	1	1
4	0	0	0	0	1	0	0	0
5	0	0	0	0	1	0	1	1
6	0	0	0	0	1	1	0	1
7	0	0	0	0	1	1	1	0
8	0	0	0	1	0	0	0	0
9	0	0	0	1	0	0	1	1
0	0	0	0	1	0	1	0	1
a	—	1	1	0	0	0	1	0
b	0	1	1	0	0	1	0	0
c	0	1	1	0	0	1	1	1
d	0	1	1	0	1	0	0	0
e	0	1	1	0	1	0	1	1
f	0	1	1	0	1	1	0	1
g	0	1	1	0	1	1	1	0
h	0	1	1	1	0	0	0	0
i	0	1	1	1	0	0	1	1
j	0	1	0	0	0	0	1	1
k	0	1	0	0	0	1	0	1
l	0	1	0	0	0	1	1	0
m	0	1	0	0	1	0	0	1
n	0	1	0	0	1	0	1	0
o	0	1	0	0	1	1	0	0
p	0	1	0	0	1	1	1	1
q	0	1	0	1	0	0	0	1
r	0	1	0	1	0	0	1	0
s	0	0	1	0	0	1	0	1
t	0	0	1	0	0	1	1	0
u	0	0	1	0	1	0	0	1
v	0	0	1	0	1	0	1	0
w	0	0	1	0	1	1	0	0
x	0	0	1	0	1	1	1	1
y	0	0	1	1	0	0	0	1
z	0	0	1	1	0	0	1	0
.	0	1	1	1	0	1	1	0
$	0	1	0	1	0	1	1	1
,	0	0	1	1	0	1	1	1
/	0	0	1	0	0	0	1	1
'	0	0	0	1	0	1	1	0
&	0	1	1	0	0	0	0	1
—	0	1	0	0	0	0	0	0
@	0	0	1	0	0	0	0	0

Upper shift

Character:	S	B	A	8	4	2	1	Parity
=	1	0	0	0	0	0	1	
c	1	0	0	0	0	1	0	1
;	1	0	0	0	0	1	1	1
:	1	0	0	0	1	0	0	0
°C	1	0	0	0	1	0	1	1
'	1	0	0	0	1	1	0	0
—	1	0	0	0	1	1	1	0
+	1	0	0	1	0	0	0	1
(	1	0	0	1	0	0	1	0
)	1	0	0	1	0	1	0	0
A	1	1	1	0	0	0	1	1
B	1	1	1	0	0	1	0	1
C	1	1	1	0	0	1	1	0
D	1	1	1	0	1	0	0	1
E	1	1	1	0	1	0	1	0
F	1	1	1	0	1	1	0	0
G	1	1	1	0	1	1	1	1
H	1	1	1	1	0	0	0	1
I	1	1	1	1	0	0	1	0
J	1	1	0	0	0	0	1	0
K	1	1	0	0	0	1	0	0
L	1	1	0	0	0	1	1	1
M	1	1	0	0	1	0	0	0
N	1	1	0	0	1	0	1	1
O	1	1	0	0	1	1	0	1
P	1	1	0	0	1	1	1	0
Q	1	1	0	1	0	0	0	0
R	1	1	0	1	0	0	1	1
S	1	0	1	0	0	1	0	0
T	1	0	1	0	0	1	1	1
U	1	0	1	0	1	0	0	0
V	1	0	1	0	1	0	1	1
W	1	0	1	0	1	1	0	1
X	1	0	1	0	1	1	1	0
Y	1	0	1	1	0	0	0	0
Z	1	0	1	1	0	0	1	1
.	1	1	1	1	0	1	1	1
!	1	1	0	1	0	1	1	0
,	1	1	1	0	1	1	1	0
?	1	0	1	0	0	0	1	0
±	1	0	0	1	0	1	1	1
+	1	1	1	1	0	0	0	0
—	1	1	0	0	0	0	0	1
*	1	0	1	0	0	0	0	1

CONTROL CHARACTERS
(Either shift)
(Either setting of S bit)

Backspace		1	0	1	1	1	0	
End of transfer		0	0	1	1	1	1	
Delete		1	1	1	1	1	1	
Down-shift		1	1	1	1	1	0	
Carriage return		1	0	1	1	0	1	
Prefix		0	1	1	1	1	1	
Idle		1	0	1	1	1	1	
Reader stop		0	0	1	1	0	1	
Space		0	0	0	0	0	0	
End of block		0	1	1	1	1	0	
Up-shift		0	0	1	1	1	0	
Line feed		0	1	1	1	0	1	
Tab		1	1	1	1	0	1	
Restore		1	0	1	1	0	0	
Bypass		0	1	1	1	0	0	
End of heading		0	0	1	0	1	1	
Punch on		0	0	1	1	0	0	
Punch off		1	1	1	1	0	0	

Fig. 2.3. Eight-bit BCD coding as on the IBM 1050 series.

30 INFORMATION CODING

and here 18 control characters are listed, which again must be recognizable in either shift. The number is often greater than this. Figure 2.8 lists 33 control characters.

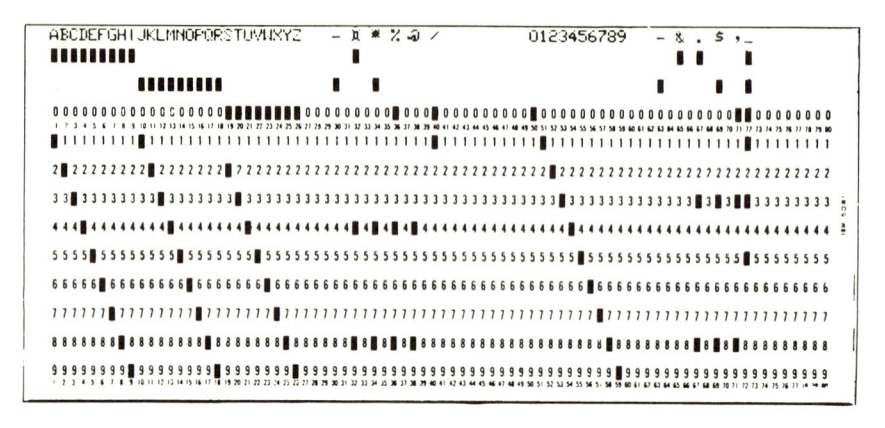

(A)

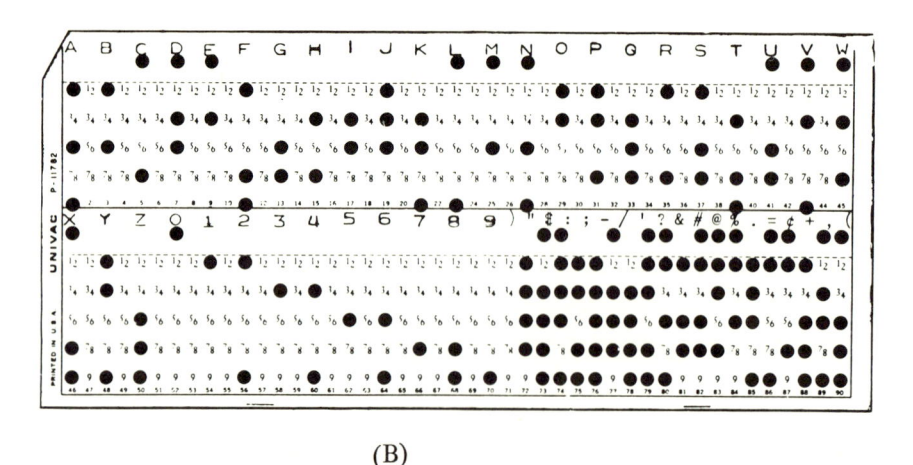

(B)

Fig. 2.4. Examples of coding on punched cards. (A) Hollerith coding on an IBM 80-column card. (B) Sperry-Rand 90-column card.

The above examples have used a special *character* for control. In some codes a special *bit* is used. This is the case in the Fieldata code of Fig. 2.2. In this the first six bits are normally data bits. In the standard code, however, the seventh bit is used for control purposes. If it is a 1, the first six bits *are* data bits. If it is a 0, then the whole character becomes our control character. Thus in Fig. 2.2 all of the characters except the last two in the list have a 1 as the seventh bit of the standard code (not the paper tape code). They are therefore data characters. The last two are control characters because their seventh bit is 0 (otherwise they would be "u" and "Idle," respectively). In the paper tape version of this code the convention is somewhat different. Here the characters are control characters if and only if the sixth and seventh bits are the same. Thus all of the paper tape code characters in Fig. 2.2, except the last two, have a different sixth and seventh bit. In the last two characters these are the same.

BINARY-CODED DECIMAL

In the early days of data processing the main medium for storing away data was punched cards, and a number of punched card codes became standardized. Two of these are shown in Fig. 2.4.

When the card is transmitted, it is desirable to make the transmitting machine as inexpensive as possible, and so a code has been used that is close to that on the card. Codes referred to as *binary-coded decimal* are used to represent in six or sometimes seven bits the data on the card.

The Hollerith card, a widely accepted standard, uses, in effect, twelve bits to code one character that could be coded with six bits. The coding is therefore condensed in producing the binary-coded decimal (BCD) form. This code commonly has six bits referred to as B, A, 8, 4, 2, 1. The zero overpunch in coding letters S to Z on the Hollerith card becomes an A bit; the 11 overpunch used in coding letters J to R becomes a B bit; the 12 over-punch of letters A to I becomes an A and B combination. Similarly, a 1 or 2 on the Hollerith card becomes a 1 or 2, respectively, in the BCD code. A 3 in the Hollerith card becomes a 1 and 2 combination. A 7 becomes a 1, 2, and 4 combination, and so on. Sixty-four possible card characters are thus coded into the six-bit BCD code.

	00	01	10	11
0000	5 i X	E r /	N t V	V Σ
0001	6 Δ	,	O " ;	W β Y
0010	— –	.	P | ;	: ;
0011	) 0	: ;	) ;	@ +
0100	½ 1	A	J	? /
0101	" 2	B	K	S
0110	# 3	C	L	T
0111	$ 4	D	M	U -
1000	% 5	E	N	V
1001	* 6	F	O	W
1010	& 7	G	P	X
1011	' 8	H	Q	Y
1100	(9	I	R	Z
1101	' .	# .	$ Z	% –
1110	& ;	C ¢ ;	* ;	T = ;
1111	(;	D @ ;	M ? 8	NOT USED

FIELDS

2		3
	1	
4		

HIGH-SPEED PRINTER

Normal: prints 1 except for characters with an entry in 4 which are non printing. **Computer Digit**: prints 1 for printing characters, 4 for non printing characters. The digit 8 prints in place of suppressed zeros.

SUPERVISORY CONTROL PRINTER

Normal: prints 1 except for characters with an entry in 3 which are non printing. If 3 is a ; then a ; is printed, if not, typewriter action occurs. 2 is upper case where different from lower. **Computer Digit**: prints 1 for printing characters, 3 for non printing.

UNITYPER

Code impulses designated by 1.

HIGH SPEED PRINTER ACTION

@ Fast Feed Symbol 1
| Fast Feed Symbol 2
? Fast Feed Symbol 3
= Fast Feed Symbol 4
r Multi Line Symbol

SUPERVISORY CONTROL PRINTER ACTION

$ Shiftlock
% Single Shift
? Unshift
r Carriage Return
t Tabulator

BOTH PRINTERS

| Printer Ignore
Δ Printer Space
β Breakpoint Stop
Σ Printer Stop

Fig. 2.5. BCD codes used by Sperry-Rand.

32 INFORMATION CODING

This may be seen by comparing the normal shift of the BCD code in Fig. 2.3 with the coding on the Hollerith card in Fig. 2.4. The BCD code in Fig. 2.3 is, in fact, an eight-bit extension of the above six-bit code. The eighth bit is an odd parity bit. The first decides which shift the transmission is in. The second shift gives an extra set of characters—fewer than 64, because some are used for control characters. The upper shift characters of Fig. 2.3 therefore have the same B, A, 8, 4, 2, 1 bits as the normal shift. Only the S bit (and hence the parity bit) is different.

It should be noted that one computer manufacturer's BCD code can be entirely different from others. The Sperry Rand code of Fig. 2.5 is entirely different from the IBM code of Fig. 2.3.

Bit positions 1,2,3,4

	0000	0001	0010	0011	0100	0101	0110	0111	1000	1001	1010	1011	1100	1101	1110	1111
0000	NUL				Blank	B	—						>	<	t	0
0001							/		a	j			A	J		1
0010									b	k	s		B	K	S	2
0011									c	l	t		C	L	T	3
0100	PF	RES	BYP	PN					d	m	u		D	M	U	4
0101	HT	NL	LF	RS					e	n	v		E	N	V	5
0110	LC	BS	EOB	UC					f	o	w		F	O	W	6
0111	DEL	IDL	PRE	EOT					g	p	x		G	P	X	7
1000									h	q	y		H	Q	Y	8
1001							/	"	i	r	z		I	R	Z	9
1010					?	!		:								
1011					-	S	,	#								
1100					←	*	%	@								
1101					(	)	ˇ	'								
1110					+	;	—	=								
1111					‡	¢	±	√								

Bit positions 5,6,7,8

Fig. 2.6. An extended BCD code with eight bits and plenty of space to spare for special characters. This code is used to transmit the eight-bit bytes of computers using this code, or to operate a printing device with up to 256 graphics.

Figure 2.6 shows an extension of the BCD code to use eight bits. This has plenty of space to spare for special characters. It is used to transmit the eight-bit bytes of computers such as the IBM 360, or to operate a printing device with up to 256 graphics.

N-OUT-OF-M CODES

It was mentioned above that a parity bit used for error detection, as in Figs. 2.2 and 2.3, is not as effective as one would like because transmission noise often changes more than one bit in a character. This noise can be counteracted in two ways: first, by using a separate error-checking pattern at the end of a record or message, as will be discussed in Chapter 4, and second, by coding self-checking characters in a more effective way than those above.

One way of coding characters to detect the loss or addition of a small group of bits is to use an N-out-of-M code. A number, M, of bits are used to transmit each character. N of these M bits will always be 1's [and $(M - N)$ will always be 0's] if the character is received correctly. A common example of this is a 4-out-of-8 code. Here eight bit characters are coded in such a way that four bits are always 1's and four bits are always 0's. If then any group of bits is lost from or added to a character, the error detection circuit will spot this (unless the character is completely wiped out). It is, of course, possible that one noise pulse could add a bit and another delete a bit.

The right-hand side of Fig. 2.7 illustrates the coding of characters using a 4-out-of-8 code. Seventy combinations are possible. This is far fewer than the 128 combinations of an eight-bit code with parity, and that is the price one pays for the added safety. As will be seen in Chapter 5, error-checking patterns can be devised which give both a lower level of redundancy and a higher level of protection than N-out-of-M codes.

Suppose that a data-processing machine that employs seven-bit BCD coding transmits using the 4-out-of-8 codes. This situation is illustrated in Fig. 2.7. The 4-out-of-8 code has six more characters than the BCD code. These are used for transmission control characters which govern the sending of data but which do not enter into the data processing.

The machines at both ends must translate from one code to the other. This is sometimes done by using a specially wired core plane so that characters read into it in one code can be read out in the other. In computers the code conversion may be done under program control, possibly by using special instructions wired or microprogrammed into the machine.

AMERICAN STANDARDS ASSOCIATION CODE

A profusion of different data transmission codes has been growing up. There are many others beside those discussed in this chapter. Individual machine designers often find their own reason for employing yet another code. A further transmission code, for example, will appear in the next chapter.

As the usage of data transmission grows, and particularly as more and more machines are used to dial-up other machines on the public network,

34 INFORMATION CODING

Character	BCD Code C B A 8 4 2 1	4-of-8 Code 1 2 4 8 R O X N
64 DATA CHARACTERS		
Space	C	2 4 8 O N
0	C 8 2	2 8 R N
1	1	1 O X N
2	2	2 O X N
3	C 2 1	1 2 R N
4	4	4 O X N
5	C 4 1	1 4 R N
6	C 4 2	2 4 R N
7	4 2 1	1 2 4 R N
8	8	8 O X N
9	C 8 1	1 8 R N
A	B A 1	1 R O X
B	B A 2	2 R O X
C	C B A 2 1	1 2 O X
D	B A 4	4 R O X
E	C B A 4 1	1 4 O X
F	C B A 4 2	2 4 O X
G	B A 4 2 1	1 2 4 N
H	B A 8	8 R O X
I	C B A 8 1	1 8 O X
J	C B 1	1 R X N
K	C B 2	2 R X N
L	B 2 1	1 2 X N
M	C B 4	4 R X N
N	B 4 1	1 4 X N
O	B 4 2	2 4 X N
P	C B 4 2 1	1 2 4 X
Q	C B 8	8 R X N
R	B 8 1	1 8 X N
S	C A 2	2 R O N
T	A 2 1	1 2 O N
U	C A 4	4 R O N
V	A 4 1	1 4 O N
W	A 4 2	2 4 O N
X	C A 4 2 1	1 2 4 O
Y	C A 8	8 R O N
Z	A 8 1	1 8 O N
/	C A 1	1 R O N
#	2 1	1 2 8 R
.	B A 8 2 1	1 2 8 N
$	C B 8 2 1	1 2 8 X
@	C A 8 2 1	1 2 8 O
☐	C 8 4	4 8 R N
•	C B A 8 4	4 8 O X
	B 8 4	4 8 X N
&	C B A	2 4 8 N
%	A 8 4	4 8 O N
—	B	2 4 8 X
⌐	C B A 8 2	2 8 O X
⌐	B 2	2 8 X N
RM	A 2	2 8 O N
GM	C B A 8 4 2 1	1 4 8 N
Delta	B 8 4 2 1	1 4 8 X
SM	A 8 4 2 1	1 4 8 O
WS	C A 8 4 1	2 8 R X
(	B A 8 4 1	2 4 R X
?	B A 8 4 2	2 4 R O
)	C B 8 4 1	4 8 R O X
:	C B 8 4 2	4 8 R O
,	C A 8 4 2	2 8 R O
"	8 4 2	2 4 8 R
;	8 4 1	1 2 4 8 R
TM	8 4 2 1	4 8 R

6 Control Characters:

Idle		1 8 R O
Inquiry/Error		1 8 R X
Transmit Leader		1 4 R O
Control Leader		1 4 R X
SOR1/ACK1 (Start of Record/Ack)		1 2 R X
SOR2/ACK2 (Start of Record/Ack)		1 2 R O

} These cannot be translated into the 6-bit BCD code

Dual Characters:

| TEL (Telephone) | 4 8 R   X |
| EOF (End of File) | 2   8 R   X |

Note:
    The 4 out of 8 code
    permits 70 combinations

    The 6-bit BCD code
    permits 64 combinations

Fig. 2.7. An example of coding in a 4-out-of-8 transmission code to transmit data coded in a six-bit BCD code. A specially wired core plane might do the translation.

so the need for standardization of transmission codes becomes very great. The American Standards Association, CCITT, and other national bodies have given much thought to this. The requirements of different users conflict considerably, so there are difficulties in agreeing upon standards.

The American Standards Association, however, has standardized the U.S. ASCII shown in Fig. 2.8. (ASCII stands for American Standard Code for Information Interchange.) This is a seven-bit code that permits double shift printing and has enough control characters for most purposes. An eighth bit can be used if desired as a parity check.

	000	100	010	110	001	101	011	111
0000	NUL	DLE	SPACE	0	@	P	`	p
1000	SOH	DC1	!	1	A	Q	a	q
0100	STX	DC2	"	2	B	R	b	r
1100	ETX	DC3	#	3	C	S	c	s
0010	EOT	DC4	$	4	D	T	d	t
1010	ENQ	NAK	%	5	E	U	e	u
0110	ACK	SYN	&	6	F	V	f	v
1110	BEL	ETB	'	7	G	W	g	w
0001	BS	CAN	(	8	H	X	h	x
1001	HT	EM	)	9	I	Y	i	y
0101	LF	SUB	*	:	J	Z	j	z
1101	VT	ESC	+	;	K	[	k	{
0011	FF	FS	,	<	L	\	l	|
1011	CR	GS	-	=	M	]	m	}
0111	SO	RS	.	>	N	^	n	~
1111	SI	US	/	?	O	_	o	DEL

Bit positions 5,6,7 (columns); Bit positions 1,2,3,4 (rows)

NUL = All zeros
SOH = Start of heading
STX = Start of text
ETX = End of text
EOT = End of transmission
ENQ = Enquiry
ACK = Acknowledgement
BEL = Bell or attention signal
BS = Back space
HT = Horizontal tabulation
LF = Line feed

VT = Vertical tabulation
FF = Form feed
CR = Carriage return
SO = Shift out
SI = Shift in
DLE = Data link escape
DC 1 = Device control 1
DC 2 = Device control 2
DC 3 = Device control 3
DC 4 = Device control 4
NAK = Negative acknowledgement

SYN = Synchronous/idle
ETB = End of transmitted block
CAN = Cancel (error in data)
EM = End of medium
SUB = Start of special sequence
ESC = Escape
FS = Information file separator
GS = Information group separator
RS = Information record separator
US = Information unit separator
DEL = Delete

Fig. 2.8. American Standard Code for Information Interchange (ASCII).

36 INFORMATION CODING

This code has come into wide acceptance in telegraphy and data transmission in the United States. Several of the major computer manufacturers are using it in their equipment, both in the United States and other countries.

The code in Fig. 2.8 is a seven-bit code. It would be of value to have also a six-bit and eight-bit standard code. The American Standards Association is working on this, but no such codes are accepted at the time of writing. Figure 2.9 shows a six-bit code used by IBM that contains a subset of the characters in the seven-bit ASCII code. IBM's "Binary Synchronous" range of transmission machines use the six-bit, seven-bit, and eight-bit codes of Figs. 2.9, 2.8, and 2.6.

Bit positions 1, 2

Bit positions 3, 4, 5, 6	00	01	10	11
0000	SOH	&	—	0
0001	A	J	/	1
0010	B	K	S	2
0011	C	L	T	3
0100	D	M	U	4
0101	E	N	V	5
0110	F	O	W	6
0111	G	P	X	7
1000	H	Q	Y	8
1001	I	R	Z	9
1010	STX	SPACE	ESC	SYN
1011		$	'	
1100	<	*	%	@
1101	BEL	US	ENQ	NAK
1110	SUB	EOT	ETX	EM
1111	ETB	DLE	HT	DEL

Fig. 2.9. A six-bit code used by IBM containing a subset of the characters in the seven-bit American standard code, ASCII. The control characters have the same meanings as those listed in Fig. 2.8.

PSEUDO-BAUDOT CHARACTERS

Many channels exist which are designed for five-bit telegraph transmission only. In fact, multimillion-dollar, highly reliable networks exist for five-bit transmission. ASCII or eight-bit codes are sent over these by splitting their characters into two. The first four bits of an eight-bit code have a 0 bit added, and the second four have a 1 bit added. These form two pseudo-Baudot five-bit characters, and can be transmitted over a Baudot network.

TRANSPARENT CODES

All of the above codes contain control characters. The transmission cannot take place without some of these. However, it is often desirable to transmit *all* of the six-, seven-, or eight-bit combinations that a computer or its peripheral devices can store. A conflict therefore arises here.

The conflict is resolved by using a pair of characters for the control character instead of one. For example, the DLE character (Data Link Escape) of the ASCII and similar codes may precede any control character, and this tells the receiving machine that the control character has its control meaning. The DLE character is regarded as not being part of the data. To transmit a DLE character and have the receiving machine accept it, it must itself be preceded by a DLE character.

This type of transmission is sometimes referred to as a *transparent code* or transmission in *transparent text mode*.

Some machines can switch backwards and forwards between transparent and normal text. Sequences of characters are needed to do this, such as the following:

DLE STX: Initiate transparent text mode.
DLE ETB: Terminate transparent transmission.
DLE ITB: Terminate transparent text mode but continue transmission in normal mode.

3 MODES OF TRANSMISSION

Over a given transmission line there is a variety of ways in which digital data can be transmitted; in other words, a variety of methods of organizing the signals sent so that they convey the information in question. For each different type of transmission mode there are families of input-output machines, computer transmission line adapters and so on, built to operate in this manner.

FULL DUPLEX VS. HALF DUPLEX   Over a given physical line, the terminal equipment may be designed so that it can either transmit in both directions at once—*full-duplex* transmission—or else so that it can transmit in either direction but not both at the same time—*half duplex*.

An input/output terminal or a computer line adapter works in somewhat different fashions depending upon which of these possibilities is used. Where full-duplex transmission is employed, it may be used either to send data streams in both directions at the same time or to send data in one direction and control signals in the other. The control signals govern the flow of data and are used for error control. Data at the transmitting end are held until the receiving end indicates that the data have been received correctly. If the data are not received correctly, the control signal indicates this, and the data are retransmitted. Control signals ensure that no two terminals transmit at once on a line with many terminals, and organize the sequence of transmission. This will be described in subsequent chapters.

Simultaneous transmission in two directions can be obtained on a two-wire line by using two separate frequency bands. One is used for transmission in one direction and the other for the opposite direction. By keeping the signals strictly separated in frequency, they can be prevented from interfering with each other.

The two bands may not be of the same bandwidth. A much larger channel capacity is needed for sending data than for sending the return signals that control the flow of data. If, therefore, data are to be sent in one direction only, the majority of the line bandwidth can be used for data. Figure 3.1 illustrates a possible division of line frequencies for this purpose.

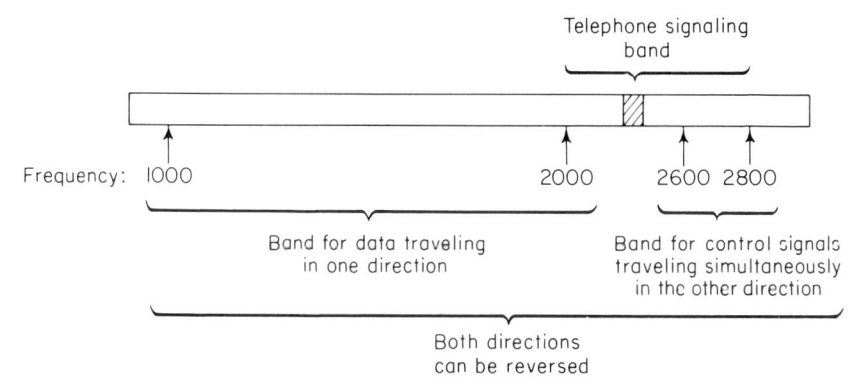

Fig. 3.1. The bandwidth of public telephone lines (Great Britain) split into a band for data traveling in one direction and a band for control signals traveling in the other.

The band used for data is much wider than that used for control signals traveling in the opposite direction. On a line such as this, the data and control signal directions may reverse together when information is to be sent in the opposite direction. The bands used in Fig. 3.1 avoid the line's signaling frequency. A more efficient, but probably more expensive, scheme would use the whole bandwidth, randomizing the data transmitted so that they would not interfere with the line's signaling system. One modem (not available commercially at the time of writing) permits transmission of data at 3600 bits per second in one direction and provides a simultaneous return path for control signals at 150 bits per second.

Many data-processing situations are not able to take advantage of the facility to transmit streams of data in both directions at the same time. Consequently, where full-duplex transmission is used, it is often with data traveling in one direction only, the other direction being used for control signals. Later chapters will illustrate this in detail.

PARALLEL VS. SERIAL TRANSMISSION

Digital data can be sent over communication lines either in a serial mode or a parallel mode. The stream of data is often divided into characters (as with the characters printed by a teleprinter). The characters are composed of bits. This stream may be sent either serial-by-character, serial-by-bit; or serial-by-character, parallel-by-bit.

Let us suppose that the characters are composed of six bits each. The serial-by-character, parallel-by-bit system must then transmit six bits at once. This is done on some terminals by using six separate communication paths, usually with a seventh for control purposes. It may, for example, require eight separate wires. This is not likely to be done on long-distance low-speed lines, as it would be more expensive than other means. A brief look at the line costs in the first chapter will indicate that several low-speed lines are more expensive than one higher-speed line. This form of parallel transmission is occasionally used between machines some miles apart when it is not possible to obtain a broadband link such as an AT&T series 8000 line.

Parallel wire transmission does, however, have the advantage that it can lower the terminal cost, as we will illustrate in Chapter 7. No circuitry is needed in the terminals for deciding which bits are which in a character. Parallel wire transmission is, therefore, commonly used over short distances where the wires are laid down by the user. For data collection terminals in a factory, for example, which connect to a computer or other machine within that factory, bunches of wires are often installed to connect the machines in a parallel fashion. The terms "in-plant" and "out-plant" system are used. The former means that the communications links are within one plant or localized area, and it implies that common carrier lines are not used. "Out-plant" implies that common carrier lines are used. The economics of what line connections are employed changes when the lines are laid down by the user.

Some machines are designed for parallel-by-bit transmission, but separate parallel wires are not used to connect them. Instead, the bits travel simultaneously, using different frequency bands on the same wire. This is a form of *frequency-division multiplexing*, which will be discussed in Chapter 11. One physical channel is split up into several effective channels, each operating on a different frequency band. As will be discussed in Chapter 11, these separate effective channels could carry signals from different machines.

MULTITONE TRANSMISSION

Another form of parallel transmission uses tones such as those generated by a push-button telephone. A Bell System touch-tone telephone keyboard can transmit eight possible audible frequencies: 697, 770, 852,

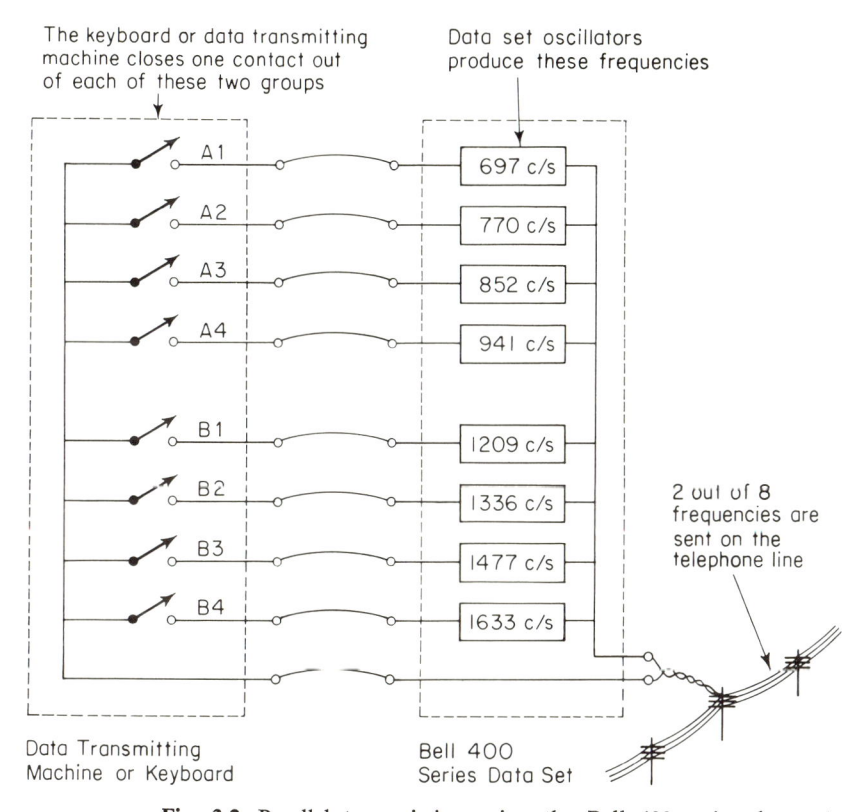

Fig. 3.2. Parallel transmission using the Bell 400 series data sets.

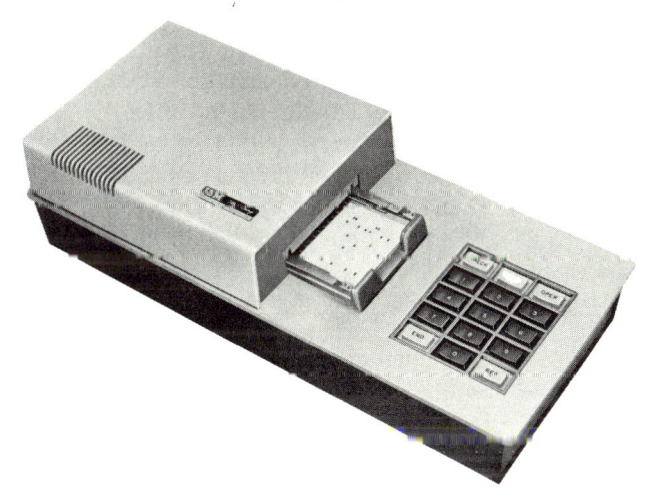

Fig. 3.3. Picture of IBM 1001.

42   MODES OF TRANSMISSION

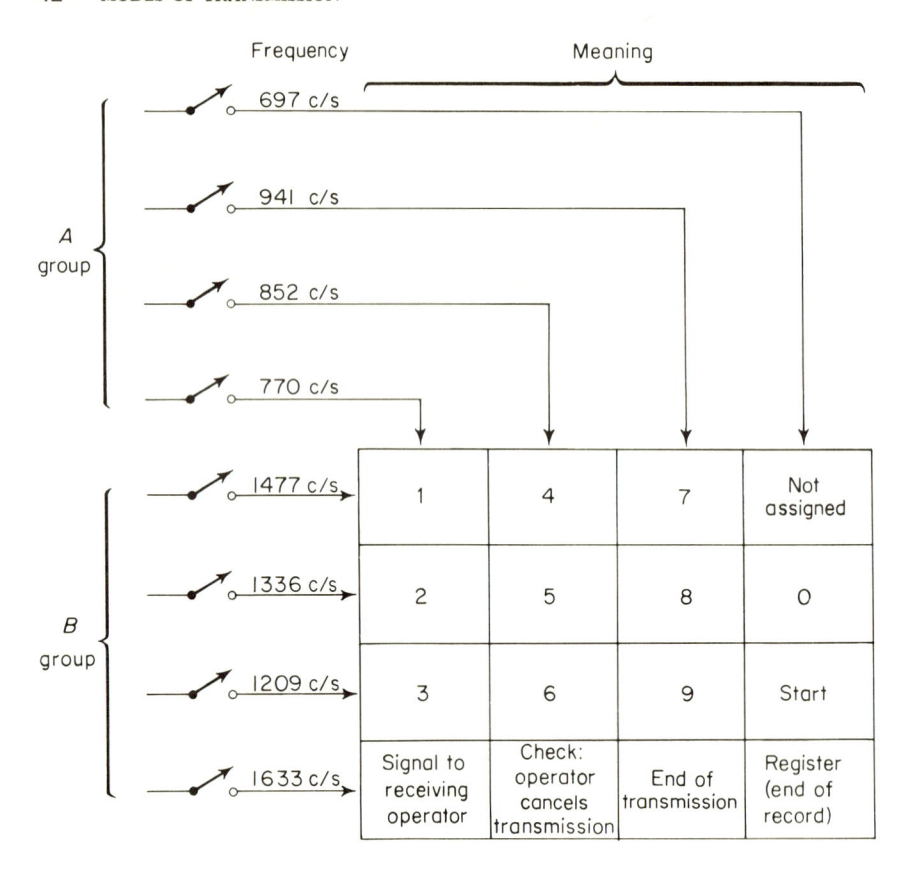

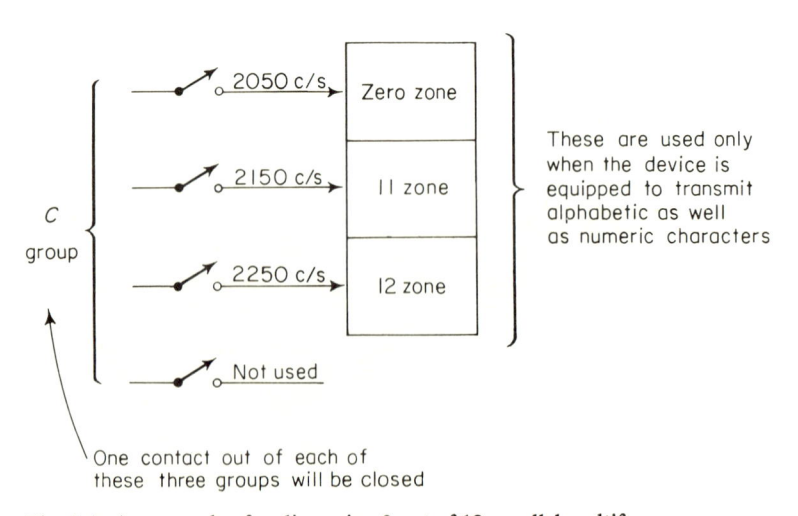

		Frequency	1	4	7	Not assigned
A group		697 c/s				
		941 c/s				
		852 c/s				
		770 c/s				
B group		1477 c/s	1	4	7	Not assigned
		1336 c/s	2	5	8	0
		1209 c/s	3	6	9	Start
		1633 c/s	Signal to receiving operator	Check: operator cancels transmission	End of transmission	Register (end of record)
C group		2050 c/s	Zero zone			
		2150 c/s	11 zone			These are used only when the device is equipped to transmit alphabetic as well as numeric characters
		2250 c/s	12 zone			
		Not used				

One contact out of each of these three groups will be closed

Fig. 3.4. An example of coding using 3 out of 12 parallel multifrequency transmissions.

941, 1209, 1336, 1477, and 1633 hertz (cycles per second). The pressing of any one key produces a discordant combination of two of these frequencies, one from the first four and one from the second. The Bell System 400 series data sets use the same eight frequencies plus others, and to these a data transmission device operating at 10 characters per second, or less, may be connected. This is illustrated in Fig. 3.2. The IBM 1001 shown in Fig. 3.3 operates in this manner. A code is used in which two frequencies out of the eight possible are transmitted at any one instant. This gives 16 possible combinations that can be transmitted. It gives some measure of transmission error detection, in that a fault causing only one, or more than two, frequencies to be received will be noted as an error, but this is not comprehensive error detection as in a 4-out-of-8 code.

This code does not have enough combinations for alphabetic transmission. When cards with alphanumeric punches are to be transmitted on the IBM 1001, three more frequencies are needed, one for each alphabetic "zone" punch. The coding remains the same as we have discussed, but now an extra (third) frequency can be sent. A frequency from the third (C) group can also be transmitted by itself. The way the characters are coded is shown in Fig. 3.4. The C-group frequencies are used when an alphabetic character is to be sent. The 12 "zone" combined with a digit gives the letters A to I, as on the punched cards. The 11 "zone" gives J to R, and the zero "zone", S to Z and a special character composed of a 0 and 1 punch.

SYNCHRONOUS VS. ASYNCHRONOUS TRANSMISSION

Data transmission can be either *synchronous* or *asynchronous*. Asynchronous transmission is often referred to as *start-stop*. With synchronous transmission, characters are sent in a continuous stream. A block of perhaps 100 characters or more may be sent at one time, and for the duration of that block the receiving terminal must be exactly in phase with the transmitting terminal. With asynchronous transmission one character is sent at a time. The character is initialized by a START signal, shown in Fig. 3.5 as a "0" condition on the line, and terminated by a STOP signal, here a "1" condition on the line. The pulses between these two give the bits of which the character is composed. Between characters the line is in a "1" condition. As a START bit switches it to "0," the receiving machine starts sampling the bits.

ASYNCHRONOUS (START-STOP) TRANSMISSION

Figure 3.5 shows the form in which a character is sent with start-stop transmission. The two most common types of character coding are illustrated, Baudot code (Fig. 2.1) and ASCII code (Fig. 2.8). Most non-American teleprinters transmit Baudot characters with five data bits plus the START and STOP elements, as shown at the

44 MODES OF TRANSMISSION

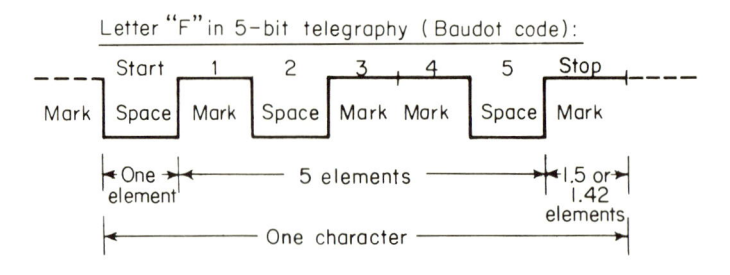

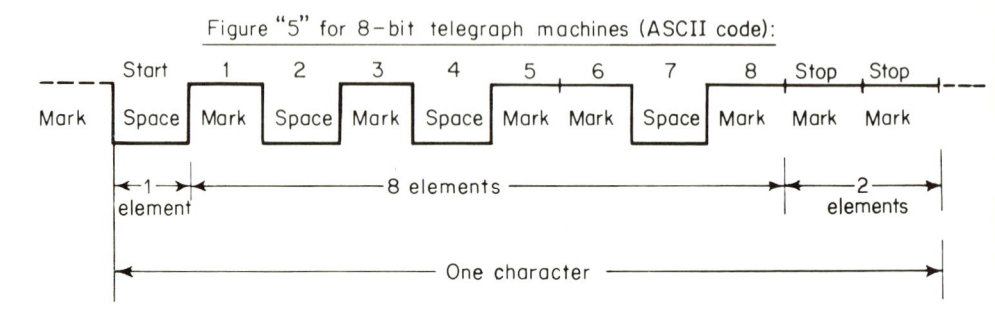

Letter "F" in 5-bit telegraphy (Baudot code):

[diagram showing start bit, 5 data bits, and stop bit with Mark/Space levels; labels: One element, 5 elements, 1.5 or 1.42 elements, One character]

Figure "5" for 8-bit telegraph machines (ASCII code):

[diagram showing start bit, 8 data bits, and 2 stop bits with Mark/Space levels; labels: 1 element, 8 elements, 2 elements, One character]

Fig. 3.5. Typical character structures for start-stop (asynchronous) transmission, used by machines such as those in Fig. 3.6.

top of Fig. 3.5. Most American teleprinters designed in recent years transmit ASCII characters, shown at the bottom of Fig. 3.5, with eight data bits (of which one is often unused) plus the START and STOP elements.

Start-stop transmission is usually used on keyboard devices which do not have a buffer, and on which the operator sends characters along the line at more or less random intervals as she happens to press the keys. The START pulse initiates the sampling, and thus there can be an indeterminate interval between the characters. Characters are transmitted when the operator's finger presses the keys. If the operator pauses for several seconds between one keystroke and the next, the line will remain in the "1" condition for this period of time.

Start-stop machines are generally less expensive to produce than synchronous machines, and for this reason many machines that transmit card-to-card or paper-tape-to-printer, card-to-computer, and so on, are also start-stop, although the character stream does not have the pauses between characters a keyboard transmission has. Figure 3.6 shows some typical start-stop machines.

The receiving machine has, in essence, a clocking device that starts when the START element is detected and operates for as many bits as there are in a character. With this, the receiving machine can distinguish which bit is which. The STOP element was made 1.5 or so times longer than the data bits in case the receiver clock was not operating at quite the same speed as the transmitter.

When this start-stop transmission is used, there can be an indeterminate period between one character and the next. When one character ends, the receiving device waits idly for the start of the next character. The transmitter and the receiver are then exactly in phase, and remain in phase while the character is sent. The receiver thus is able to attach the correct meaning to each bit it receives.

When an automatic machine such as a paper tape reader is sending START/STOP signals, the length of the STOP condition is governed by the sending machine. It is short, always 1.42 (1.5 or 2) times the other bits, so as to obtain the maximum transmission rate. When a typist uses the keyboard of a start-stop machine, on the other hand, the duration between her keystrokes varies. The transmission occurs when she presses each key, so the stop condition varies in length considerably. When a scientist uses a teleprinter on a time-sharing system, he may be doing work that involves a great amount of thinking. Between one character and another there may occasionally be a very long pause while he thinks or makes notes. The STOP "bit" will last for the duration of this.

Figure 3.7 illustrates the principle of operation of mechanical start-stop instruments. The sending device and the receiving device each have an armature A, which can rotate at a constant speed when a clutch connects it to an electric motor in these machines. Brush contacts on the armatures connect an outer ring of contacts B to an inner ring C. In Fig. 3.7, the armature is in its "stop" position. Consequently, current from the battery in the sender flows via the outer contact labeled STOP through the armature and onto the line. This current causes the contact of a receiver relay D to be in the position shown, and thus no current from the battery in the receiver flows to any of the operating magnets shown.

Now suppose that a girl operating the sending device presses the **H** key on the keyboard. In accordance with the code in Fig. 2.1 the **3** and **5** data contacts now close. Also the **START** contact closes, and the armature of the sender begins to rotate counterclockwise. The armature connects the contacts on the outer ring to the line in sequence: **START, 1, 2, 3, 4, 5, STOP**. The **3, 5,** and **STOP** contacts convey current; the others do not. As soon as the armature of the sender travels to its **START** contact, the positive currents on the line cease. The contact of the relay D flips across, and positive current from the battery of the receiver flows through the inner ring C, armature A, and outer ring B, to the start magnet E. The start magnet causes the armature of the receiver to rotate counterclockwise at the same speed as that of the receiver.

46   MODES OF TRANSMISSION

A typical Telex installation. The user has a paper tape reader, punch, a keyboard, and printer, and may dial other Telex subscribers throughout the world. The machines use Baudot code at 50 bits per second (66 words per minute). *Courtesy G.P.O., London*

A typical TWX installation, similar to the Telex installation above, but operating in North America (not worldwide like Telex) at speeds up to 150 bits per second. The U.S. ASCII code is used. This Teletype 35 Automatic Send Receive machine is also used on leased lines. *Photograph courtesy A.T. & T.*

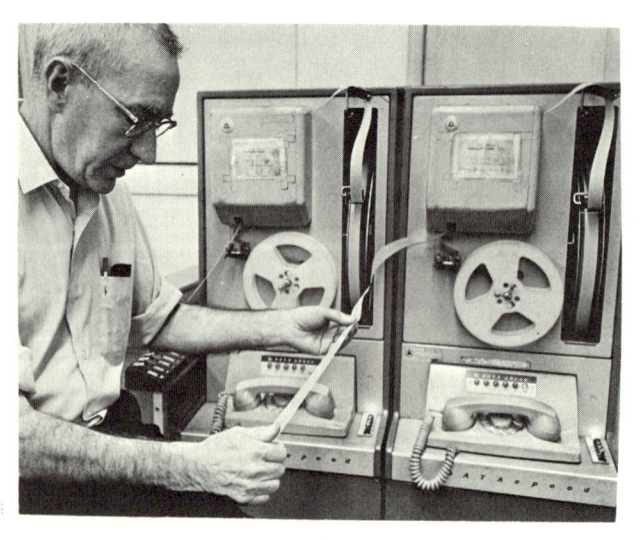

A teletype DATASPEED paper-tape transmitter. This transmits over dial-up or private voice lines at 1050 words per minute, to DATASPEED paper-tape receivers or printers. *Photograph courtesy A.T. & T.*

Fig. 3.6. Typical start-stop machines.

MODES OF TRANSMISSION 47

Typical terminals on a time-sharing system—IBM 2740's operating at 14.8 bits per second on a dial-up or leased subvoice grade lines. Sometimes a toll telephone line is used because of the convenience of the public telephone network. BCD coding is used. *Photograph courtesy IBM.*

Special-purpose terminals like the American Airlines SABRE reservations set are often geared to electronic typewriter speed. It transmits BCD code asynchronously over a subvoice grade line (e.g., Bell Type 1006) to a concentrator (Fig. 12.2) which sends the data onwards in a synchronous fashion. *Photograph courtesy American Airlines.*

Fig. 3.6.—continued.

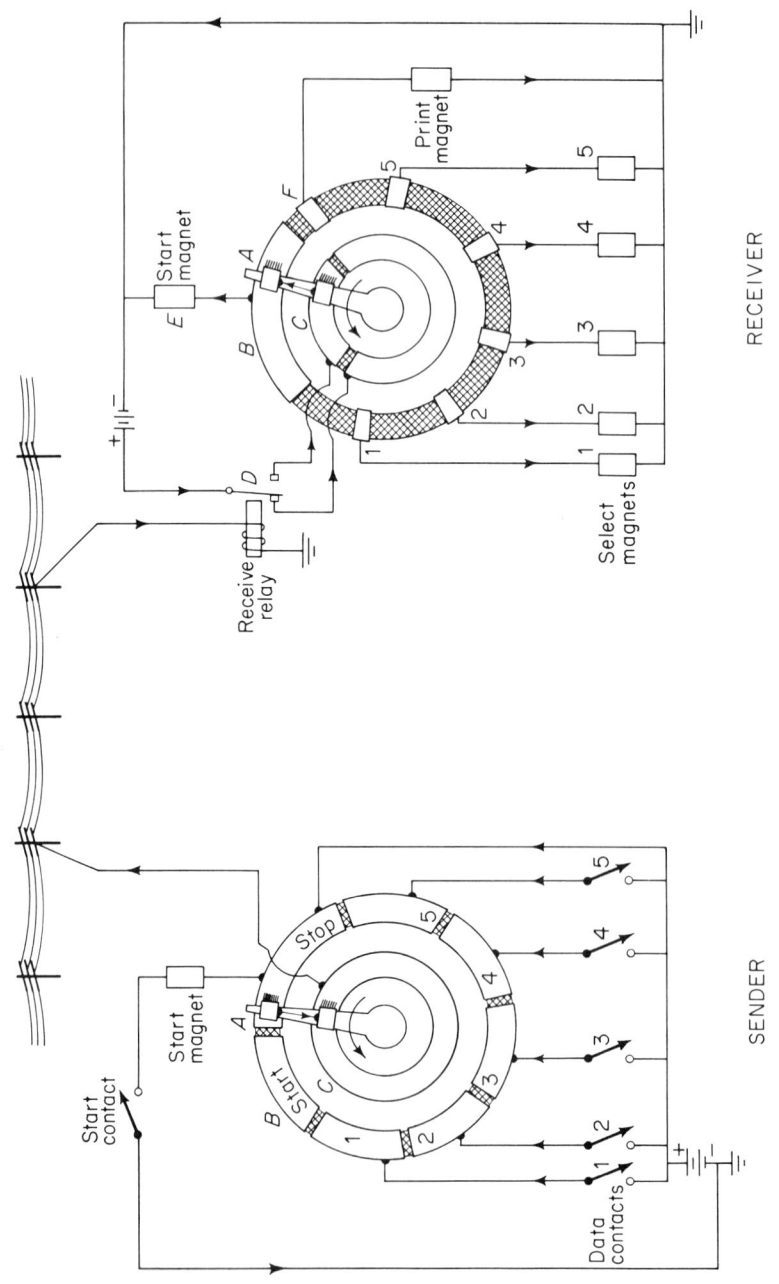

Fig. 3.7. The principle of start-stop telegraph equipment.

When the receiver armature is over contact **3**, current again flows on the line. Receiver relay *D* again moves to the position shown in the figure, and so as the armature of the receiver passes over the **3** contact on the outer ring, current flows to the select magnet **3**. Similarly, select magnet **5** is operated. These cause the letter **H** to be selected on the type mechanism of the receiver. When the sender armature reaches its **STOP** contact, the receiver armature passes over the contact **F,** and current flows to the print magnet. The letter **H** is printed. Both armatures then come to rest in the position shown in the figure, unless the sender is ready to transmit another character immediately.

There are several variations of this basic operation principle, but the speeds and timing have become standardized. This means that a variety of different telegraphy machines can communicate with one another. A computer can send data to a telegraph machine by sending pulses with the same timing as the sender in Fig. 3.7.

Teletype speeds in common use are listed in Table 3.1. Such speeds are often quoted in "words per minute." An average teletype word is considered to be five characters long. As there is a space character between words, there are, then, six characters per word, and x words per minute = $x/10$ characters per second.

Table 3.1

Speed in Bauds (bits per second)	Number of Bits in Character	Stop Bit Duration (in bits)	Information Bit Duration	Characters per Second	Words per Minute (nominal)
45.5	7.42	1.42	21.97	6.13	60
50	7.42	1.42	20	6.74	66
50	7.50	1.50	20	6.67	66
74.2	7.42	1.42	13.48	10	100
75	7.50	1.50	13.33	10	100
75	10	1.00	13.33	7.5	75
75	11	2.00	13.33	6.82	68
150	10	1.00	6.67	15	150

A considerable amount of distortion can occur in the length and positioning of the start-stop pulses without the receiver's misinterpreting them. As will be seen in Fig. 3.7, the contacts **1, 2, 3, 4, 5,** and **F** on the receiver are much narrower than their counterparts on the sender. The sender's pulse can therefore become shortened or lengthened, or it can be late or early, by an amount somewhat less than half the pulse width, and the receiver will still interpret it correctly. Of course, if it is delayed by one pulse width, then it will be wrongly interpreted. However, the margin for distortion that does exist is useful because, as we will see later, pulses do become shortened, lengthened, or delayed slightly if the equipment on the line is not in perfect adjustment.

SYNCHRONOUS TRANSMISSION When machines transmit to each other continuously, with regular timing, *synchronous* transmission can give the most efficient line utilization. Here the bits of one character are followed immediately by those of the next. Between characters there are no START or STOP bits and no pauses. The stream of characters of this type is divided into blocks. All the bits in the block are transmitted at equal time intervals. The transmitting and receiving machines must be exactly in synchronization for the duration of the block, so that if the receiving machine knows which is the first bit it will be able to tell which are the bits of each character (or words).

Devices using synchronous transmission have a wide variety of block lengths. The block size may vary from a few characters to many hundreds of characters. Often it relates to the physical nature of the data medium. For example, in the transmission of punched cards it is convenient to use 80 characters as the maximum block length, as there are that many characters per card. Similarly, the length of print lines, the size of buffers, the number of characters in records, or some such system consideration may determine the block size. Some time is taken up between the transmission of one block and the next, so the larger the block length, in general, the faster the overall transmission. Figure 3.8 on pages 52 and 53 shows some typical machines that use synchronous transmission.

With asynchronous transmission, the unit of transmission is normally the character. The operator of a teletype machine presses a key on her keyboard and one character is sent, complete with its START and STOP bits. It is independent in time of any other character. With synchronous transmission the characters are stored until a complete block is ready to be sent. The block is sent from a buffer at the maximum speed of the line and its modems. There are no gaps between characters as there are when a teletype operator taps at her keyboard. Synchronous transmission is therefore of value when one communication line has several different terminals operating on it. In order to permit synchronous transmission, however, terminals must have buffers, and hence are more expensive than asynchronous devices.

On many systems the synchronization of the transmitting and receiving machines is controlled by oscillators. Before a block is sent, the oscillator of the receiving machine must be brought exactly into phase with the oscillator of the transmitting machine. This is done by sending a synchronization pattern or character at the start of the block. If this were not done the receiving device would not be able to tell which bit received was the first bit in a character, which the second, and so on. Once the oscillators at each end are synchronized, they will remain so until the end of the block. Oscillators do, however, drift apart very slightly in frequency. This drift is very low if highly stable oscillators are used, but with those low enough in price to be used in quantity in input-output units it is significant. Oscillators in common use in these machines are likely to be accurate to about one part in 100,000. If they are sampling the transmission 2500 times per second, say, then they are likely to stay in synchronization for a time of the order of 20 seconds. Most data-processing machines resynchronize their oscillators every one or two seconds for safety. Synchronization can also be maintained by "framing" blocks and carrying timing information in the frames.

On some systems, this places an upper bound on the block length, though this is not always so because resynchronization characters may be sent in the middle of a block. The IBM range of "binary synchronous" equipment, for example, inserts two synchronization characters into the text at one-second intervals. In the U.S. ASCII code with parity checking (Fig. 2.8) these would be coded 01101000 01101000. The receiving station is constantly looking for these and ensures that the transmitter and receiver are in step.

In addition to giving faster transmission because no START and STOP bits are needed between characters, synchronous transmission permits higher-speed modulation techniques to be used. Multilevel signaling can be used (Fig. 1.9), whereas this is not normally used in start-stop transmission. The signal can be randomized to prevent repetitive codes' causing harmful interference with other telephone users, and hence a higher transmission level can be tolerated, giving a better signal-to-noise ratio. (Many synchronous data sets do not do this as yet.) On high-speed circuits, synchronous transmission can tolerate a higher degree of jitter and distortion than start-stop transmission, and for this reason it is usually used.

Synchronous transmission can give better protection from errors. At the end of each block an error-checking pattern is transmitted. The coding of this pattern is selected to give the maximum protection from noise errors on the line. As well as the error code at the end of the block, each character may also have parity bit for checking (Fig. 3.5, bottom). This, however, is often not done, and an end-of-block check is used alone. As discussed earlier, a parity bit is not too useful as a protection against communication line errors. It is an extremely useful protection against loss of bits in the core of a computer, because there it is likely that only one bit will be lost at a time. On a communication line, however, several bits are often lost at once because of a noise impulse or dropout. Where two, four, or six bits are changed in a character, a parity check will not detect this.

52   MODES OF TRANSMISSION

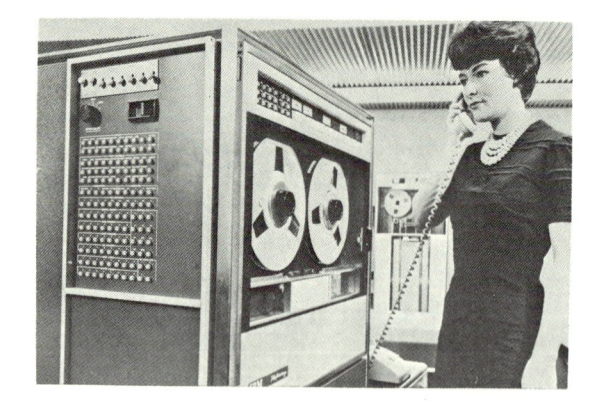

The IBM 7702 Magnetic Tape Transmission Terminal transmits the contents of magnetic tapes to another 7702 or other type of compatible device. Synchronous Transmission over a voice line is used at speeds up to 300 characters per second.

The Ultronic Time Share Multiplex Unit which combines the transmissions on many low-speed lines and sends them together, synchronously, over a voice line. Multiplexing is discussed in Chapter 11.

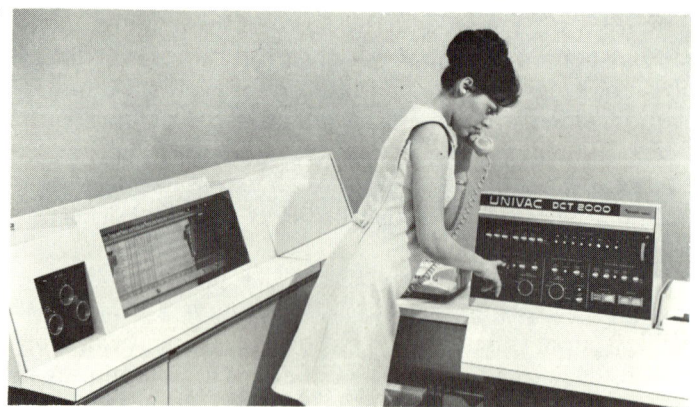

The Univac DCT 200 Data Communication Terminal consists of a 250-line-per-minute printer, a 200-card-per-minute card reader, a 75-card-per-minute punch and their control unit. These transmit or receive synchronously over a voice line.

Fig. 3.8. **Typical synchronous machines.**

MODES OF TRANSMISSION

The IBM 2770 data communications system. Many different devices can be attached to the control unit, including punched card machines, paper tape machines, a visual display screen, a manual keyboard, magnetic tape cartridges, a magnetic character document reader, and a medium speed printer. Transmission to and from these is synchronous over a voice line.

The Sanders 620 stand-alone data display system is designed for transmission either synchronously or in a start-stop fashion like the machines in Fig. 3.6.

Fig. 3.8.—continued.

Some form of block check is therefore desirable that can detect the loss of several consecutive bits. The faster the transmission, the more likely is the loss of more than one adjacent bit, and thus the block check becomes more important with faster transmission. Start-stop transmission may also have a block check.

BLOCK STRUCTURE

A block of bits sent by synchronous transmission must have certain features. It must, for example, start with the *synchronization pattern* or character. It will normally end with an *error-checking pattern* or character. The block length, as with other data records used by computers, may be of fixed length or variable length. It is often the latter, as this usually allows better line utilization. On most systems it would be necessary to pad many blocks with blank characters if fixed-length blocks were used. If the block is of variable length, an end-of-block pattern must be used to tell the receiving machine to begin the actions needed when a block ends. This pattern will normally be sent immediately prior to the error-checking pattern.

Often data are sent in the form of characters, or groups of (usually) six, seven, or eight bits. The above patterns can be one, two, or more characters. One transmission scheme, for example, uses six-bit characters. These are transmitted without parity checking, so the whole block is divided up into groups of six bits. The block must start with the following characters: 111111, 111110 in that sequence. This constitutes the synchronization pattern. A circuit in the receiving machine spends its life scanning the input for this pattern. When it finds it, then the receiving device knows that next bit it receives is the first data bit. The synchronization pattern is unique. The coding of characters must be such that it could not occur anywhere else in the transmission.

The block ends with a six-bit error-checking pattern (one character) and immediately preceding that is the end-of-message character. When the text is being transmitted, the receiving device is generating its own error-checking pattern, computed from the characters received. At the same time it is examining each character received to see whether it is the end-of-message character. When this is received the machine knows that the next character is to be the transmitted error-checking pattern, and so it compares that with the pattern it has generated itself. If there is a difference, the receiving machine sends a message to the transmitting machine to demand a retransmission of that message.

Figure 3.9 shows the format of a block of text transmitted in this manner. It is designed for a line to which many input-output machines are attached. These are arranged in groups, and each group is connected to a control unit, which itself is connected to the line transmitting data to and from the computer. After the synchronization pattern in each block comes the address of the control unit (one character) and the address of the input-output machine (one character) to which the message is going or from which the message has come. It is possible that messages transmitted to the computer may be longer than the maximum length of a block. In this case

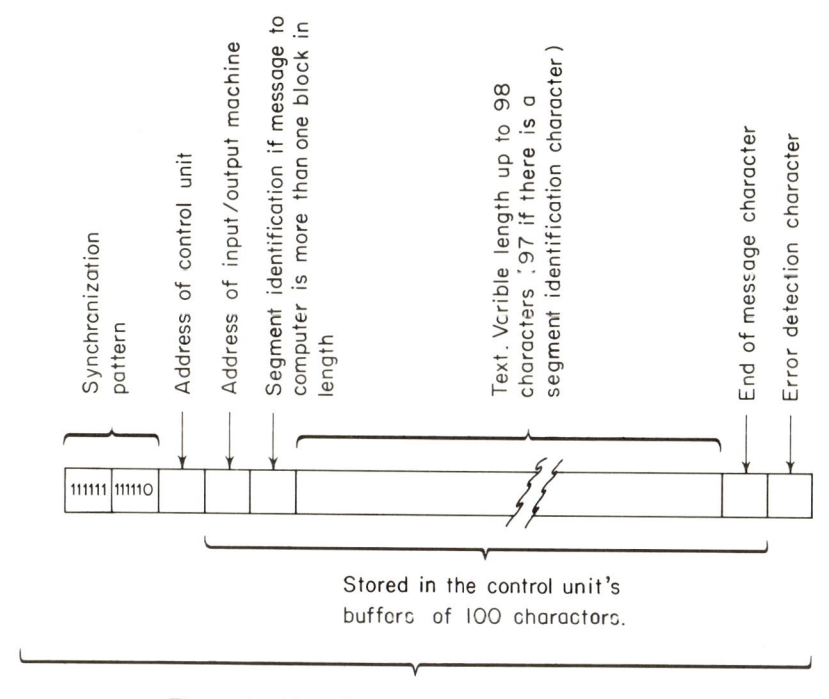

[Figure: Block format diagram with labeled fields — Synchronization pattern (111111 111110), Address of control unit, Address of input/output machine, Segment identification if message to computer is more than one block in length, Text. Variable length up to 98 characters; 97 if there is a segment identification character, End of message character, Error detection character. Stored in the control unit's buffers of 100 characters. The entire block is composed of six-bit characters.]

Fig. 3.9. Typical format of a block of data for synchronous transmission on a heavily loaded line with many terminals where efficient line loading is important.

they are divided into as many blocks as are necessary, and to link them a character is used as a segment identifier. The control unit places this, if it is needed, in the block immediately before the text. The text itself is again in six-bit characters and can be of any length up to 98 characters. This maximum is imposed by the size of the buffers, 100 characters, used in the control units.

There are many variations of this type of format. Sometimes one character is designated as the "synchronization character," and a stream of these characters is sent continuously between messages when the line would otherwise be idle. At least two such characters are necessary prior to a message to establish synchronization.

Figure 3.10 shows a synchronous block structure composed of eight-bit characters. The error-checking facilities in this example are better than in the former, more redundancy being built into the block. First, each

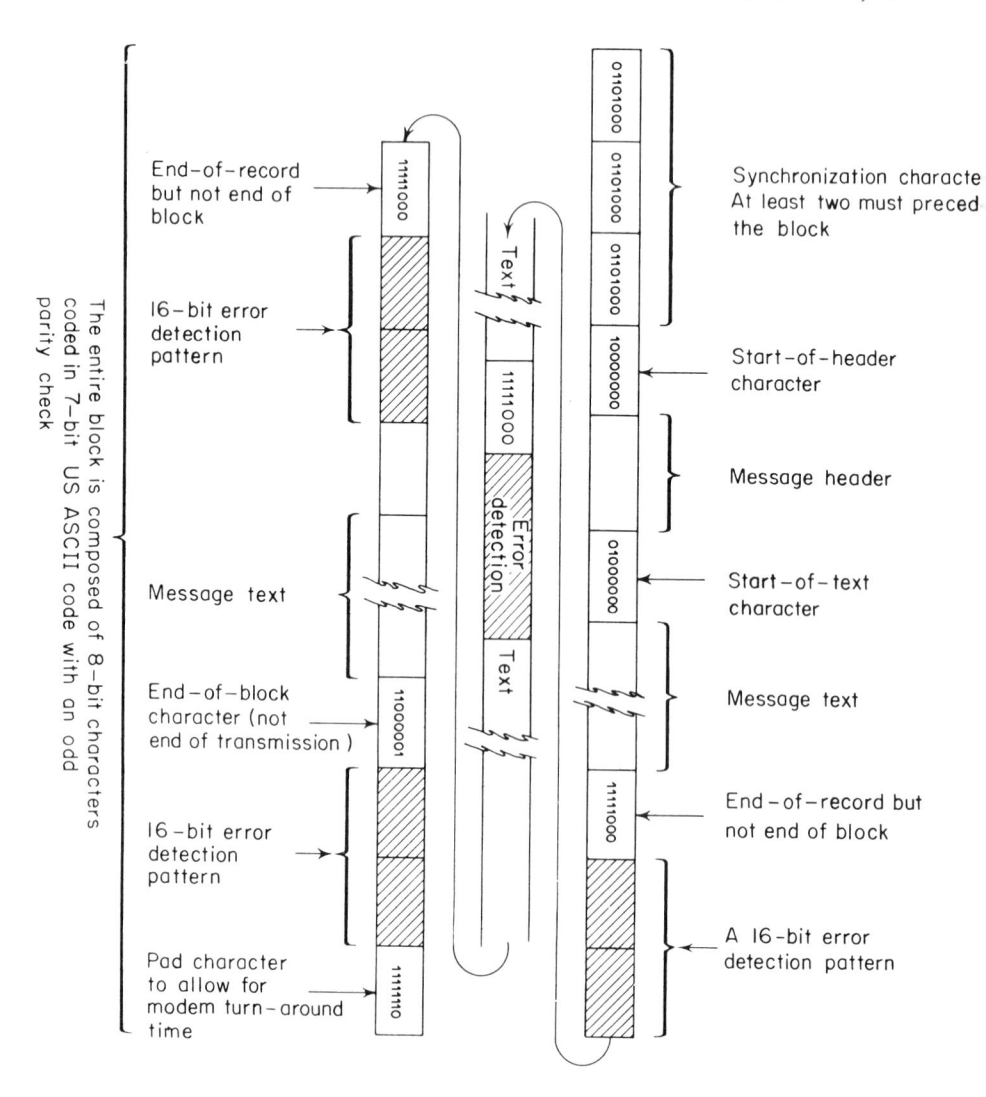

Fig. 3.10. Typical format of a block of data for synchronous transmission in which several records may be included in one block, and good error-detection coding is needed.

character has its own parity check. The characters are seven-bit U.S. ASCII code with a parity check bit added. Second, *two* error-checking characters are used at the end of each record, giving a 16-bit error-checking bit as opposed to six bits in the former example. Third, this pattern is used at the end of each record within the block, not merely at the end of the block.

The block in this example is subdivided into separate records and a header. These are all of variable length, so special characters must be inserted to say where these items start and finish. In this illustration *start-of-header*, *start-of-text*, *end-of-record*, and *end-of-block* characters are used. The end of block is not necessarily the end of transmission. The sending device may have more data yet to come, and so an *end-of-transmission* character is also needed in the repertoire. This could have replaced the *end-of-block* character here.

The last character in Fig. 3.10 is a *pad* character. This does nothing. It is not recognized by the receiving machine. Its purpose is to give a pause, a brief moment in which the modems can reverse their direction of transmission.

SYNCHRONIZATION REQUIREMENTS

There are basically three levels of synchronization needed in digital data transmission. These are needed regardless of the transmission mode. They are *bit synchronization*, *character synchronization*, and *message synchronization*.

Bit synchronization ensures that the receiving machine knows at what instant a bit starts and ends. The receiving machine must "sample" the bit at its center, not during a period of transition. *Character synchronization* ensures that the receiving machine knows which bit is which in a character. Without this synchronization the receiving machine might think the second bit of a character transmitted was in fact the beginning of the character, for example, so the characters would be misinterpreted. *Message synchronization* is needed to ensure that the receiving device knows which characters are the starting and ending characters of records or messages.

Bit synchronization may be all taken care of in the modem. Some modems are described as being "self-clocked" and automatically establish the "bit phase." The modem tells the data-processing machine when to sample the bit. This is not always the case, however. Some modems merely send the data-processing machine a pulse train, and this machine must decide when to sample the pulses. In this case a synchronization pattern is needed to establish the bit phase as well as the character synchronization. Figure 3.10 shows a minimum of two synchronization characters being needed for character synchronization; on some systems as many as four such characters are needed before the two for character synchronization.

This is, in effect, saying that three types of timing information are needed by the receiving machine: timing information telling it the exact position of a bit, information as to which is the first bit of a character, and information as to what is the start and end of a message, giving a total of six synchronization characters needed. In other systems a separate bit synchronization pattern is sent; for example, in one, 16 consecutive bits are needed consisting of alternating 0's and 1's (0101010101010101).

The character synchronization pattern generally needs to be sent along with the bit synchronization pattern, if one is needed, at the start of each new message or data block.

One alternative to this type of technique is to use a third signal level in addition to the 1 and 0 levels conventionally used. This third level would carry the timing information needed for synchronization and so alleviate need for synchronization characters. Extra bandwidth is needed for this, however, and it does not permit the maximum amount of data to be sent over a given bandwidth. Where *parallel* data transmission is used, one of the paths is sometimes used exclusively to carry synchronization information.

With start-stop information each character in effect carries its own synchronization information. When the transition to the start bit is detected, this causes the receiving machine to sample the data bits at the correct intervals afterward. The starting and stopping gives a form of transmission less resiliant to noise and jitter than synchronous transmission. This is increasingly so at higher speeds. Noise pulses, for example, can cause false starts. Some start-stop schemes of greater resilience have been built, for example, using more complex start patterns, or making the stop pulse time an integral multiple of the other bit times, and then using an oscillator in the receiving machine as in a synchronous system. This loses the advantage of simplicity and low cost that start-stop machines have without fully gaining the speed advantage of synchronous transmission.

Message synchronization is needed on start-stop or synchronous devices, and is likely to take the same form with either of these. The *start-of-header, start-of-text, end-of-record, end-of-block*, and *end-of-transmission* characters discussed above, or their equivalents, may frame asynchronous as well as synchronous messages. The use of these will be discussed further in the chapters on line control.

HIGH-SPEED PULSE STREAM

A technique that might be regarded as a hybrid between synchronous and asynchronous transmission is one that uses a high-speed pulse stream traveling in a constant and unbroken fashion, to carry data in which the characters originate at random. Such a scheme is used in some in-plant systems. It has the advantage that both the terminals and the in-plant communication links can be relatively inexpensive.

Binary pulses can be carried at a very high rate over simple pairs of copper wires for a distance of a few thousand feet. If sent further than this, they become too distorted to be recognizable. Small solid state repeaters placed in the line at intervals, perhaps as close as 1000 feet, can reconstruct the pulse train, however. The repeater catches a pulse before it becomes too distorted, recognizes whether it is a 1 or 0, and retransmits a clean square-edged pulse down the wire pair. In this way a single pair of wires laid down in a factory, say, can carry half a million binary pulses per second.

The wires may be arranged in a *loop* that originates at a computer or controlling machine, wanders around various data-handling machines, and then returns to the computer. It may carry a train of small fixed-length blocks that convey characters to or from the terminal devices. Each block is able to carry one character. If a terminal at a particular instant has a character to transmit, its controller waits for an empty block, then places the character in it along with the terminal address. Similarly, it receives characters being sent *to* the devices.

The terminals can be very simple, as they transmit one character at a time, and the mass of wiring that is usually necessary on in-plant systems with many terminals is avoided.

RELATIVE ADVANTAGES AND DISADVANTAGES   The advantages and disadvantages of the various transmission modes discussed are listed in Table 3.2 on pages 60 and 61.

Table 3.2. Advantages and Disadvantages of The Various Transmission Modes

Mode of Transmission	Advantages and Disadvantages
Four-wire	Permits full-duplex transmission.
Two-wire	Full-duplex transmission still possible with separate frequency bands for the two directions.
Simplex	Rarely used for data transmission, as there is no return path for control, or error signals.
Half duplex (Data in both directions at once)	Commonly used for data transmission, though a full-duplex line may cost little more. (Often 10% more in the United States.) System sometimes cannot take advantage of this, as data cannot be made available for transmitting in both directions simultaneously. Can substantially reduce the response time, however, on a conversational multidrop line. Often requires a more expensive terminal. Commonly used on a link between concentrator and computer.
Full duplex (Data in one direction; control information in the other)	A common arrangement, though, as data are still only being sent in one direction at a time, half-duplex transmission may give better value for money at low character rates. With high character rates the line turnaround time may be long compared to the character time and full-duplex operation may eliminate most turnaround delay.

Mode of Transmission		Advantages and Disadvantages
Serial-by-character Parallel-by-bit	Separate wires	Low transmitter cost, but high line cost. Economical for in-plant use. Line costs too expensive for long distances.
	Separate frequencies	Used on voice lines to give a slow but inexpensive terminal. For efficient line utilization, however, data set costs are high, and receiver cost can be high.
Serial-by-character Serial-by-bit		The most common system, especially on long lines.
	Start-stop transmission	Inexpensive terminal, e.g., telegraph machines. Only one character lost if synchronization fails. Not too resilient to distortion at high speeds.
	Synchronous transmission	More expensive terminal. Block lost if synchronization fails. Efficient line utilization. High ratio of data to control bits. More resilient to noise and jitter than start-stop transmission, especially at high transmission speeds. The most common system on lines of 600 bits per second and faster.
	High-speed pulse train	In-plant or private wiring only at present. Low wiring cost with low terminal cost. High accuracy.

4 ERRORS AND THEIR TREATMENT

On many data transmission systems the control of errors is of vital importance. On some, however, it is not of great significance. The number of bits that are incorrect after data transmission (with no error correction) is typically in the range of one in 50,000 to one in 500,000. (Short-wave or long-wave radio links are much worse than this.) A variety of error correction procedures are in use, but most of those in common use still leave a number of undetected errors that is greater than the number that would be expected on other components of a computer system. Tape and file channels on a computer, for example, have an error rate very much lower than those on the telecommunication links in conventional use today.

In addition to transmission errors, the errors made by operators constitute a significant problem in the design of on-line systems. The number of errors made by the operators of the input/output devices on a large system usually far exceeds the number of errors caused by noise or distortion on the transmission lines. It is usually important that accuracy controls be devised for the human input. On many systems a tight network of controls is necessary to stop abuse or embezzlement. It is also important to ensure that nothing is lost or double-entered when hardware failures occur on the system or when switchover takes place. These systems design considerations are beyond the scope of this book, and in this chapter we will discuss the treatment of errors arising from noise and distortion on the communication lines.

NUMBERS OF LINE ERRORS The number and pattern of errors that can be expected on various types of communication lines is discussed in Chapter 21 of *Telecommunications and the Computer* (James Martin, Prentice-Hall, Inc., 1969). It is useful for the systems analyst to have a rule of thumb for estimating the numbers of errors that he is likely to encounter. He will use these figures for doing probability calculations that indicate to him the viability of different systems approaches. With them he may answer such questions as what degree of error checking is needed and what block length he should use for transmission.

The figures in Table 4.1 are typical average error rates, and might be used as the rule of thumb if the systems analyst has no more precise figures relating to his particular lines. These rule-of-thumb figures are suggested

Table 4.1

Type of Channel	Transmission Rate (bits per second)	Bit Error Rate
50-baud Telex	50	1 in 50,000
150 or 200 band subvoice grade lines	150 or 200	1 in 100,000
Public voice lines	600	1 in 500,000
	1200	1 in 200,000
	2400	1 in 100,000

by surveying many reports published by the International Telecommunications Union and individual common carriers. Lines of the same type in most countries appear to have about the same error rate when they are working correctly with well-adjusted data sets. Certain individual lines, however, have a large amount of noise on them that considerably exceeds the above figures. This is usually caused by an old plant, ill-adjusted equipment, or workmen on the lines.

It is likely that transmission over a voice line at 4800 bits per second will give a higher error rate than those above.

POSSIBLE APPROACHES There is a variety of approaches that we can take to the dealing with noise on transmission lines. All of the approaches discussed below are found on data transmission systems in use today.

The first, and easiest approach, is to ignore the noise, and this is often done. The majority of telegraph links in operation today, for example, have no error-checking facilities at all. Part of the reason for this is that they normally transmit English language text that will be read by human beings. Errors in English language caused by the changing of a bit or of a small group of bits are usually obvious to the human eye, and we correct these in the mind as we read the material. Telegrams that have figures as well as text in them commonly repeat the figures. This inexpensive approach is also taken on computer systems where the transmission handles verbal text. For example, on administrative message-switching systems, it is usually acceptable to have transmission to and from unchecked telegraph machines. If the text turns out to be unintelligible, the user can always ask for a retransmission.

An error rate of one bit in 10^5 is possibly not quite so bad as it sounds. Suppose that we considered transmitting the text of this book, for example, and coded it in five-bit Baudot code. If one bit in 10^5 were in error, in the entire book there would be about 40 letters that were wrong. The book would certainly still be readable, and the majority of its readers would not notice most of these errors. The human eye has a habit of passing unperturbed over minor errors in text. This book was first set in galley proofs by the publishers. The text was then scrutinized and checked by a team of professional editors. It was then divided into pages and page proofs were produced. By now, most of the errors should have been removed from the text, but, in fact, those remaining corresponded to an error rate of one bit in 2×10^4, an error level higher than that which would be found on unchecked telegraph transmission.

In any case, on most systems *some* of the errors are ignored. An error-detection procedure that catches all of them has been too expensive. Many systems in current use might raise the level of undetected errors from one bit in 10^5 to one bit in 10^7 or 10^8. It is possible to devise a coding scheme that gives very much better protection than this. In fact, one that is on the market gives an undetected error rate of one bit in 10^{14}, but this is expensive.

An undetected error rate of one bit in 10^{14} is much lower than is needed for most practical purposes. If one had transmitted nonstop at that rate over a voice line at 2400 bits per second, for a normal working week (no vacations) since the time of Christ, one would probably not have had an error yet!

By using sufficiently powerful error-detecting codes virtually any measure of protection from transmission errors can be achieved, as we will see in the next chapter. It is hoped that some future terminals will give a *very* high measure of protection at reasonable cost, using "large-scale integration" circuits for encoding and error detection.

CALCULATIONS OF THE EFFECT OF ERRORS

In designing a computer system it is important to know what error rate is expected. Calculations should then be done to estimate the effect of this error rate on the system as a whole. On some systems the effect of infrequently occurring errors is cumulative, and it is in situations such as these that special care is needed in eliminating errors. For example, if messages cause the updating of files, and an error in the message causes an error to be recorded on the file, then it is possible on some systems that as the months pass by the file will accumulate a greater and greater number of inaccuracies.

ERRORS AND THEIR TREATMENT   65

Suppose, for example, that teletype transactions with one bit in 10^5 in error update a file of ten thousand records. Suppose that on the average an item is updated 100 times a month, and that if any one of 20 five-bit characters is in error in the transmission then the item will be updated incorrectly. After six months of this no less than 4500 file records will be incorrect. If an error-correction procedure on the telecommunication lines reduces the rate of undetected errors to one bit in 10^7, then 60 of the file records are likely to be incorrect at the end of six months. With one bit in 10^8, six records are likely to be wrong. Probability calculations of this type need to be done on various aspects of the system when it is being designed.

DETECTION OF ERRORS

To *detect* communications errors, redundancy is built into the messages transmitted. In other words, more bits are sent than need be sent for the coding of the data alone. One of the simplest and most commonly used forms of this was the parity check that was discussed in Chapter 2. The 4-out-of-8 code also discussed in Chapter 2, and shown in Fig. 2.7, is more effective in catching multiple errors.

There is a very wide variety of error-detection codes. They differ considerably in their power and in the degree of redundancy that is needed. A discussion of these is left until the next chapter.

The probability of a transmission error remaining undetected can be made very low, either by a brute force method that gives a high degree of redundancy, or, better, by an intricate method of coding. Many systems in actual usage do not employ a *very* secure error-detection scheme, but compromise with one that is not too expensive but which will probably let one error in a thousand or so slip through.

Those illustrated in Figs. 3.9 and 3.10 are typical. In Fig. 3.9 a message that is composed of up to one hundred six-bit characters has one character at the end of the message for error detection. No parity bits or other forms of redundancy are built into the six-bit characters themselves. In Fig. 3.10 the level of redundancy is somewhat higher. First, each of the seven-bit characters in this example has an eighth bit as a parity check. Second, at the end of each record in the bulk of text, two eight-bit characters are added for checking purposes. If several records are included in one block, each of the records would have these checking characters.

DEALING WITH ERRORS

Once the errors have been detected, the question arises: what should the system do about them? It is generally desirable that it should take some automatic action to correct the fault. Some data transmission systems, however, do not do this and leave the fault to be corrected by human means at a later time. For example, one system that transmits data to be punched into cards causes a card to be offset in its position in the stacker when an error is detected. The offset cards are later picked out by the operator, who then arranges for retransmission. In general, it is much better to have some means of automatic retransmission rather than a manual procedure, and this is usually less expensive than employing an operator for this. In some real-time terminal systems, however, automatic retransmission has not been used because it is easy for the terminal operator to re-enter a message or request retransmission.

Again, on some systems it is possible to ignore incorrect data, but it is important to know that it is incorrect. On such systems error detection takes place with no attempt to correct the errors. This could be the case with statistical data where erroneous samples can be discarded without distortion. It is used on systems for reading remote instruments where the readings are changing slowly and an occasional missed reading does not matter. The advantage of a detection-only scheme is that it requires a channel in only one direction. In systems with telephone and telegraph lines this is not a worthwhile advantage, because such channels are half- or full-duplex. However, it is a great advantage with certain tracking and telemetry systems, and here we find detection-only schemes. Clearly, for most commercial applications they will not suffice.

ERROR-CORRECTING CODES

Automatic correction can take a number of forms. First, sufficient redundancy can be built into the transmission code so that the code itself permits *automatic error correction* as well as detection. As no return path is needed, this is sometimes referred to as *forward error correction*. To do this effectively in the presence of bursts of noise can require a large proportion of redundancy bits. Codes that give safe forward error correction are, therefore, inefficient in their use of communication line capacity. If the communication line permitted the transfer of information in one direction only, then they would be extremely valuable. However, again, most earthbound systems use half-duplex or full-duplex communication links. In general, error-*correcting* codes used on voice-grade or subvoice-grade lines do not give us as good value for money, or value for bandwidth, as error-*detecting* codes coupled with the ability to retransmit automatically data that are found to contain an error.

On higher speed links the argument for forward error correction becomes stronger because the time taken for reversing the direction of transmission is equivalent to many bits of transmission. This time is relatively high on wideband links and on half-duplex voice-grade links with high-speed modems.

LOOP CHECK

One method of detecting errors does not use a code at all. Instead, all of the bits received are retransmitted back to their sender, and the sending machine checks that they are still intact. If not, then the item in error is retransmitted. Sometimes referred to as a *loop check* or *echo check*, this scheme is normally used on a full-duplex line. Again, it uses the channel capacity less efficiently than would be possible with an error-detection code, although often the return path of a full duplex is underutilized in a system, because the system does not produce enough data to keep the channel loaded with data in both directions. A loop check is most commonly found on short lines and in-plant lines where the wastage of channel capacity is less costly. It gives a degree of protection that is more certain than most other methods.

RETRANSMISSION OF DATA IN ERROR

Many different forms of error *detection* and retransmission are built into data-handling equipment. In a typical high-speed paper tape transmission system, a "vertical" parity check, that is, a parity check on each character, is used along with a "horizontal" checking character at the end of a block of characters. At the end of each block the receiving station sends a signal to the transmitting station saying whether the block has been received correctly or whether an error has been detected. If any error is found, both the transmitting tape reader and the receiving tape punch go into reverse and run backwards to the beginning of that block. The punch then erases the incorrect data by punching a hole into every position—the *delete* code (shown in Figs. 2.2, 2.3, 2.8, and 2.9). The block is then retransmitted. If transmission of the same block is attempted several times (four times on much equipment), and is still incorrect, then the equipment will stop and notify its operator by means of a warning light and bell or buzzer. Other automatic facilities are usually used to detect broken or jammed paper tape, and to warn when the punch is running short of tape.

Where data are being transmitted to a computer, automatic retransmission is sometimes handled under control of the program, and sometimes by circuitry external to the main computer. More detailed examples of this will be given later.

ERROR CONTROL ON RADIO CIRCUITS

When high frequency radio is used for telegraphy, the mutilation of bits is normally much worse than with land-based telegraph circuits. It is subject to severe fading and distortion, especially in times of high sun-spot activity. Because of its high error rate and general unreliability its use is avoided as far as possible for transmission of computer data.

However, it is still used in some more isolated areas and in ship-to-shore links.

The system of error detection and retransmission most commonly in use for radio telegraphy is the van Duuren ARQ system. This transmits full duplex synchronously, the characters being sent in blocks or "words," a 3-out-of-7 code being used (which permits 35 different combinations, as opposed to 32 with the five bits of normal telegraphy). The start and stop bits of the Baudot code are stripped off, and the remaining five are recorded into 3-out-of-7 code and transmitted.

If the receiving equipment detects more or less than three bits in any character, transmission of data in the opposite direction is interrupted. An error signal is sent back to the transmitter of the data now in error. This transmitter then interrupts *its* sending, returns to the invalid word, and retransmits it. On a long radio link one or more words may have been sent after the message that had the error, depending upon the duration of the transmission path. These are discarded by the receiver. When the transmitter receives the error indication, it stops what it is transmitting, backtracks to the word in error, and retransmits that and all following words.

High-frequency radio links can be expected to have an error rate before correction of one character in 1000, and sometimes much worse than this. Most of these errors will be detected with the 3-out-of-7 code, but there is a certain probability of a double mutilation that makes a character incorrect while still leaving it with three 1 bits. The number of *undetected* errors in this case is one character in 10,000,000, approximately. This mutilation rate can rise as high as one character in 40 or even as high as one character in 4 on bad links and at certain bad points of time. If the mutilation rate is one character in 40, the undetected error rate rises to one character in 16,000, and the effective speed of the link would drop to a speed of perhaps 90 percent of the nominal speed, depending upon the word size and link retransmission time. If the mutilation rate rises to one character in four, the rate after error detection and retransmission is about one character in 160, and the effective speed is likely to drop to about half the nominal speed. This is still usable for human language messages, because we can apply our own error-correction thinking.

HOW MUCH IS RETRANSMITTED? Systems differ in how much they require to be retransmitted when an error is detected. Some retransmit only one character when a character error is found. Others retransmit many characters or even many messages.

There are two possible advantages in retransmitting a *small* quantity of data. First, it saves time. It is quicker to retransmit five characters than

500 when an error is found. However, if the error rate is one character error in 20,000 (a typical figure for Telex and telegraph lines), the percentage loss in speed does not differ greatly between these two cases. It *would* be significant if a block of 5000 had to be retransmitted.

Second, when a large block is retransmitted, it has to be stored somewhere until the receiving machine has confirmed that the transmission was correct. This is often no problem. In transmitting from paper tape, for example, the tape reader merely reverses to the beginning of that block. The paper tape is its own message storage. The same is true with transmission from magnetic tape or disk. With transmission from a keyboard, however, an auxiliary storage, or *buffer*, is needed if there is a chance that the message may have to be retransmitted automatically. On some input devices a small core storage unit constitutes the buffer. On others the keys themselves are the storage. They remain locked down until successful transmission is acknowledged. Again, several input devices may share a common control unit, and this contains the buffer storage. Buffer storage in quantity can be fairly expensive, so it may be better to check and retransmit only a small number of characters at a time. Again, when data on punched cards are transmitted, a buffer is needed for retransmission unless the machine is designed in an ingenious manner so that the card can be reread if required.

The *disadvantages* of using small blocks for retransmission are first that the error-detection codes can be more efficient on a large block of data. In other words, the ratio of the number of bits *with* the error-detection code to the number if the data were sent *without* protection is smaller for a given degree of protection if the quantity of data is large. Second, where blocks of data are sent synchronously, a period of time is taken up between blocks in control characters and line turnaround procedures. The longer the block, the less significant this wasted time.

The well-designed transmission system achieves the best compromise between these factors. Let us consider some examples of different retransmission quantities:

1. *Retransmission of One Character.* Characters are individually error checked, as with a 4 out of 8 code. As soon as a character error is detected, retransmission of that character is requested. This is likely to be used only on a very slow link.
2. *Retransmission of One Word.* The British ICL type 7000 series equipment uses blocks of nine characters. Figure 4.1 shows such a block. The block has a parity bit for each "row" or character, and also for each column. A transmitting terminal has buffers that hold two such blocks. Should the receiver detect an error in row or column parity, retransmission of that block is requested.

70 ERRORS AND THEIR TREATMENT

Some systems use smaller blocks than this. The Western Union EDAC (Error Detection Automatic Correction) System uses four-character blocks in one of its models. Four five-bit teletype-code characters are transmitted synchronously, with a five-bit error checking pattern, from a buffer.

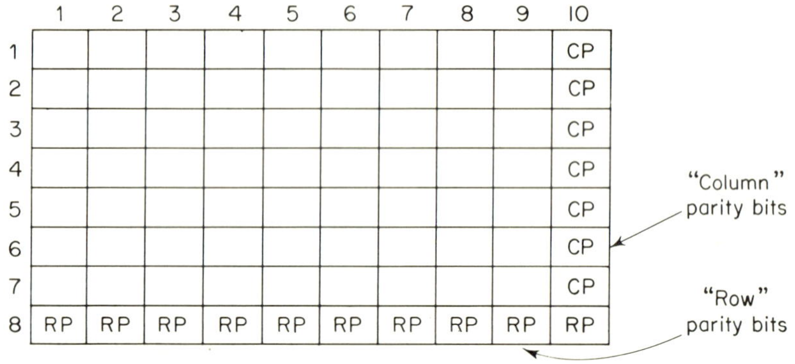

	1	2	3	4	5	6	7	8	9	10
1										CP
2										CP
3										CP
4										CP
5										CP
6										CP
7										CP
8	RP	RP	RP	RP	RP	RP	RP	RP	RP	RP

"Column" parity bits — column 10
"Row" parity bits — row 8

Fig. 4.1. ICL type 7000 series block of data; nine seven-bit characters.

A three-bit control signal is returned by the receiver to indicate correct reception. If this signal is mutilated, the transmitting machine repeats the transmission. It does this until it receives confirmation of correct receipt, and then the four characters in storage can be erased.

3. *Retransmission of a Message or Record.* The above "words" were both short and fixed in length. Many machines retransmit complete messages or records at one time, which are much longer than the above words and more often than not are variable in length. They may have a format such as that in Fig. 3.9, in which an end-of-transmission character terminates the text and this is followed by the error-checking characters. The record may be retransmitted from back-spaced tape from a variable-length area in the core of a computer or from a buffer in a control unit attached to the transmitting device. If a buffer is used, there may be a maximum size for the amount that can be retransmitted. If the message exceeds that size, it is broken into separate messages that are linked together with a control character indicating that a given transmission has not completed the message.

4. *Retransmission of a Block of Several Messages or Records.* When the transmission speed is high, it becomes economical on many systems to transmit the data synchronously in large blocks. These may be longer than one "message" or "record." A typical format for a multirecord block is that in Fig. 3.10. On most systems when

an error occurs in any of the records, the whole block of records is retransmitted. By using machines with good logic capabilities it would be possible to resend only the faulty record and not all of the other records in the block.

5. *Retransmission of a Batch of Separate Records.* Sometimes a control is placed on a whole batch of records. Like the controls conventionally used in batch data processing, the computer adds up account numbers and/or certain data fields from each record to produce *hash totals.* These totals are accumulated at the sending and the receiving end and are then compared. At one of these stations such a control might, in some cases, be produced manually on an adding machine, and it is often used to detect not only the errors in transmission, but also errors in manual preparation of data. When one computer sends a program to another computer, it is vital that there should be no undetected error in the program, so the words or groups of characters are added up into an otherwise meaningless hash total. This hash total is transmitted with the program, and only if the receiving computer obtains the same total in *its* addition is the program accepted.

Some form of batch control of this type is often used, where applicable, *as well as other* automatic transmission controls overriding safety precaution. Its use is entirely in the hands of the systems analyst and can be made as comprehensive and secure as he feels necessary.

A typical application of batch totals might be on the network shown in Fig. 4.2. In the figure, transactions are to be sent from many branch locations to the head office computer. The transactions are punched into paper tape at the branches on a telegraph. The transactions are grouped into batches of about fifty and these are added up on an adding machine before punching. The totals obtained are punched as the last transaction of the batch.

The batches are sent on telegraph lines to a programmed hold-and-forward concentrator (discussed in detail in Chapter 12). This machine handles all the data from one area and forwards it periodically to the distant head office computer on a dial-up voice line using synchronous transmission. The main purpose of this is to reduce the cost of the communication lines.

As the concentrator receives a batch, it adds up the batch totals and compares them with the totals received. It prints a message at the branch, saying whether the totals were correct or not. If this were not done, the telegraph transmission would be unchecked. Here the batch totals are being used to check the accuracy of punching as well as the accuracy of transmission. The concentrator stores the data on its file and later dials up the head office computer to transmit the batches it has collected. This link has other error-detection facilities, but for additional safety, the batch totals are again checked by the head office computer.

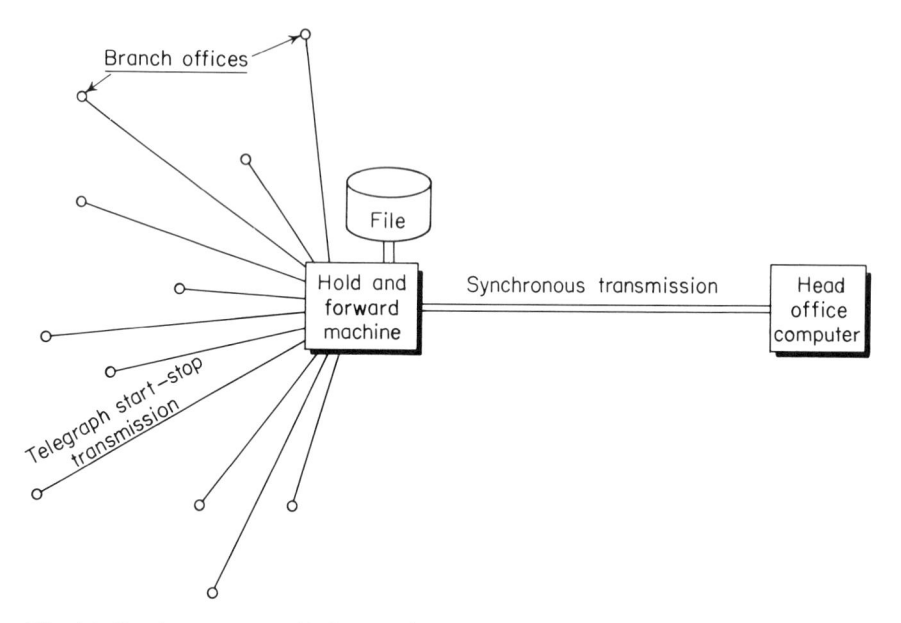

Fig. 4.2. Batch error control is here performed at the hold and forward machine.

6. *Retransmission at a Later Time.* The batch totals in the above illustration were checked immediately; the transmission was complete, and the sender was notified whether they were correct or not. Some forms of validity check might not be capable of being used until the items are processed. They may, for example, necessitate comparing transactions with a master tape. They are, nevertheless, valuable error controls, and an originating computer might keep the data in its files until a receiving computer has confirmed this validation.

TRANSMISSION ERROR-CONTROL CHARACTERS In order to govern the automatic retransmission of information in which an error has been detected, a number of special characters are used —sometimes sequences of characters. Typical *control characters* for this purpose were listed in the code tables of Chapter 2. The ASCII code, for example, shown in Fig. 2.8 uses the codes ACK, NAK, CAN, and DEL.

The ACK code is used by the receiver to signal the transmitter that a block of code has been received correctly. Similarly, the NAK code is sent by the receiving terminal to tell the transmitter that a block of code received had an error in it. When the transmitter sends a block of code on most systems, it waits before it sends the next one until the ACK or NAK control character is received from the transmitter. If ACK is received, it proceeds normally; if NAK, it resends the block in error.

The transmitter itself commonly does some error checking on what it sends. It is possible that the circuits doing this may detect an error in a message on which transmission has already begun. The transmitter must then cancel the message, and so it sends a CAN character. The DEL character is normally a character that should be totally ignored. It is used to delete characters already punched in paper tape, by punching a hole in every position.

The transmitting and receiving machines have circuits designed to detect these special characters. Sometimes the special character combination of bits might itself have a meaning in the data to be sent. The data may, for example, contain the bit combination 0001100, which in the ASCII code is a CAN character. If this is a possibility, then clearly 0001100 cannot be used by itself to mean "cancel." In this case, each of the control signals is composed of two characters, the first of which is DLE (0000100 in ASCII code). DLE CAN (0000100 0001100) is then interpreted as "cancel," and similarly with other control characters. If DLE itself is to be sent as a valid *data* character, it must be sent as DLE DLE (0000100 0000100). In this way any combination of data bits can be sent without being confused with the control signals.

ODD-EVEN
RECORD COUNT

It is possible that the control characters themselves or end-of-transmission characters could be invalidated by a noise error. If this happens, then there is a danger that a complete message might be lost or two messages inadvertently joined together. It is possible that during the automatic retransmission process a message could be erroneously sent twice. To prevent these errors, an odd-even count may be kept of the records transmitted.

Sometimes at the start of a block a control character is sent to indicate whether this is an odd-numbered or even-numbered block. On some systems, two alternative *start-of-transmission* characters are used. With other schemes it is the ACK characters that contain this odd-even check. Two different ACK-type signals may be sent: ACK 0 and ACK 1. On the ASCII code there is only one ACK character, so if this is used a two-character sequence may be employed.

If an odd-numbered block does not follow an even-numbered block, then the block following the last correct block is retransmitted. It is very improbable indeed that *two* blocks could be lost together, or two blocks transmitted twice in such a manner that the odd-even count would not detect the error. However, some systems use a *serial number* to check that this has not happened, instead of the odd-even count. The serial number may be examined by hardware, but often it is a check that is applied by a computer program; one of its main values is to bridge the continuity gap when a terminal computer, or other hardware, failure occurs. During the period of recovery from such a failure there is a danger of losing or double-processing a transaction.

AN EXAMPLE OF A TYPICAL ERROR-CONTROL SYSTEM

Finally, let us examine the error-control circuits on a typical machine. Consider the transmission between two computers both of which use BCD characters. A line control machine is attached to the computer channels and transmits with a 4-out-of-8 code. It uses a checking character at the end of each block. The sequence of events is likely to be as follows:

1. The computer passes the data to its line control machine a character at a time. The line control machine checks that it has received each character with valid parity.
2. The line control machine translates the character into 4-out-of-8 code, and then a circuit checks that there are in fact four 1 bits in the newly formed character.

If either of the above checks detect an invalid character, the line control machine signals its computer. Either the line control machine or the computer program terminates the transmission and sends a signal to the receiving computer saying that this has been done. This is likely to consist of a CAN control character (0001100 in ASCII code).

Another attempt is made to transmit the message. If either of the above checks fails repeatedly, the line control machine normally sounds an alarm to notify the operator.

3. The receiving line control machine checks that the characters received have four 1 bits. If an error is detected in a character, the receiving device registers this fact for later transmission of a NAK character. Where a half-duplex line is used, the NAK character cannot be sent until the end of the block has been transmitted. The receiving device also informs its computer of the error.
4. The receiving device converts each character to BCD and checks that the character it has formed has the correct parity.

5. Both the receiving and transmitting machines keep an odd-even count of the number of blocks. At the beginning of each block transmitted, the transmitting device sends a start-of-transmission character that indicates whether its block count is odd or even.
6. As the transmitting machine sends its characters, it accumulates the longitudinal checking character that is to be sent at the end of the block. When the data in the block have all been sent, the end-of-transmission character is sent and then the longitudinal check character. Sometimes this checking character is sent in the middle of a record, as in Fig. 3.10.
7. The receiving device also accumulates a longitudinal checking character as the data characters are being received. When the end-of-transmission character is detected, it compares the next character with its longitudinal check. If the two do not agree, it registers this fact.
8. When the receiving device notifies the computer that a complete block has been received, it also sends it a signal (normally on parallel lines), saying whether any error has occurred. The computer then decides whether to send back an ACK or NAK character. Before it does this it may attempt to store the block, perhaps on tape or disk. If all goes well it instructs its line control machine to send an ACK signal. If not, a NAK.
9. The transmitting computer receives this signal and then decides whether to transmit the *next* message or retransmit the previous one.

5 ERROR-DETECTING CODES

A large variety of error-detecting and -correcting codes have been invented. There are too many to give details of them all; the bibliography at the end of the chapter will direct the reader to descriptions of specific codes. This chapter discusses the codes that the reader is most likely to meet in planning computer transmission networks. Many codes with specific names fall into the general category of "polynomial codes." Much of the chapter discusses these. As we will see, data can be protected very well indeed on a noisy transmission line by using high-order polynomial codes.

CRITERIA FOR CHOICE OF CODE   The merit of any scheme for correcting transmission errors is a function of three properties:

1. What is its efficiency in detecting errors? How many incorrect messages does it let through? Ideally, we would like a scheme which catches *all* errors.
2. How much does it reduce the line throughput? Both redundant bits and retransmission lessen the total data throughput on the line.
3. How much does it cost?

We can swing the design between these parameters by the choice of different coding methods. In some systems high accuracy is all-important, and a high price will be paid for it if necessary. It is very rare, however, to find those codes that give the highest measures of protection in use today. Systems usually settle for a reasonably high measure of protection and a cost that is not too high. In some systems, however, in the author's view, the systems analysts have paid too little attention to transmission accuracy.

Line throughput is usually not too much of a worry in selecting a code. Higher-speed modems can be used if necessary, or more lines can be added. For this reason simple codes have been used which have a high degree of redundancy but which do not inflate terminal costs too much.

The balance is, however, swinging in favor of the more complex codes. In the early days of computing, the circuitry for composing the error check and for examining it after transmission was expensive. Circuit cost dominated the choice of code. Today, however, and especially in view of large-scale integration technology, the cost associated with the more complex codes is much lower. Furthermore, more systems need a high degree of protection because they are transmitting data that should not be allowed to acquire errors—programs or financial data, for example. It would seem worthwhile to mass-produce circuits that handle the high-order polynomial codes discussed at the end of this chapter, especially as these do not introduce a high level of redundancy and hence throughput degradation.

Terminal cost assumes great importance on some of today's systems with very large numbers of terminals. Therefore, one sometimes finds a cheap-to-implement code used at the terminal, but a more complex code giving better transmission throughput on the concentrators, or on control units handling many terminals.

With error-detecting codes it is not necessary, as we will see, to have a very high proportion of redundancy in order to achieve a very high measure of protection. If we transmit a long block of data, we can protect it very well with relatively few bits. The error-detecting power depends primarily on the number of checking bits in a group rather than on the percentage of redundancy. Therefore, it gives more efficient coding if long blocks are sent and protected by one group of bits so that there is a high ratio of data bits to check bits. The error-detecting power, however, is *highly* dependent on the nature of the code.

As we will see, we can transmit 100-character messages with only about ten percent redundancy and can detect virtually all errors if we use a sufficiently powerful code.

ERROR CORRECTING CODES The ability of a code to *correct* errors is related to its ability to *detect* them. A code that can detect double errors can correct *single* ones. If a single error occurs the receiving machine could attempt to correct it, for example, by changing the bits one at a time. If the receiving machine could detect all double errors, then it could detect when the erroneous bit has been corrected rather than a second error being made. Similarly, a device that can detect quadruple errors can *correct* double ones; a device that can detect $2x$ errors can correct x. Similarly, some codes can detect two error bursts of length b bits, in which bits within the burst may or may not be correct; such codes could *correct* one such burst.

In theory an error-correcting code could be formed from a set of tables. If a set of bits representing a character, word, or block is received, the receiving machine could look up this set in a table which gives the correct character, word, or block that is the most likely to be the original. To construct such a table would require a knowledge of the most probable types of errors. In practice we need something very much easier to implement than this.

Whatever way we implement an error-correcting code there is always a probability that it will mis-correct. Sometimes it will turn a correct bit into an incorrect bit because the particular error pattern that occurred was not one that the code was designed to correct. If the bits in error were randomly distributed, this would not be much of a problem. It would happen very rarely. However, they are not. The noise is likely to come in bursts in which almost any error pattern could occur.

During a period of excessive noise a good detection and retransmission scheme would increase the number of messages that were rejected, lowering the throughput until the noisy period ended. A forward-error-correction system, on the other hand, would not degrade throughput but would increase the number of erroneous messages that it let slip through. For most computer systems that latter would be undesirable. Detection and transmission is to be preferred.

Furthermore, because the bit errors tend to come in clusters, the number of messages that need retransmission is less. Suppose that we transmit blocks of 1000 bits on a line with an error rate of one bit in 10^5. If the errors were randomly distributed, one block in a hundred would need to be retransmitted. However, if the errors are clustered, we might receive one block with ten errors, and then 10^6 bits without error. The retransmission rate may thus have dropped to one block in a thousand.

The cost of schemes for forward-error correction is generally greater than that for effective detection. The number of redundant bits required for the same measure of protection as a good error-detection code is much higher. Error detection and retransmission therefore gives better value for money, and better value for bandwidth. It could be argued that retransmission necessitates the return path; however, this is almost invariably needed anyway for control of the machines and for responses.

Error-*correcting* codes are of great value in protecting records on magnetic tape, disk, or other media where an error is detected and the original is unobtainable. With data transmission, however, the original still exists in the transmitting machine, and can easily be re-sent.

With computer data transmission, over full-duplex voice-grade or subvoice-grade lines, with today's state of the art, forward-error-correction systems have less merit for all of our three criteria—reliability, throughput, and cost. On half-duplex lines, used at high speed, the line turn-around time becomes long compared to the character transmission time, and so here forward error correction is more attractive. The remainder of this chapter describes error-*detecting* codes.

PARITY CHECKS   It is common to find two kinds of check on transmitted data, one on each character, and one on each message, message segment, or block. Many systems have both. Some systems have only character checks; this is generally not very secure. Some systems have only the message or block check, and this, as we will see, can be made extremely secure, given a good code.

The simplest check, and one of the most commonly used because it is inexpensive to implement, is the parity check. However, one single parity check will fail to detect an error if that error damages an even number of bits. In data transmission a single noise pulse or drop-out (loss of signal) is frequently of greater duration than the length of one bit. This is more likely to be so when a high bit rate is used. Even at low bit rates double errors are common. A CCITT study[1] of 50 band telegraph lines gave the following figures:

Isolated single-bit errors	50–60%
Error bursts with two erroneous bits	10–20%
Error bursts with three erroneous bits	3–10%
Error bursts with four erroneous bits	2– 6%

A burst was defined here as bits in error separated by less than ten nonerroneous bits.

Curves published by AT&T showing errors on their switched public network[2] indicate that when 1200 bit-per-second transmission is used, about 49 percent of the error bits have another error following within seven bits. If the error bit in question is the first one of an eight-bit character, there is a 49 percent chance that the parity check will fail to detect the error (if the possibility of having three or more bits in error is ignored). For the character as a whole, the curves can be used to calculate that there is about a 30 percent chance that the single parity check will fail. Such figures are very approximate estimates of code performance. Tests made on transmission lines with parity-checked characters have confirmed that this can

---

[1] CCITT Special Study Group A (Data Transmission), Contribution 92, Annex XIII, p. 131 (October 18, 1963).

[2] "Error Distribution and Error Control Evaluation," Extracts from Contribution GT. 43, No. 13, February 1960, CCITT Red Book, Vol. VII, The International Telecommunication Union, Geneva, 1961.

hardly be regarded as a satisfactory way to protect data (although, surprisingly, it is used as such on some machines).

A parity check on a character is sometimes referred to as a "vertical" parity check. A parity bit checking all of the equivalent bits in a message is referred to as a "horizontal" or "longitudinal" check. They are also referred to as "row" and "column" parity checks. Used in conjunction with each other, they provide a measure of protection much greater than either "vertical" or "horizontal" parity checks alone. Fig. 4.1 illustrates their use.

THE U.S. ASCII RECOMMENDATION The American Standards Institute recommendation for checking on the ASCII code is to use horizontal and vertical parity bits, like Fig. 4.1 but on variable length messages.

The level of redundancy needed is fairly high. If x characters are sent, the ratio of check bits to data bits is

$$\frac{x + 8}{7x}$$

Thus for a 20-character message, one-fifth as many check bits as data bits are needed. A very long message needs about one-seventh. This is a higher ratio than with the polynomial codes, which will be discussed shortly.

For an undetected error to occur, the bits changed must be self-compensating, as shown in Fig. 5.1. This type of check can detect all messages with one-, two-, or three-bit errors, all with an odd number of errors and some with an even number of errors. The probability of self-compensating errors occurring is low.

Where the transmission system is prone to double errors there will be a higher probability of self-compensating errors occurring. Some transmission schemes are designed in such a way that double-bit errors can occur. In some modulation schemes the bits are represented by a *change* in state of the carrier rather than by the state of the carrier itself. This is referred to as *transition coding* rather than *state coding*. For example, a change of phase in one direction represents a 1 bit, and a change of phase in the opposite direction represents a 0 bit. Here an erroneous phase change on the line may cause two errors rather than one. A second similar occurrence shortly afterwards may give the compensating error that is undetectable by the vertical and horizontal parity check.

Again, on some modems, data are coded into pairs of bits (di-bits)[3] rather than single bits. The carrier may be in one of four possible states at

---

[3] *Telecommunications and the Computer*, James Martin, Prentice-Hall, Inc., 1969, Chapter 13.

one instant, conveying one of four possible bit pairs. A noise impulse is then likely to change a pair of bits. Two such impulses could give a compensating error.

We see, then, that the error-detecting power of this checking is dependent on the method of modulation being used. It will be lower with transition coding than with state coding, and lower if di-bits rather than single bits are encoded.

	Data character 1	Data character 2	Data character 3	Longitudinal check character
Bit position 1	1	0	1	0
Bit position 2	0	0	0	0
Bit position 3	0	0	1	1
Bit position 4	0	0	0	0
Bit position 5	0	1	0	1
Bit position 6	0	0	0	0
Bit position 7	1	1	1	1
Parity bit	1	1	0	0

A message of three data characters in ASCII code, with ASCII checking

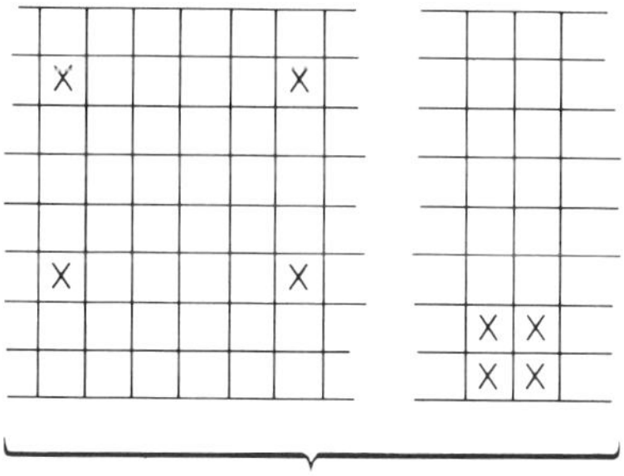

For an undetectable error to occur an even number of bits, greater than three, must be changed, in compensating positions

Figure 5.1

Measurements of the effectiveness of coding with a vertical and horizontal parity check indicate that it lessens the number of undetected errors by a factor ranging from 100 to 10,000. A telephone line with an error rate of 1 to 10^5 might have an undetected error rate of from 1 in 10^7 to 1 in 10^9.

M-OUT-OF-N CODES

Many transmission schemes have coded characters so that a fixed number of bits per character must be used. If a number of bits other than this is received, then an error is recognized.

Figure 2.7 illustrates one such code that has been extensively used—the 4-out-of-8 code. Eight-bit characters are transmitted, each of which must have four 1 bits and four 0 bits. This gives 70 possible combinations, much fewer than the 256 combinations when all bits are used or the 128 combinations when one bit is used to check parity.

A similar system used extensively in radio telegraph circuits (the van Duuren ARQ system) uses a 3-out-of-7 code.

In general, an M-out-of-N code permits

$$\frac{N!}{M!(N-M)!}$$

combinations out of 2^N possible ones.

If all errors in a noise burst were the same type of change—for example, if they were all 0's changed to 1's—then the M-out-of-N code would be very secure. It might be easy to imagine that this is the case, if we visualize noise impulses as always an increase in voltage on the line, for example, which would always change 0's to 1's when amplitude modulation or baseband signaling was in use.

Unfortunately, however, noise is often of an oscillating nature, a peak in voltage being followed by a dip in voltage. It is not uncommon that when a 0 is changed to a 1, a nearby 1 is changed to a 0. If this happens the M-out-of-N code loses its effectiveness. Similarly, where transition coding is used, or di-bits rather than bits are encoded, an undetectable compensating error can occur.

Experiments were performed by IBM using a 4-out-of-8 code transmitted over a voice line at 1200 bits per second. By comparing what was received with the original, the number of errors that were undetected by this code were counted. The results were compared with those for characters protected by a parity check, transmitted over the same line. It was found that the percentage of undetected errors for the parity-checked characters was about 1.9 times greater than with the 4-out-of-8 code. This ratio was approximately the same with both state coding and transition coding. The 4-out-of-8 code was, then, an improvement over the parity check, but, in view of the extra redundancy needed, was not a spectacular improvement. Two-phase modulation was used. The error rate would probably have been worse if four-phase modulation had been used (transmitting dibits). With amplitude modulation the 4-out-of-8 code might have improved its performance.

Clearly, it would not be secure to transmit characters coded in 4-out-of-8 code without also some form of longitudinal redundancy check. In practice a longitudinal parity check is commonly employed on such messages, as with those using character parity.

POLYNOMIAL CODES

After M-out-of-N codes and codes with parity checks, vertical and longitudinal, the next most common class of codes are specific examples of polynomial codes, including the Hamming codes, the Bose-Chaudhuri codes, the Fire codes, the codes of Melas, various interleaved codes, and, for that matter, the simple parity check. (See bibliography for detailed references to all these.)

All of these codes can be described in terms of the properties of divisions of polynomials.[4] Polynomial codes can be made to perform with very high efficiency.

Let us suppose that the data block that has to be transmitted is composed of k bits. We can represent this by a polynomial in a variable x, having k terms—a polynomial of order $(k-1)$. If we represent the bits in the data block by the terms $a_{k-1} + a_{k-2} + \cdots + a_2 + a_1 + a_0$, the polynomial is then

$$M(x) = a_{k-1}x^{k-1} + a_{k-2}x^{k-2} + \cdots + a_2x^2 + a_1x + a_0$$

As an example, if the data message being sent is 1010001101, the polynomial representing it is

$$x^9 + 0 \cdot x^8 + 1 \cdot x^7 + 0 \cdot x^6 + 0 \cdot x^5 + 0 \cdot x^4 + 1 \cdot x^3 + 1 \cdot x^2 + 0 \cdot x + 1$$
$$= x^9 + x^7 + x^3 + x^2 + 1$$

The high-order term of the polynomial is the bit that is transmitted first.

This is simply a convenient mathematical way of expressing the message to be sent. We will manipulate this using the laws of ordinary algebra, except that modulo 2 addition must be employed. This uses binary addition with no carries, as follows:

Example

ADDITION IN MODULO 2 ARITHMETIC:

$x^7 + x^6 + x^5 +$	$x^2 + 1$	1 1 1 0 0 1 0 1 +
$x^7 +$	$x^5 + x^4 + x^3 + x^2$	1 0 1 1 1 1 0 0 =
x^6	$+ x^4 + x^3$ 1	0 1 0 1 1 0 0 1

---

[4] "Cyclic Codes for Error Detection," by W. W. Peterson and D. T. Brown, *Proceedings of the IRE*, January 1961.

84   ERROR-DETECTING CODES

MULTIPLICATION IN MODULO 2 ARITHMETIC:

$$(x^7 + x^6 + x^5 + x^2 + 1)(x + 1)$$

$$\begin{array}{r} x^8 + x^7 + x^6 + x^3 + x \\ + x^7 + x^6 + x^5 + x^2 + 1 \\ \hline x^8 + x^5 + x^3 + x^2 + x + 1 \end{array}$$

$$\begin{array}{r} 1\,1\,1\,0\,0\,1\,0\,1 \quad \times 1\,1 = \\ 1\,1\,1\,0\,0\,1\,0\,1\,0\,+ \\ 0\,1\,1\,1\,0\,0\,1\,0\,1 \\ \hline 1\,0\,0\,1\,0\,1\,1\,1\,1 \end{array}$$

To transmit the data block we need a second polynomial, referred to as the *generating polynomial*, $P(x)$. $P(x)$ is of degree r, where this is less than the degree of the message polynomial $M(x)$, but is greater than zero. $P(x)$ has a unity coefficient on the x^0 term (i.e., the lowest-order term is 1).

Thus to transmit the above message,

$$M(x) = x^9 + x^7 + x^3 + x^2 + 1$$

we might use a generating polynomial:

$$P(x) = x^5 + x^4 + x^2 + 1$$

The steps involved in the transmission are, in effect, as follows:

Step 1: The data message $M(x)$ is multiplied by x^r, giving r 0's in the low-order positions.

Step 2: The result is divided by $P(x)$. This gives a unique quotient $Q(x)$ and remainder $R(x)$:

$$\frac{x^r \cdot M(x)}{P(x)} = Q(x) \oplus \frac{R(x)}{P(x)}$$

($\oplus$ is the sign for addition in modulo 2 arithmetic.)

Step 3: The remainder is added to the message, thus placing up to r terms in the r lower-order positions.

This is the message that is transmitted. Let us call it $T(x)$.

$$T(x) = x^r M(x) \oplus R(x)$$

As an example, suppose that the generating polynomial $P(x) = x^5 + x^4 + x^2 + 1$ is used, for which $r = 5$. The data block to be sent is the above 1010001101. We have

Step 1: $x^r M(x) = x^5(x^9 + x^7 + x^3 + x^2 + 1)$

$$= x^{14} + x^{12} + x^8 + x^7 + x^5$$

which is equivalent to 101000110100000.

Step 2: This is divided by $P(x) = x^5 + x^4 + x^2 + 1$, which gives $x^9 + x^8 + x^6 + x^4 + x^2 + x$ and a remainder of $x^3 + x^2 + x$, which is equivalent to 01110. Figure 5.2 shows the division.

Step 3: The remainder is added to $x^r M(x)$. This gives the bit pattern 101000110101110, which is the message that is transmitted.

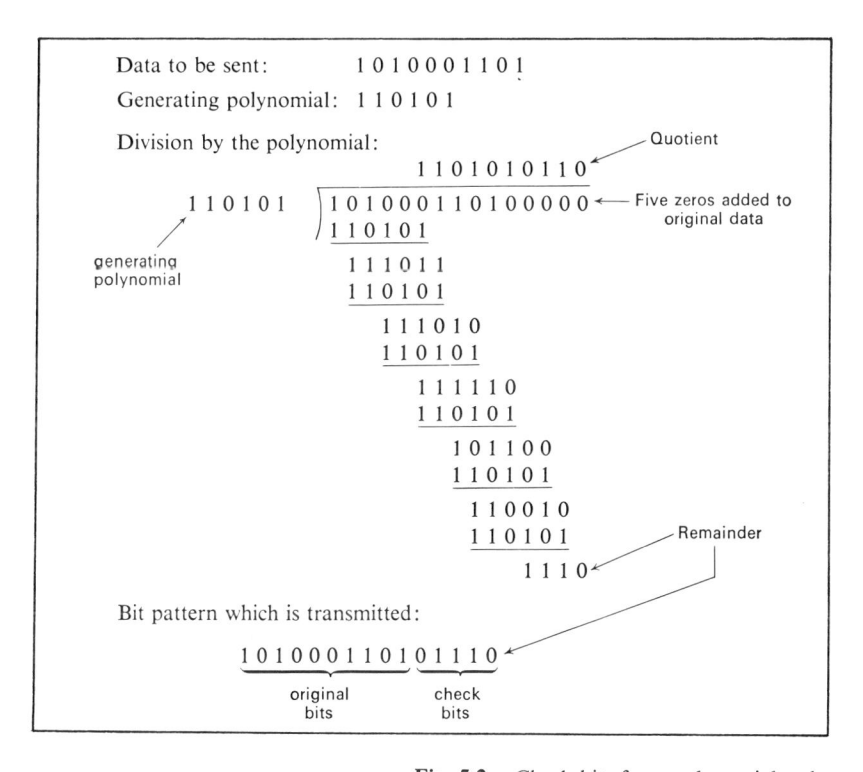

```
Data to be sent:         1 0 1 0 0 0 1 1 0 1
Generating polynomial:   1 1 0 1 0 1
Division by the polynomial:                    Quotient
                                    1 1 0 1 0 1 0 1 1 0
           1 1 0 1 0 1  ) 1 0 1 0 0 0 1 1 0 1 0 0 0 0 0  ← Five zeros added to
                          1 1 0 1 0 1                       original data
  generating
  polynomial              1 1 1 0 1 1
                          1 1 0 1 0 1
                            1 1 1 0 1 0
                            1 1 0 1 0 1
                              1 1 1 1 1 0
                              1 1 0 1 0 1
                                1 0 1 1 0 0
                                1 1 0 1 0 1
                                  1 1 0 0 1 0
                                  1 1 0 1 0 1          Remainder
                                      1 1 1 0

Bit pattern which is transmitted:
         1 0 1 0 0 0 1 1 0 1 0 1 1 1 0
         _____/ _____/
          original     check
            bits        bits
```

Fig. 5.2. Check bits for a polynomial code.

We are thus sending the original bit pattern with five bits accompanying it for error detection. The bits are transmitted from left to right, the five check bits last.

The division is represented by the equation

$$\frac{x^r \cdot M(x)}{P(x)} = Q(x) \oplus \frac{R(x)}{P(x)}$$

Therefore,

$$x^r \cdot M(x) = Q(x) \cdot P(x) \oplus R(x)$$

Subtraction is the same as addition in modulo 2 arithmetic (no carries); therefore,

$$x^r \cdot M(x) \oplus R(x) = Q(x) \cdot P(x)$$

Hence the message transmitted is given by

$$T(x) = x^r \cdot M(x) \oplus R(x) = Q(x) \cdot P(x)$$

The message transmitted is therefore exactly divisible by the generating polynomial P(x). It is this property that we will check in attempting to see whether an error has occurred. The receiving machine, in effect, divides the message polynomial it receives by $P(x)$. If the remainder is nonzero, then an error has occurred. If it *is* zero, then either there is no error or an undetectable error has occurred.

When the message is transmitted, a number of bits may be changed by noise. We may refer to this pattern of error bits by another polynomial $E(x)$. Thus for a message in error $T(x) + E(x)$ will be received. If $T(x) + E(x)$ is exactly divisible by $P(x)$, then the error will not be detected.

It follows that if $E(x)$ is divisible by $P(x)$, then the error will not be detected. On the other hand, if $E(x)$ is not divisible by $P(x)$, we will detect it. Knowing the characteristics of the communication lines, we must, therefore, pick our generating polynomial $P(x)$ such that it is very improbable that the pattern of error bits will be divisible by it.

ERROR-DETECTION PROBABILITIES

The choice of generating polynomial should be dependent on a knowledge of the error patterns that are likely to occur on the channel in question. There are certain error characteristics that we can be sure to protect the data from:

1. *Single Bit Errors*

If the data message or block protected by our polynomial code has one single bit in error, this can be represented by $E(x) = x^i$, where i is less than the total number of bits in the message, n.

If we give our generating polynomial more than one term, then x^i cannot be divided by it exactly. All single bit errors will be detected.

2. *Double Bit Errors*

Double bit errors can be represented by the polynomial $E(x) = x^i + x^j$, where i and j are both less than n. If $i < j$, we can write $E(x) = x^i(1 + x^{j-i})$. For the error to be detected, neither x^i nor $(1 + x^{j-i})$ may be divisible by the generating polynomial. If this polynomial has a factor with three terms, then this will be so and all double errors will be detected.

3. Odd Numbers of Errors

If the error message contains an *odd* number of bits in error, then the polynomial that represents it is not divisible by $(x + 1)$.

This can be proved as follows: Suppose that a message is represented by a polynomial $E(x)$, which is divisible by $(x + 1)$.

We can write $E(x) = (x + 1)Q(x)$. Substituting $x = 1$ into this, we have

$$E(1) = (1 + 1)Q(x)$$

Therefore, $E(1) = 0$. $(1 + 1 = 0$ in binary arithmetic with no carries) Therefore, $E(x)$ must contain an even number of terms.

Hence, if we employ a generating polynomial $P(x)$ with a factor $(x + 1)$, then *any* message with an odd number of errors will be caught.

Any polynomial of the form $(x^c + 1)$ contains a factor $(x + 1)$, since $(x^c + 1) = (x + 1)(x^{c-1} + x^{c-2} + \cdots + 1)$. Therefore, any generating polynomial of the form $(x^c + 1)$ will detect all errors with an odd number of bits incorrect.

4. Bursts of Errors

A burst of errors refers to a group of incorrect bits within one data message or block. We will define the length of a burst b as being the number of bits in a group having at least its first and last bits in error. Thus if $E(x)$ represents the error pattern 0000010100110000, this contains an error burst of length $b = 7$.

We can factorize $E(x)$ as follows:

$$E(x) = x^i E_1(x)$$

where i is less than the number of bits in the message.

Thus the above error pattern is represented by

$$E(x) = x^{10} + x^8 + x^5 + x^4$$

and would be written $x^4(x^6 + x^4 + x^1 + 1)$. x^i is not divisible by $P(x)$ because it is a single term. Therefore, the error will go undetected only if $E_1(x)$ is exactly divisible by $P(x)$.

When the length of the burst b is less than the length of $(r + 1)$ of $P(x)$, the generating polynomial $E_1(x)$ will be detected. Thus if we use a generating polynomial of 13 bits, all bursts of length 12 bits or less will be detected. To achieve this we will have to use 12 redundant bits in the message [the maximum size of the remainder $R(x)$].

When the number of bits in the burst b is equal to the number of bits in the generating polynomial $(r + 1)$, (13 in the above example), then the error will go undetected if and only if the burst is identical to the generating polynomial. The first and last bits in the burst are error bits, by definition. Therefore, the remaining $(r - 1)$ bits must be identical. If we regarded all combinations of bits as equally possible, the probability of an error being undetected would be the probability that $(r - 1)$ independent bits are identical with the generating polynomial. This is $(\frac{1}{2})^{(r-1)}$. In the above case $r = 12$, and so the probability of an undetected error is $1/2^{11} = 0.00049$, given that the block contains a burst of length 13 bits (a very rare event).

When the number of bits in the burst b is greater than $(r + 1)$, there is a variety of possible error patterns that are divisible by $P(x)$. If $E_1(x)$ is divisible by $P(x)$, then we can write:

$$E_1(x) = Q_1(x)P(x)$$

where Q_1 is the quotient obtained by dividing $E_1(x)$ by $P(x)$.

$E_1(x)$ is a polynomial of degree $(b - 1)$; $P(x)$ is a polynomial of degree r; therefore, the degree of the polynomial $Q_1(x)$ must be $(b - 1) - r$.

The number of bits represented by $Q_1(x)$ is, therefore, $(b - 1 - r) + 1 = b - r$. The first and last terms of $E_1(x)$ are always 1, and this causes the first and last terms of $Q_1(x)$ to be always 1. There are, therefore, $b - r - 2$ terms in $Q_1(x)$ which can alternate in value. This means that there are $b - r - 2$ ways in which $E_1(x)$ is divisible by $P(x)$.

As above, there are $2^{(b-2)}$ possible combinations of $E_1(x)$. If all combinations are equally probable, then the probability of an error being undetected in this case is

$$\frac{2^{(b-2-r)}}{2^{(b-2)}} = 2^{-r}$$

In the above case in which $r = 12$ the probability of an undetected error is $2^{-12} = 0.00024$, given that the block contains a burst of length greater than 13 bits (again, a rare occurrence).

To Summarize: If we choose a polynomial having $(x + 1)$ as a factor, and one factor with three or more terms, then the following protection will be given:

Single bit errors : 100% protection.
Two bits in error (separate or not) : 100% protection.
An odd number of bits in error : 100% protection.
An error burst of length less than $(r + 1)$ bits : 100% protection.
An error burst of exactly $(r + 1)$ bits in length : $[1 - (\frac{1}{2})^{(r-1)}]$ probability of detection.
An error burst of length greater than $(r + 1)$ bits: $[1 - (\frac{1}{2})^r]$ probability of detection.

The latter two terms assume an equal probability of any error pattern. In practice some error patterns are more prevalent than others, so some generating polynomials of a given r are better than others.

It will be seen that polynomial codes can provide a high measure of protection. As r becomes larger, so the measure of protection against bursts becomes greater. This is especially so as long bursts are rarer than short bursts on telephone and telegraph lines.

Figure 5.3 shows measurements of burst length made on a long-distance leased line in Europe. In these measurements the transmission rate was 2000 bits per second using binary phase modulation, and the data were sent in blocks of 792 bits. Of these blocks, 36 percent of those in error had only one bit in error; 81 percent had less than 10 bits, but 15 percent had large numbers of errors. The upper curve in Fig. 5.3 shows this distribution.

The lower curve in Fig. 5.3 shows the distribution of burst lengths. This is defined here as being the distance in bits between the first bit error and last bit error in the blocks. This curve states again that 36 percent of the error blocks have a burst length of only one bit. Thirty-four percent of the bursts are from two to eight bits in length, and 30 percent are longer than eight bits.

Unfortunately, there *is* a fairly high proportion of long bursts—greater than 35 bits in length, for example, in the lower curve of Fig. 5.3. This means that *some* errors are not going to be caught by the polynomial checks, or for that matter by any other checking schemes that are reasonable to implement.

We could, however, if it were necessary, make r very large; in other words, we could use a high-order generating polynomial. In this way we could produce a *very* high measure of protection indeed.

90 ERROR-DETECTING CODES

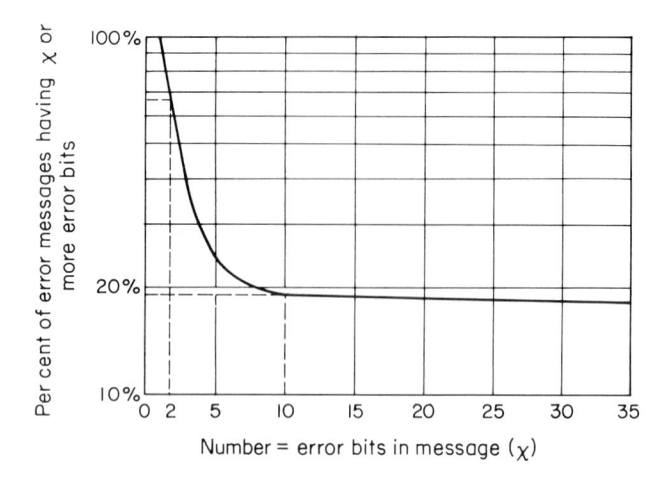

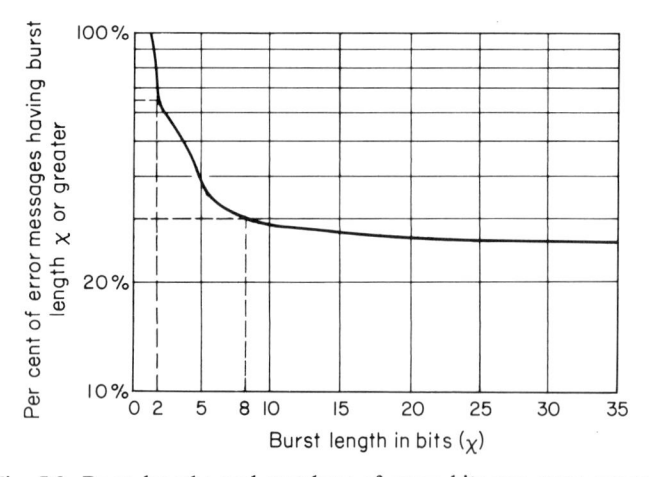

Fig. 5.3. Burst lengths and numbers of error bits per error message encountered on data transmission tests, at a 2000 bit/sec transmission rate, on a multipoint leased voice line from London to Rome.* Fixed message lengths of 792 bits were used. **Data Transmission Test on a Multipoint Telephone Network in Europe*, CCITT Blue Book, Supplement No. 37. Published by the International Telecommunication Union, Geneva, November 1964.

RESULTS OBTAINED IN PRACTICE

Results obtained in practice with polynomial codes on telephone lines have been reasonably close to the theoretical prediction. Often the number of undetected errors has been slightly lower than predicted. To gather statistics on undetected errors, a large quantity of random data is transmitted, and the data received are compared with the original. In this way the effectiveness of different polynomials has been tested.

Table 5.1 shows a typical set of results. These measurements were made on a leased voice line from London to Rome transmitting random bit messages in blocks of 729 bits. The transmission speed was 2000 bits per second:[5]

Table 5.1

Generating Polynomial Used	Fraction of Undetected Error Messages	
	Expected	Actual
$x^6 + 1$	0.0156	0.0107
$x^6 + x^5 + 1$	0.0156	0.0075
$x^6 + x + 1$	0.0156	0.0066
$x^7 + x^3 + 1$	0.0078	0.0035
$x^{12} + 1$	0.0003	0.0021
$x^{12} + x^{11} + 1$	0.0003	0.0003
$x^{12} + x + 1$	0.0003	0.0004

ENCODING AND DECODING CIRCUITS

Polynomial coding and decoding requires circuitry more complex than the use of vertical and horizontal parity bits. It is still not very complex, and, indeed, one of the advantages claimed for these codes is that of ease of coding and decoding.

The necessary division, such as that in Fig. 5.2, can be performed with a series of 1 bit shift registers and modulo 2 adders (exclusive OR circuits). The number of shift register positions is the same as the degree of the divisor, $P(x)$—five for the division in Fig. 5.2. The number of exclusive OR circuits is equal to (the number of 1 bits in the divisor − 1)—three for the divisor in Fig. 5.2.

---

[5] Data taken with permission from "Data Transmission Test on a Multipoint Telephone Network in Europe," *CCITT Blue Book*, Volume VIII, Data Transmission, Third Plenary Assembly, Supplement No. 37. Published by the International Telecommunication Union, Geneva, 1964.

92   ERROR-DETECTING CODES

Figure 5.4 shows such a circuit and gives the contents of the shift registers as the division in Fig. 5.2 is performed. The message to be coded, 1010001101, enters one bit at a time, followed by five 0 bits. When this is complete the shift register positions contain the requisite remainder, as shown.

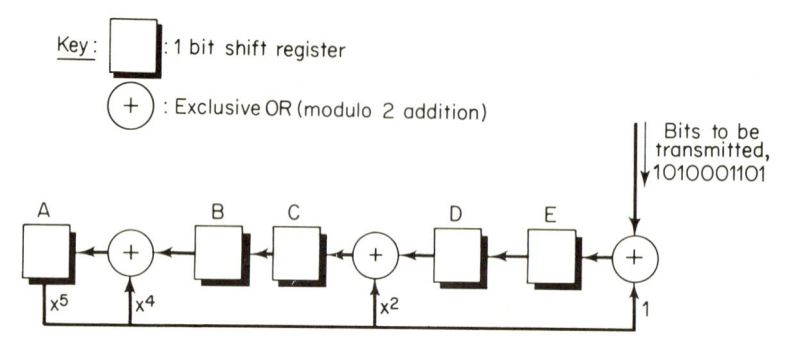

Key: ☐ : 1 bit shift register

⊕ : Exclusive OR (modulo 2 addition)

Bits to be transmitted, 1010001101

Contents of shift registers:

	A	B	C	D	E	Input bit	
Initial contents:	0	0	0	0	0		
Step 1	0	0	0	0	1	1	
Step 2	0	0	0	1	0	0	
Step 3	0	0	1	0	1	1	
Step 4	0	1	0	1	0	0	
Step 5	1	0	1	0	0	0	
Step 6	1	1	1	0	1	0	Message to be sent
Step 7	0	1	1	1	0	1	
Step 8	1	1	1	0	1	1	
Step 9	0	1	1	1	1	0	
Step 10	1	1	1	1	1	1	
Step 11	0	1	0	1	1	0	
Step 12	1	0	1	1	0	0	
Step 13	1	1	0	0	1	0	Five 0's added
Step 14	0	0	1	1	1	0	
Step 15	0	1	1	1	0	0	

Remainder (which is sent as the five check bits)

Fig. 5.4. Circuit with shift registers for dividing by the polynomial $(x^5 + x^4 + x^2 + 1)$. The division is shown in Fig. 5.2.

A disadvantage of this circuit is that there will be a delay between the message bits entering the circuit and the check bits being available to be sent. Figure 5.5 shows a refinement of the circuit that avoids this. The data message bits, now without any 0's following them, are fed in with switch 2 closed and switch 1 open. As soon as the last data bit is received, switch 2 is opened and switch 1 closed, and the shifting process continues. The five bits which then have the shift registers are five remainder bits, the required check bits. These follow the data bits with no time gap.

The circuit in Fig. 5.5 has the great advantage that it can be used for checking the received message as well as for generating the check bits for transmission. Only one such circuit is needed in a machine that transmits and receives.

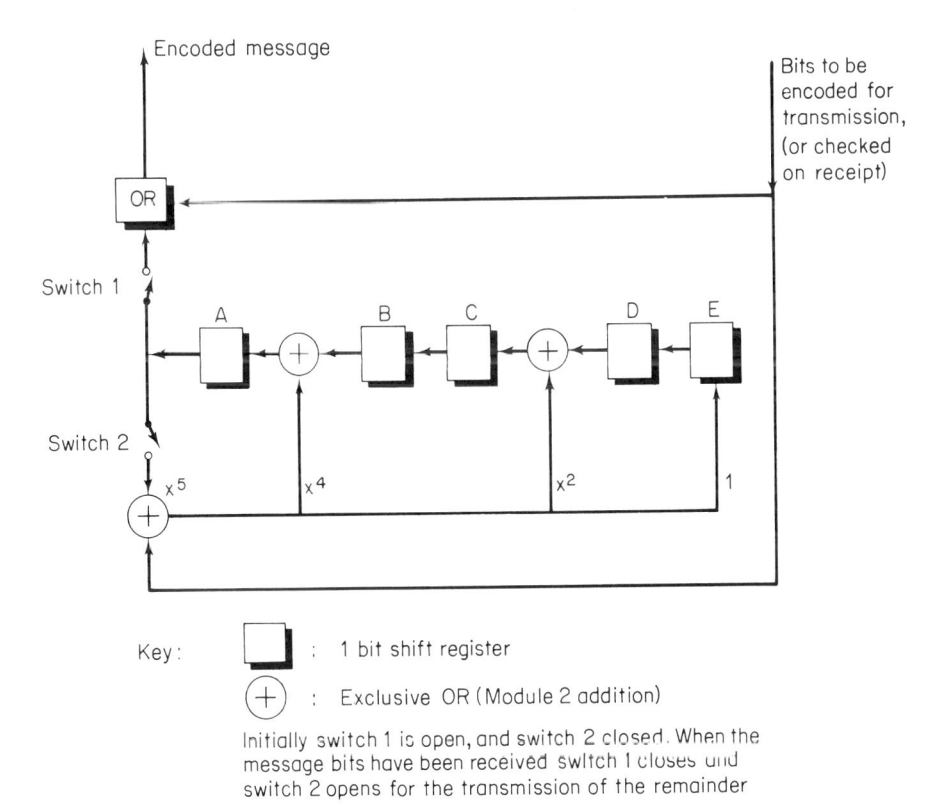

Fig. 5.5. A circuit that carries out the same operation as that in Fig. 5.4 but gives no delay between message bits and check bits. The same circuit also checks the received message. *Diagram redrawn with permission from "Cyclic Codes for Error Detection," by W. W. Peterson and D. T. Brown, Proceedings of the IRE, January 1961.*

POLYNOMIAL CHECKING ON VARIABLE-LENGTH MESSAGES

When polynomial checking is used, the messages checked can be of fully variable length. The checking characters shown in the messages in Figs. 3.9 and 3.10 may be polynomial checking characters and the messages they check may be of variable length.

IBM's "binary synchronous" mode of data transmission, used on a variety of different equipment, has checking characters shown in Fig. 3.10. This range of equipment can either use the ASCII code with its recommended vertical and horizontal parity checks, or it can use polynomial checking. When six-bit data characters are sent, two six-bit checking characters are used, giving 12 check bits. A generating polynomial of order 12 will, therefore, be used ($T = 12$). When eight-bit data are sent, two eight-bit checking characters enable a generating polynomial of order 16 to be used.

The polynomials used are

$$x^{16} + x^{15} + x^2 + 1 = (x + 1)(x^{15} + x + 1)$$

and

$$x^{12} + x^{11} + x^3 + x^2 + x + 1 = (x + 1)(x^{11} + x^2 + 1)$$

It will be seen from the above theory that these will catch all messages with one or two errors, all with an odd number of errors, all with single bursts of less than 16 and 12 bits respectively, and most of the few messages with larger bursts. In practice they have been found to perform slightly better than the above theory.

TO ACHIEVE A *VERY* HIGH MEASURE OF PROTECTION

We commented that it is possible to achieve a *very* high measure of protection with polynomial codes if a generating polynomial of sufficiently high order is used. Let us look at this a little more closely.

For the sake of discussion suppose that when one is transmitting fixed-length blocks of 100 data characters (800 bits) over a certain telephone line, there is a probability of 10^{-3} that a block will be perturbed by an error burst of length greater than 17 bits (pessimistic). If we make $r = 16$ and use 16 redundant bits for protection, then the probability that the burst greater than 17 bits is undetected is $[1 - (\frac{1}{2})^{16}] = (1 - 1.5 \times 10^{-5})$. The probability of having an undetected error is, then, theoretically of the order of $10^{-3} \times 10^{-5} = 10^{-8}$.

If we make $r = 80$, then the probability that bursts of length greater than 81 bits will be undetected is $[1 - (\frac{1}{2})^{80}] = (1 - 0.83 \times 10^{-24})$. Bursts of less than 80 bits will always be detected. So now our probability of undetected error is at least $10^{-3} \times 10^{-24} = 10^{-27}$.

This is a much higher degree of protection than is needed for most practical purposes. To repeat our example in the last chapter, if we had transmitted data protected in this way from all of the locations in the world where there is now a telephone, transmitting at the maximum speed of a voice line, and if we had been transmitting nonstop since the time of Christ (with no equipment failures), it is unlikely that there would yet have been an undetected error, anywhere in the world!

Furthermore, we had to add only 10 redundant characters to a message of 100 data characters. So the transmission efficiency is quite high—higher, indeed, than using the ASCII horizontal and vertical parity checking. The cost of the encoding and decoding equipment would have been higher. However, if it were mass-produced in great quantities in *large-scale-integration* circuitry, it might not be significantly higher.

BIBLIOGRAPHY

1. W. W. Peterson, *Error Correcting Codes*, The M.I.T. Press, Massachusetts Institute of Technology; and John Wiley & Sons, Inc., New York and London, 1961.
2. F. F. Sellers, Jr., M-Y. Hsiao, and L. W. Bearnson, *Error Detecting Logic for Digital Computers*, McGraw-Hill, Inc., New York, 1968.
3. D. T. Tang and R. T. Chien: "Coding for Error Control," *IBM Systems Journal*, Vol. 8, Number 1, 1969.
4. R. W. Hamming, "Error Detecting and Error Correcting Codes," *Bell System Tech. Journal*, April, 1950.
5. P. Fire, "A Class of Multiple-Error-Correcting Binary Codes for Non-Independent Errors," Stanford Electronics Laboratories, Technical Report No. 55, April 24, 1959.
6. N. Abramson, "A Class of Systematic Codes for Non-Independent Errors," *IRE Trans. on Information Theory*, IT-5 150 (1959).
7. L. H. Zetterberg, "Cyclic Codes from Irreducible Polynomials for Correction of Multiple Errors," *IRE Trans. on Information Theory*, IT-8, 13 (1962).
8. S. H. Rieger, "Codes for the Correction of 'Clustered' Errors," *IRE Trans. on Information Theory*, IT-6, 16 (1960).
9. M. Melas, "A New Group of Codes for Correction of Dependent Errors in Data Transmission," *IBM Journal*, **4**, 58 (1960).
10. R. Bose and D. Ray-Chaudhuri, "A Class of Error-Correcting Binary Group Codes," *Inf. and Control*, Vol. 3, March, 1960.
11. D. Hagelbarger, "Error Detection Using Recurrent Codes." Presented at the AIEE Winter General Meeting, February, 1960.
12. W. Peterson and D. Brown, "Cyclic Codes for Error Detection," *Proceedings of the IRE*, January, 1961.

6 POINT-TO-POINT LINE CONTROL

When transmission devices send data to each other, a variety of control signals must pass to and fro between the devices to ensure that they are working in step with each other. For example, the sending machine must tell the receiving machine when it is about to start transmitting data. The receiving machine must tell the sending machine whether it is ready to receive. Throughout the transmission exact synchronization must be maintained, and slippages in synchronization must be corrected. When the receiving machine detects errors, it must notify the sending machine, and the erroneous data must be retransmitted.

This chapter describes the signals that pass to and fro to ensure that the machines are working together, on a point-to-point line. Chapter 9 examines the same topic for a multipoint line. It is, of course, more complicated on a multipoint line, because now a machine may be conversing with only one of many machines to which it is connected. It must address the requisite machine and the others must be kept from interfering.

The lines in question may be half duplex or full duplex. Many data transmission machines are designed so that they can operate on either half- or full-duplex lines. The technique of line control, however, varies somewhat, because on a full-duplex line, control signals can travel in one direction while data are traveling in the other.

The lines may go through a switching center, or they may be private leased lines. Private leased lines themselves may be switched through a private branch exchange. A variety of configurations with switching is shown in Chapter 8. Switched or not, all the line configurations illustrated in Chapter 8 are basically point-to-point lines. If the public network is used, the first step must be dialing the connection. This may either be done automatically, or an operator may dial a "talking" connection and then switch the modem she is using from "voice" to "data."

GETTING STARTED

Some data transmission links need an operator at each end of the link in order to establish the connection. Many machines, however, are capable of unattended operation and so, when called, the modem must turn itself and its associated data-handling machine on without human intervention.

Figure 6.1 illustrates the opening sequence of events on a typical start-stop system. The frequencies quoted in this diagram are those of AT&T's type 103A data set. This machine uses frequency-shift keying on two separate frequency bands, one for each direction of transmission. In this way it gives full duplex transmission over switched public voice lines at 300 bits per second or over TWX lines at 150. It transmits in a binary fashion, sending either a MARK or a SPACE frequency ("1" or "0"). These frequencies are audible, and if an earphone were connected the operator would be able to hear the whistle of data rushing long the line.

The first step in establishing the connection is to dial the distant machine. This might be done either by the originating operator, or in some cases automatically by an originating machine such as a computer. When the connection is made, the answering data set must be placed in its "data mode" so that it is ready to receive and transmit data. This may be done automatically or it may be done by the answering operator pressing the DATA key. It may be done entirely nonautomatically by two operators talking to each other and agreeing to press the DATA key.

One and a half seconds after the establishment of *data* mode, the answering set places its MARK frequency on the line. If the originating call came from an operator, she will hear this as a high-pitched whistle or "data tone." She will then press the DATA key on her set. She can then no longer hear the *data tone* because she has switched the set from *talk* to *data* mode and telephone is no longer connected. If the call was not originated by an operator, the data set will be automatically placed in *data* mode. The pause of 1.5 seconds before answering is used to avoid interfering with certain tone signaling actions that are used on the telephone network.

A similar pause of 1.5 seconds occurs after the originating set is placed in data mode and it then places its MARK frequency on the line (a different frequency because these modems are designed for full-duplex operation). Any echo suppressor disablers on the line are told by this signal that this is a data transmission, so they stop the action of their associated echo suppressor. The echo suppressor normally permits transmission in one direction at a time only, but here simultaneous two-way transmission is needed, so the echo suppressors are automatically disabled.

Both sets, once they receive the other set's MARK frequency, place their transmitting circuits under control of their data-handling machine. Thus the connection for full-duplex data transmission has been set up.

98 POINT-TO-POINT LINE CONTROL

	OPERATOR AT BOTH ENDS	OPERATOR AT ORIGINATING END ONLY	FULLY AUTOMATIC
1.	Operator at A dials B and talks to operator there	Operator at A dials B	Data processing machine at A dials B
2.	Operator at B presses DATA key, placing modem B in data mode	Modem B is automatically placed in data mode. B placed on "off-hook" condition on the line to A	
3.	"MARK" TONE (2025 cycles) — Modem B is now connected to the data handling machine at B	"MARK" TONE (2025 cycles)	"MARK" TONE (2025 cycles)
	1.5 seconds after the data mode is established, modem B places a MARK condition on the line			
4.	B's "MARK" TONE — Operator at A hears the 2025 cycle frequency and presses the DATA key, placing modem A in data mode	B's "MARK" TONE	B's "MARK" TONE — Modem A is automatically placed in data mode
5.	B's "MARK" TONE / A's "MARK" TONE (1070 cycles)	B's "MARK" TONE / A's "MARK" TONE (1070 cycles)	B's "MARK" TONE / A's "MARK" TONE
	1.5 seconds after the data mode is established, modem A places its MARK condition on the line			
6.	DATA / DATA	DATA / DATA	DATA / DATA
	After a slight delay whilst circuits are connected, data is transmitted (full duplex if so desired)			

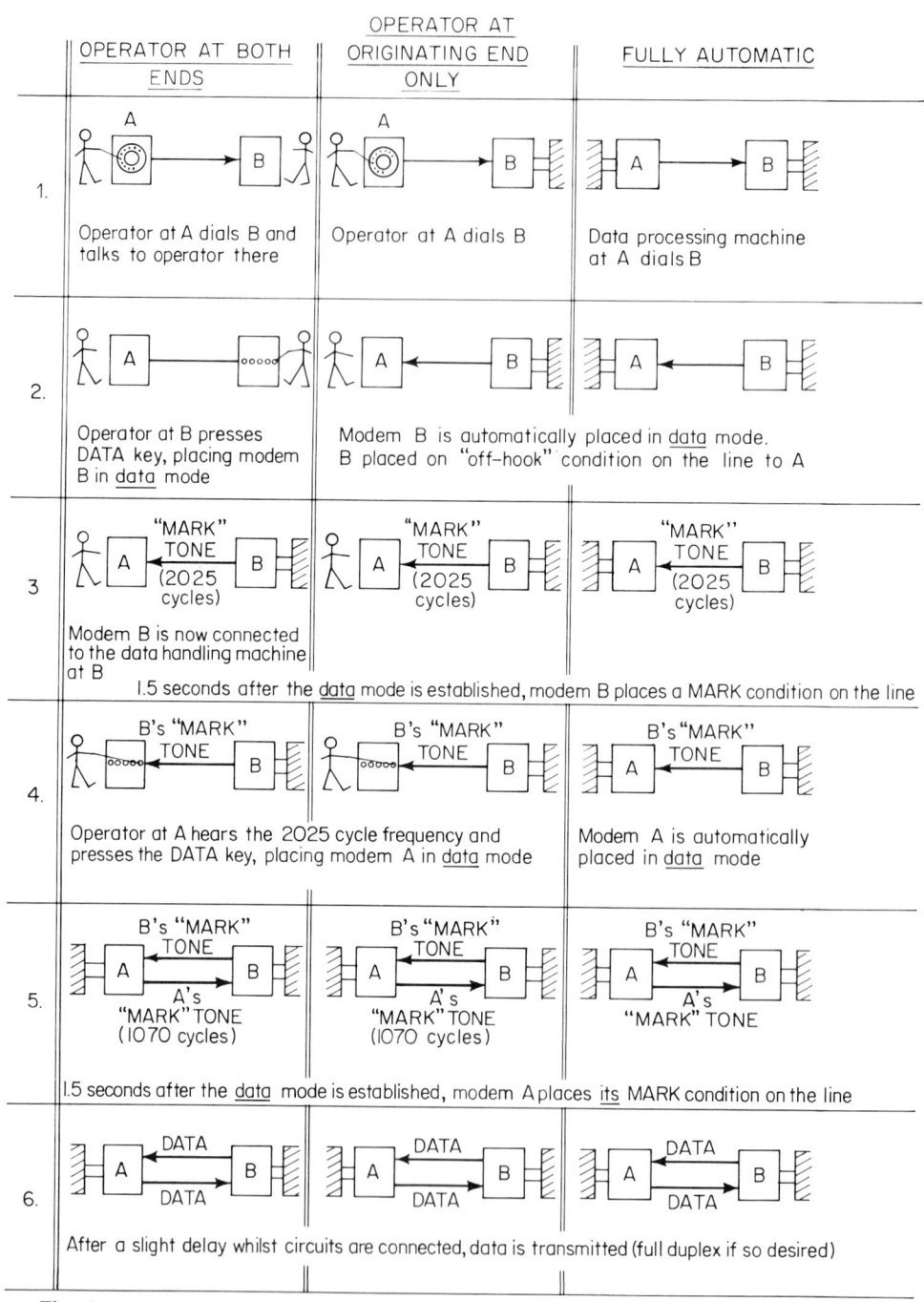

Fig. 6.1. Modem "handshaking" establishing a connection ready for transmission.

ESTABLISHING SYNCHRONIZATION

The above modem is normally used only with start-stop transmission, so no synchronization problem arises. After the connection is established, the two machines are sending a continuous MARK (or "1") line condition to each other. The first change to a SPACE (or "0") condition is interpreted as a START bit, and the bit after that as the first data bit (as was shown in Fig. 3.5).

With synchronous transmission—usually used on higher speed lines—the terminals must first establish synchronization. As was discussed in Chapter 3, this may be done by sending a special unique synchronization pattern before each message, like the 111111, 111110 of Fig. 3.9. More commonly with point-to-point transmission the machines, when they are idling, send a continuous stream of synchronization characters to each other. They are then sure to be in synchronization whenever a data message is sent.

On a full-duplex circuit the synchronization or "idle" characters travel continuously in both directions at once. On a half-duplex line each terminal may alternate between sending these characters and listening. On a typical system one terminal transmits synchronization characters for 1.5 seconds and then sends a *turnaround* character. When the other terminal receives this, it switches into transmitting mode and itself sends synchronization characters for 1.5 seconds, and then a turnaround character.

Some terminals send synchronization characters in the middle of a data record. The normal flow of data is interrupted every second or so to transmit a synchronization pattern.

CONTROL CHARACTERS

In order to control the flow of data on the line, to retransmit messages in error, and so on, a number of *control characters* are needed. The codes shown in Chapter 2 vary widely in the number of control characters they permit. Figure 2.8, for example, lists the large number of characters that are permitted in the American Standard (ASCII) code. Other codes do not have as many as this. The 4-out-of-8 code, for example, in Fig. 2.7 has only six characters that are uniquely *control* characters, though some of its 64 data characters are also designated for control functions. Baudot code in Fig. 2.1 is still worse off.

In many transmission systems, because of the shortage of allowable control characters, a combination of *two* characters has to be used. Two characters form a unique pair that is recognized by the receiving machine and which then causes some action to take place, or conveys some supervisory signal.

In the illustrations of control sequence that are to follow in Figs. 6.4 through 6.11, 4-out-of-8 code is used. This is a code that is widely used in

100   POINT-TO-POINT LINE CONTROL

practice because it gives characters in which there is a good chance of detecting communication line noise. The characters used in these illustrations are listed in Fig. 6.2, and the two-character control sequences in

CHARACTER	BIT CONFIGURATION
	1 2 4 8 R O X N
IDLE (Used in establishing synchronization)	1 0 0 1 1 1 0 0
TL	1 0 1 0 1 1 0 0
CL	1 0 1 0 1 0 1 0
INQ or ERR ⎫	1 0 0 1 1 0 1 0
SOR 1 or depending upon whether	1 1 0 0 1 0 1 0
ACK 2 TL or CL precedes it.	
SOR 3 or	1 1 0 0 1 1 0 0
ACK 4 ⎭	
EOT ⎫ These are data characters	0 1 0 1 1 0 1 0
TEL ⎭ when not preceded by CL.	0 0 1 1 1 0 1 0
LRC The longitudinal redundancy check can be *any* combination of bits.	

Fig. 6.2. Four-out-of-eight code: bit configurations of the control characters.

Fig. 6.3. These are the characters and control sequences used in IBM's Synchronous Transmit Receive (STR) range of equipment. The first character of any control sequence is **CL** or **TL**, and this conditions the meaning of the second character.

	FIRST CHARACTER	SECOND CHARACTER
End of control period	CL	IDLE
Inquiry	TL	INQ
Start of odd-numbered record	TL	SOR 1
Start of even-numbered record	TL	SOR 2
End of record	TL	LRC
Acknowledgment of odd-numbered record	CL	ACK 1
Acknowledgment of even-numbered record	CL	ACK 2
Error detected in message received	CL	ERR
End of transmission	CL	EOT
Operator request for telephone connection	CL	TEL

Fig. 6.3. Two-character control sequences.

POINT-TO-POINT LINE CONTROL 101

IDLING, WAITING FOR A SIGNAL

Figure 6.4 illustrates the flow of characters on a half-duplex line when the terminals are idling, waiting for something to happen. For 1.5 seconds Terminal A is in control. It sends **IDLE**, synchronization characters, to Terminal B, then it passes control to B by sending the **CL, IDLE** sequence. Now B is in charge. B sends **IDLE** characters to A for 1.5 seconds, and then passes control back to A. This "handshaking" goes on until either A or B has some data to send.

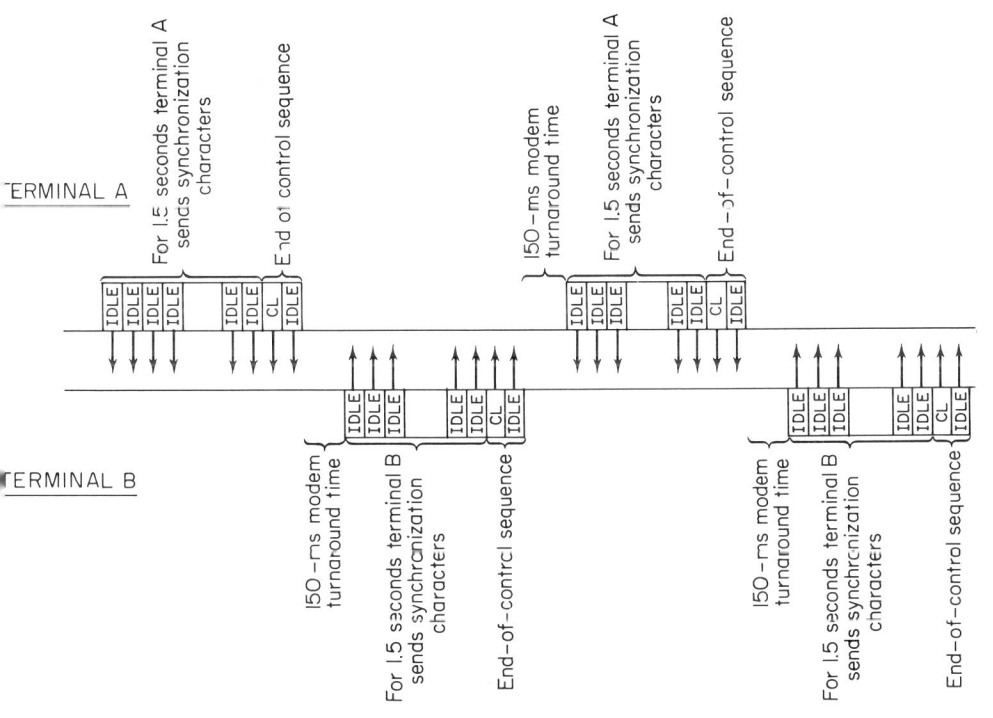

Fig. 6.4. Maintaining synchronization during an idle period ("handshaking" mode) on a typical half-duplex line.

The modems were not involved at all in the sequence in Fig. 6.4. Sometimes, particularly on higher-speed lines, the modems transmit their own synchronization characters, but on lines such as that in Fig. 6.4 they must have some other means of maintaining synchronization. Figure 6.5 shows a full-duplex line in which the modems transmit their own special character, **DSI** (Data Set Idle), during the idle period. It is possible that continuous transmission of the terminal's **IDLE** character could interfere with line equipment on high-speed lines, and the modem uses an **IDLE** character least likely to give any interference.

102 POINT-TO-POINT-LINE CONTROL

Once every half second, the terminal transmits *its* **IDLE** character so that the data-handling machines can stay in synchronization. Again there is a **CL**, **IDLE** sequence at intervals by means of which the terminals alternate in their control of the line. In Fig. 6.5, A is in control for half a second, then B for half a second, and so on. If during this period A needs

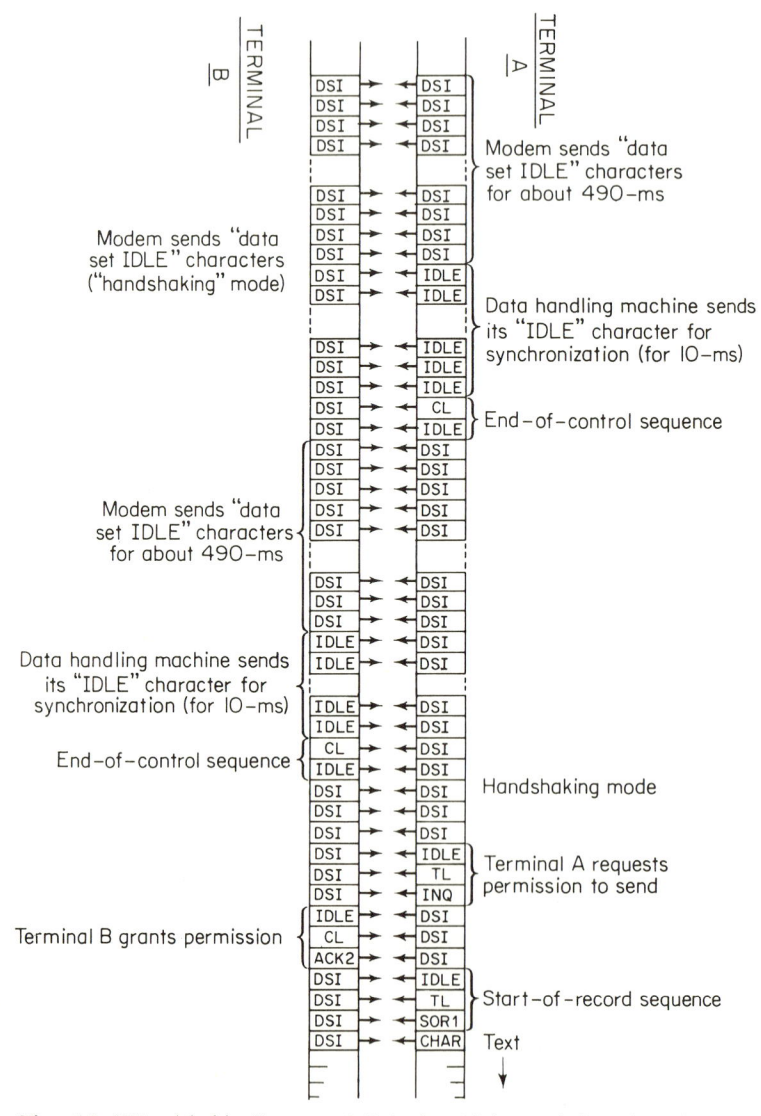

Fig. 6.5. "Handshaking" on a full-duplex high-speed line (usually higher than voice grade) on which the modems send their own synchronization characters (DSI—"Data set idle" character).

POINT-TO-POINT LINE CONTROL 103

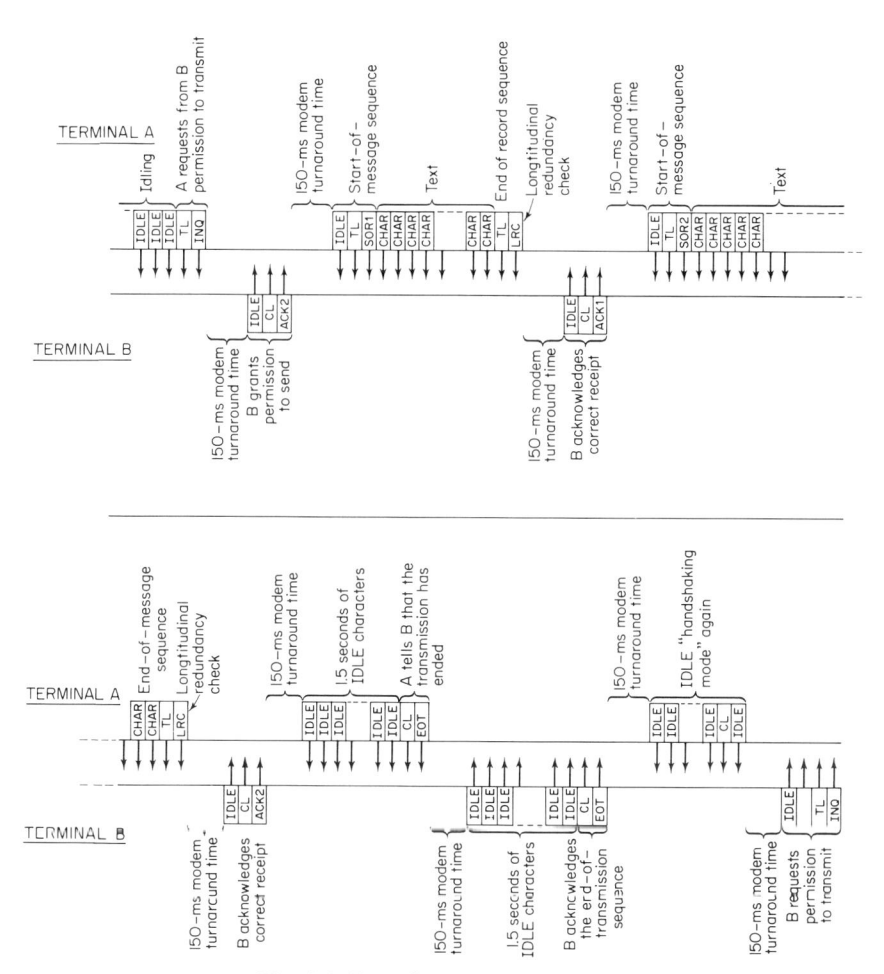

Fig. 6.6. Error-free transmission on a typical half-duplex line.

to send something, it changes the "handshaking" sequence to a request for B's permission to transmit.

REQUESTING PERMISSION TO TRANSMIT

The two characters used to request permission to transmit are **TL, INQ**. These can be seen in Fig. 6.5 for a full-duplex line and in Fig. 6.6 for a half-duplex line. During the handshaking sequence the terminal in control inserts these amid its idle characters.

The other terminal wired to recognize these characters must then reply, granting permission to send. After sending an **IDLE** character, it replies **CL, ACK 2** in these illustrations. This tells the sending terminal that it is ready to receive. The sending terminal replies (Fig. 6.6): after an **IDLE** character it sends **TL, SOR 1**, thereby telling the receiving terminal that the next character is to be the first character of data. **SOR 1** is used, not **SOR 2**, because this is the first message of what may be a group of messages transmitted by A. The next one will use **SOR 2**; after that, **SOR 1** again, and so on. This will aid the receiving terminal in ensuring that no single message has been lost. The operation on full-duplex lines shown in Fig. 6.5 is similar.

CONFIRMATION OF CORRECT RECEIPT

As discussed on p. 79, data transmitted may have two means of error detection built into them. First, checking may be built into each character with, for example, parity coding. Second, the message or record transmitted may have an error-detection pattern sent with it, often in the form of a redundant character or characters. In the examples of Figs. 6.4 through 6.11 both of these error-detection means are used. The 4-out-of-8 character coding tells the receiving machine whether bits have been lost from or added to any character, and a redundant character labled **LRC** in the figures is sent with each message. The latter is commonly assembled as the message is sent, both by the receiving machine and the sending machine. The sending machine transmits its **LRC**, and the receiving machine compares it with the one it has assembled to see whether it can detect any difference.

Some *start-of-message* character must tell the receiving machine when to begin assembling its error-detection pattern, and some *end-of-message* character must tell it when to stop, and to make the comparison. In Figs. 6.4 through 6.11 the start-of-message sequence **TL, SOR 1** or **TL, SOR 2** causes the former, and the **TL** at the end of the message stops the assembly.

If the receiving machine receives a message completely with no error detection, it may send back a signal acknowledging this success. In Figs. 6.6 through 6.10 it sends back (after an **IDLE** character) **CL, ACK 1** or **CL, ACK 2**, depending upon whether it is replying to a message started with **SOR 1** or **SOR 2**. The sending terminal receives this positive acknowledgment and so sends the next message, if it has any more to send. If it does not have any more to send, it must relinquish its control of the line. In Fig. 6.6 it idles for a time to give it data-processing machine time to transmit if this is likely, then it sends an *end-of-transmission* sequence **CL, EOT**. The receiving machine acknowledges that it has received this, by itself sending **CL, EOT**. Then the two terminals resume their "handshaking" operation again, waiting for further action.

ERROR SIGNALS

In Fig. 6.7 an error was detected by terminal B in the transmission it received. This may have been a 4-out-of-8 code error, or an error in the LRC, or even an error in a transfer of data within the machine's own circuits. Whatever the cause of the error, it signals back **CL, ERR**, instead of **CL, ACK 1** or **CL, ACK 2**.

The transmitting machine will have retained its access to the message up to this time, and will now retransmit it in its entirety. In Fig. 6.7 it is correctly received this time. The receiving machine replies with the normal **CL, ACK 1**, and transmission continues.

Suppose that the receiving terminal failed to reply at all. The sending terminal would sit waiting for the acknowledgment of success which it would not receive. It must be designed with some form of *timing mechanism* so that it does not wait too long. In Fig. 6.8 terminal B fails to reply to terminal A. Terminal A waits patiently for three seconds and then assumes that B has not received the message. Terminal A requests permission to send it again: **TL, INQ**. It is possible that the control characters themselves had become garbled in the transmission and that this is why A did not receive an intelligible reply from B. Whatever the reason was, all is well this time. B grants permission to re-send: **CL, ACK 1,** and this time sends a correct acknowledgment to A.

In Fig. 6.9 the situation is somewhat worse. Terminal B never replies. Perhaps the cleaning lady has accidentally pulled the plug out with her mop. Terminal A waits the customary three seconds and then requests permission to re-send: **TL, INQ**. B does not reply to this. A tries again, twice, in vain, and then gives up. A sends **CL, EOT** down the line, relinquishing control, and sounds an alarm to notify its operator.

The action would have been similar if there had been failure to transmit because of persistent errors on the line or in B's circuitry. After a number of sequential unsuccessful attempts to retransmit, A would have relinquished control and notified its operator. B, on most schemes, would notify its operator also.

CONVERSATIONAL TRANSMISSION

If the receiving terminal has something to transmit, itself, then it is perhaps not necessary that it should send a separate positive acknowledgment signal. The sending back of data can be taken as implying a successful receipt of the previous transmission. This is illustrated in Fig. 6.11, in this case for a full-duplex line, though it could apply equally well to a half-duplex line.

It will be seen that this speeds up the "conversation." On a half duplex line it might be of more value than on a full-duplex line, because a certain time is needed to reverse the transmission direction on the line.

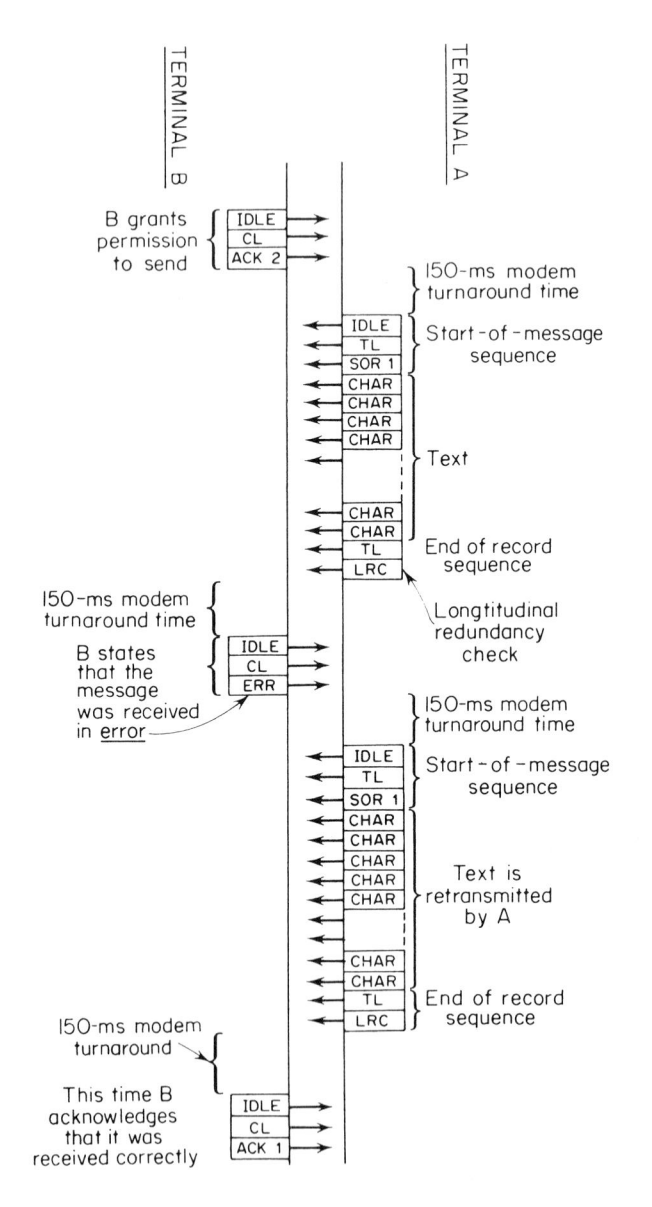

Fig. 6.7. The same transmission as in Fig. 6.6, but now an error is detected in the transmission of the first record.

Fig. 6.8. Half-duplex transmission as in Fig. 6.6, but now terminal B fails to reply. Terminal A retransmits and then B does reply.

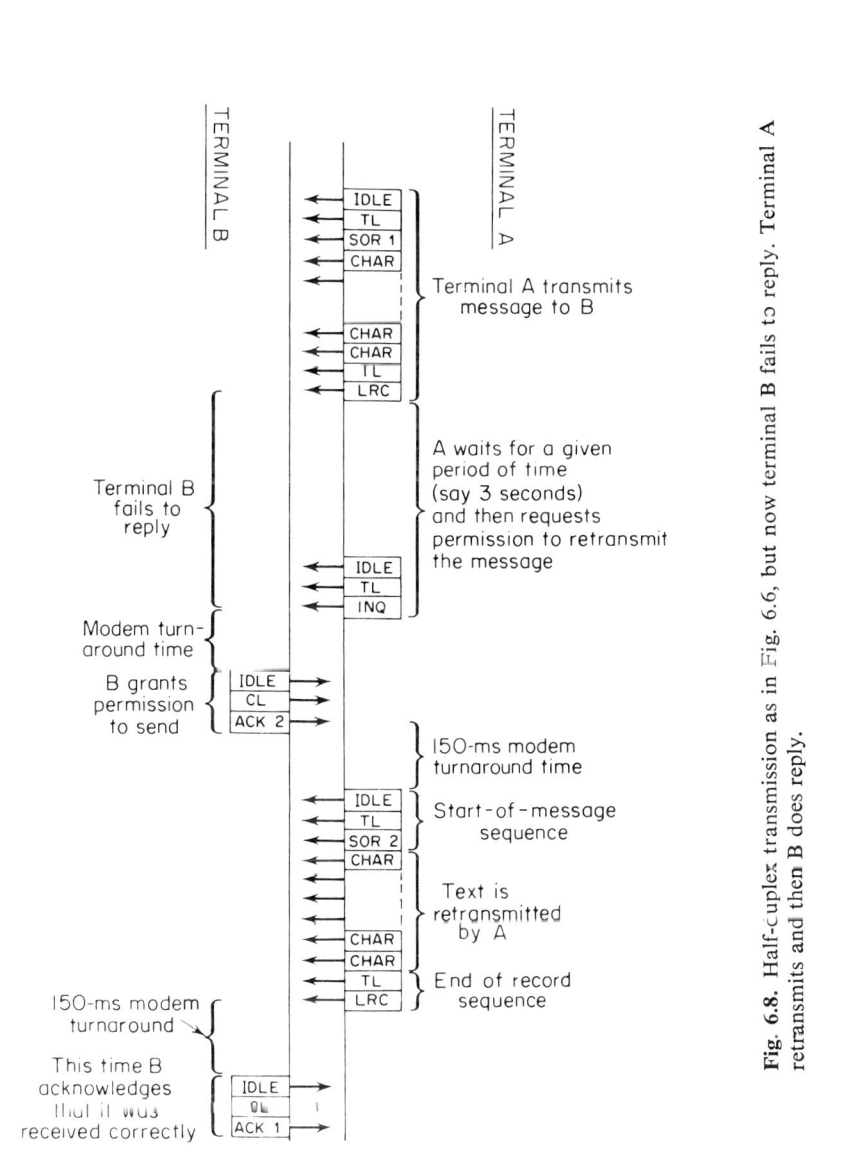

108   POINT-TO-POINT LINE CONTROL

This is shown as the 150 millisecond "modem turnaround time" in Fig. 6.6. Comparing Fig. 6.6 with Fig. 6.11, the reader will see that considerable time-saving could result from this conversational interaction. We will meet it again in talking about multipoint lines, for example, lines connecting several display-screen terminals to a computer. Here it is more often of more value than on point-to-point lines because more "conversation" takes place.

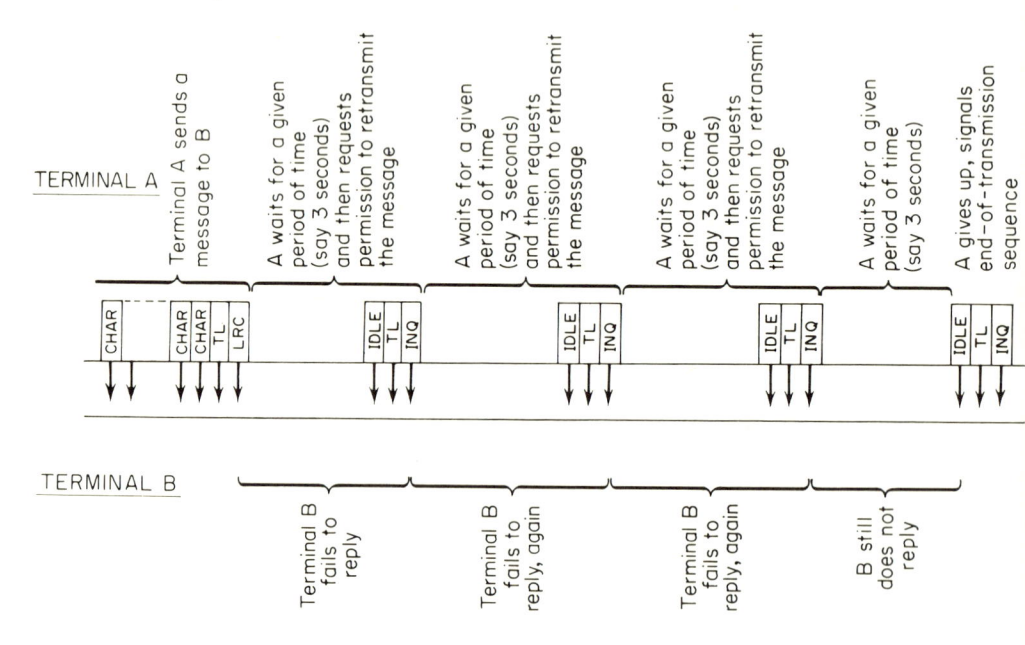

Fig. 6.9. Terminal B fails to reply to a message as in Fig. 6.8. Terminal A tries three times to contact B but without success, so A terminates the operation, sounding a buzzer to call its operator.

MODEM TURN-   It is often not practical to organize a full-duplex
AROUND TIME   line so that *data* are transmitted in both directions at once. Sometimes data can be flowing in one direction while control signals travel in the other. For example, the acknowledgment of successful receipt may be sent back at the same time as the following record is being transmitted. Many transmission devices, however, do not even do this, so there is no overlapping of either data or control characters. The flow of characters on such a full-duplex line is shown in Fig. 6.10.

Comparing Fig. 6.10 with Fig. 6.6, we see that the only savings that accrue from the full-duplex facility are the avoidance of the periods of time for modem turnaround. These are shown in the illustrations as being

POINT-TO-POINT LINE CONTROL 109

150 milliseconds in duration, and this is a typical time for a high speed line. Sometimes the time taken to reverse the direction of transmission on a half-duplex line is less than this; sometimes it is greater.

If the communication line is transmitting at a speed of 2400 bits per second, the 150 milliseconds is equivalent to the transmission of

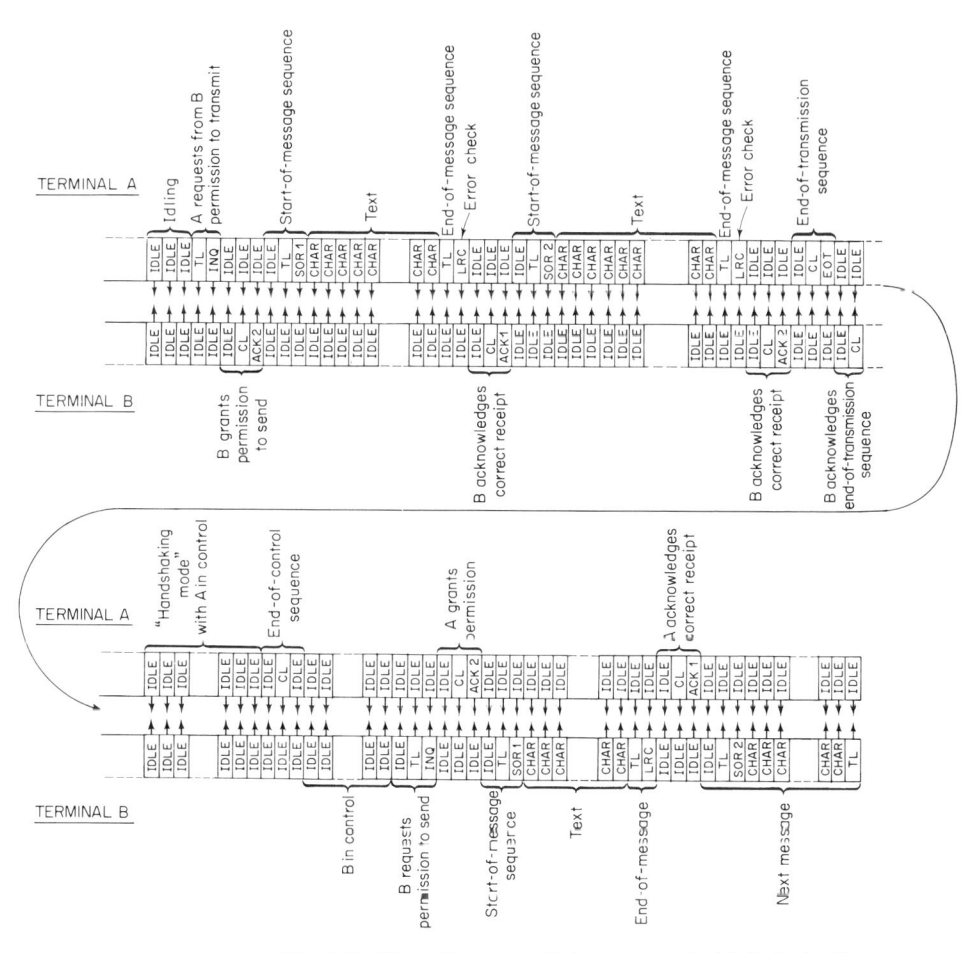

Fig. 6.10. Error-free transmission on a typical full-duplex line.

(2400 × 150) ÷ 1000 = 360 bits. For the 4-out-of-8 code in the illustrations this is equivalent to 45 characters. It will thus be seen that the modem turnaround time is not drawn to scale. If it were, on this basis, it would be almost as wide as the entire figure. For transmission at 4800 bits per second it would be twice this width.

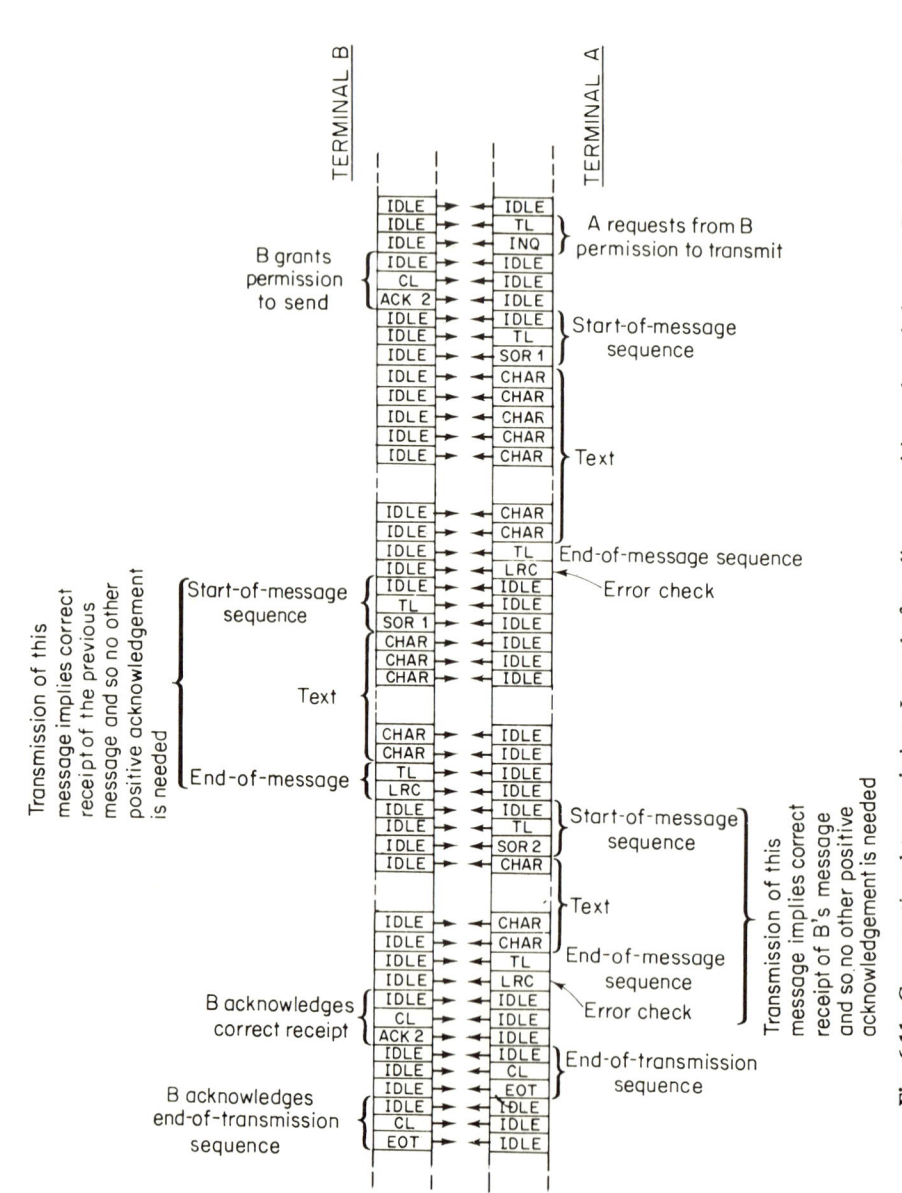

Fig. 6.11. Conversational transmission. Instead of sending a positive acknowledgment, B sends a data message to A. This implies correct receipt of A's message. Terminal A then responds similarly to B. Where this is possible, it saves time.

The sequence of operations in sending one record in Fig. 6.6 is as follows:

1. A requests permission to transmit.
2. Modem turnaround.
3. B grants permission to transmit.
4. Modem turnaround.
5. A transmits.
6. Modem turnaround.
7. B acknowledges receipt.
8. Modem turnaround.

There are four modem turnaround periods. These are, therefore, equivalent to the transmission time of 180 characters, on a 2400-bit-per-second line. Perhaps the message being sent might be only 100 or so characters itself, though sometimes much longer records are transmitted. It is clear that the turnaround time is a significant part of the whole, and it can be saved by using a full-duplex line as in Fig. 6.10.

TRANSMISSION OF DATA IN BOTH DIRECTIONS AT ONCE

Tighter organization of the full-duplex line can clearly save further time. Overlapping the message acknowledgments with the transmission of the subsequent message is a step towards this, and better still is a line on which data can be sent in both directions at once. Such a line is illustrated in Fig. 6.12.

It is difficult to organize full-duplex transmission so that fully variable-length records can be sent and acknowledged in both directions at once. Some schemes therefore transmit fixed-length data blocks from the buffers of the data-handling machines. This is the case in Fig. 6.12. A record in this illustration begins with a single start-of-record character and ends with two checking characters.

As soon as one record is transmitted the next is sent on the assumption that no error has occurred. If one should occur, and be detected, the terminal receiving that message would stop its transmission of data, and send an error character to the other terminal. When the other terminal received this, it would also stop *its* transmission, placing a cancel character in the message it is sending in this illustration. The message that was in error would then be retransmitted, and after this both terminals would continue transmission where it was interrupted.

In spite of its efficiency this duplicate transmission of data is not often used in practice. It is rare that the data-processing situation provides enough data for both directions of transmission. Perhaps in the future it will become more common.

112 POINT-TO-POINT LINE CONTROL

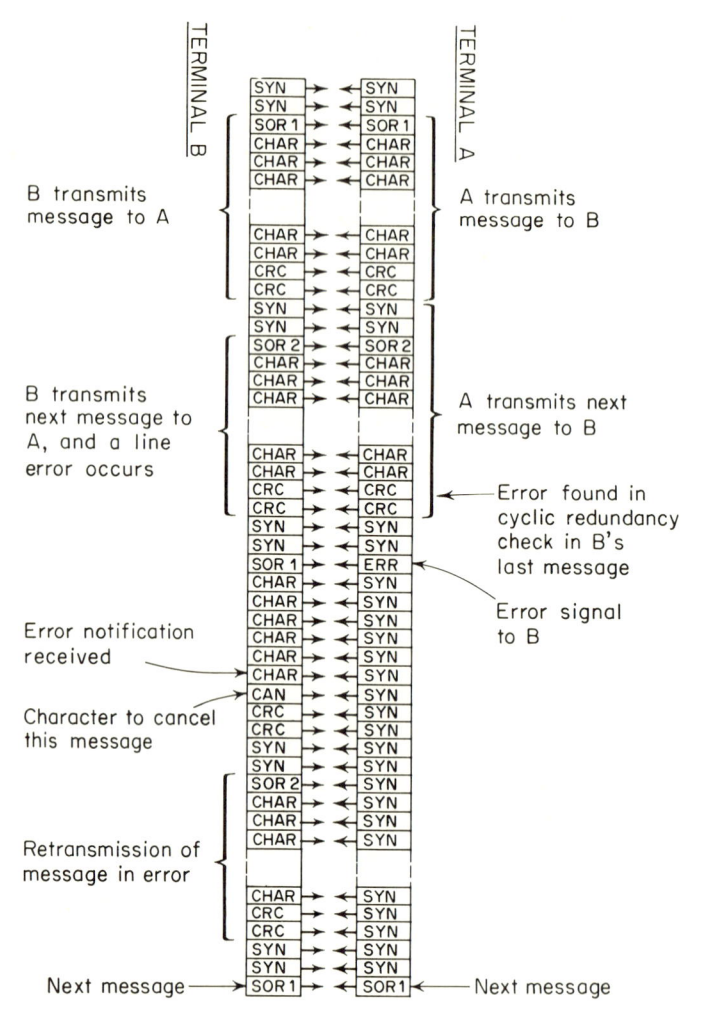

```
                    TERMINAL B                              TERMINAL A
                     SYN  →  ← SYN
                     SYN  →  ← SYN
                     SOR1 →  ← SOR1
                     CHAR →  ← CHAR
                     CHAR →  ← CHAR
                     CHAR →  ← CHAR
   B transmits                                    A transmits
   message to A                                   message to B
                     CHAR →  ← CHAR
                     CHAR →  ← CHAR
                     CRC  →  ← CRC
                     CRC  →  ← CRC
                     SYN  →  ← SYN
                     SYN  →  ← SYN
                     SOR2 →  ← SOR2
                     CHAR →  ← CHAR
                     CHAR →  ← CHAR
   B transmits       CHAR →  ← CHAR
   next message to                                A transmits next
   A, and a line                                  message to B
   error occurs
                     CHAR →  ← CHAR
                     CHAR →  ← CHAR
                     CRC  →  ← CRC  ← Error found in
                     CRC  →  ← CRC     cyclic redundancy
                     SYN  →  ← SYN     check in B's
                     SYN  →  ← SYN     last message
                     SOR1 →  ← ERR
                     CHAR →  ← SYN  ← Error signal
                     CHAR →  ← SYN     to B
                     CHAR →  ← SYN
   Error notification CHAR →  ← SYN
   received          CHAR →  ← SYN
                     CHAR →  ← SYN
                     CAN  →  ← SYN
   Character to cancel CRC →  ← SYN
   this message      CRC  →  ← SYN
                     SYN  →  ← SYN
                     SYN  →  ← SYN
                     SOR2 →  ← SYN
                     CHAR →  ← SYN
   Retransmission of CHAR →  ← SYN
   message in error  CHAR →  ← SYN

                     CHAR →  ← SYN
                     CRC  →  ← SYN
                     CRC  →  ← SYN
                     SYN  →  ← SYN
                     SYN  →  ← SYN
   Next message →    SOR1 →  ← SOR1  ← Next message
```

Fig. 6.12. Full-duplex transmission of fixed-length blocks with data being transmitted in both directions at once. Transmission of the next message acts, in effect, as an acknowledgment unless an error occurs. Few data-processing systems are able to utilize this two-way transmission of data.

HOUSEKEEPING CORE PLANES

On machines such as these, is it necessary to translate the coding used internally in the machine into the coding used on the communications line, if this is different, and vice versa. This is done by a transmit translator and a receive translator. These can be simple devices. Often a specially wired core plane is used.

Similarly the machine must recognize certain special characters. **ACK, NAK, ETX,** and the other control characters on the communication line must be recognized, as well as synchronization characters and nondata characters generated by the data-processing machine. Again, a specially wired core plane is commonly used to assist in these housekeeping functions.

Figures 6.14 and 6.15 illustrate the use of such core planes. In Fig. 6.14 each vertical line represents *one* ferrite core. This is a tiny doughnut-shaped ring of ferrite that has two stable magnetic states. It can be in a state of magnetization in either one direction or the opposite, as shown in Fig. 6.13. If the magnetization is in one direction, the core is referred to as being "set"; if in the other, it is "reset."

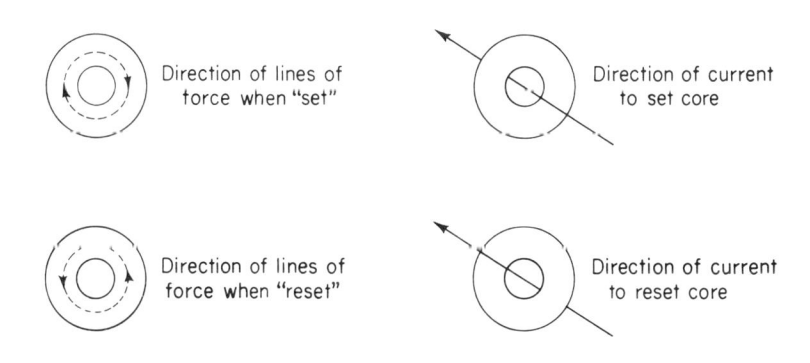

Figure 6.13

A number of wires may pass through each core; these are shown as horizontal lines in Fig. 6.14. As shown on the right-hand side of this diagram, the wires may pass through the cores once, in either direction, or twice by being looped around the core. The symbols at the intersections of the horizontal and vertical lines indicate whether the wire passes through the core once, twice, or not at all. It will be observed that several wires pass through each core.

CODE TRANSLATION Let us consider a machine which uses BCD coding internally, but which transmits and receives characters in 4-out-of-8 code. These two codes were shown in Figs. 2.5 and 2.7. Before the BCD characters are transmitted, they must be converted to 4-out-of-8 code. Figure 6.14 is used to do this. The characters received must be converted to BCD with a similar core plane. A variety of different core mechanisms are in use for this. We will illustrate a typical one.

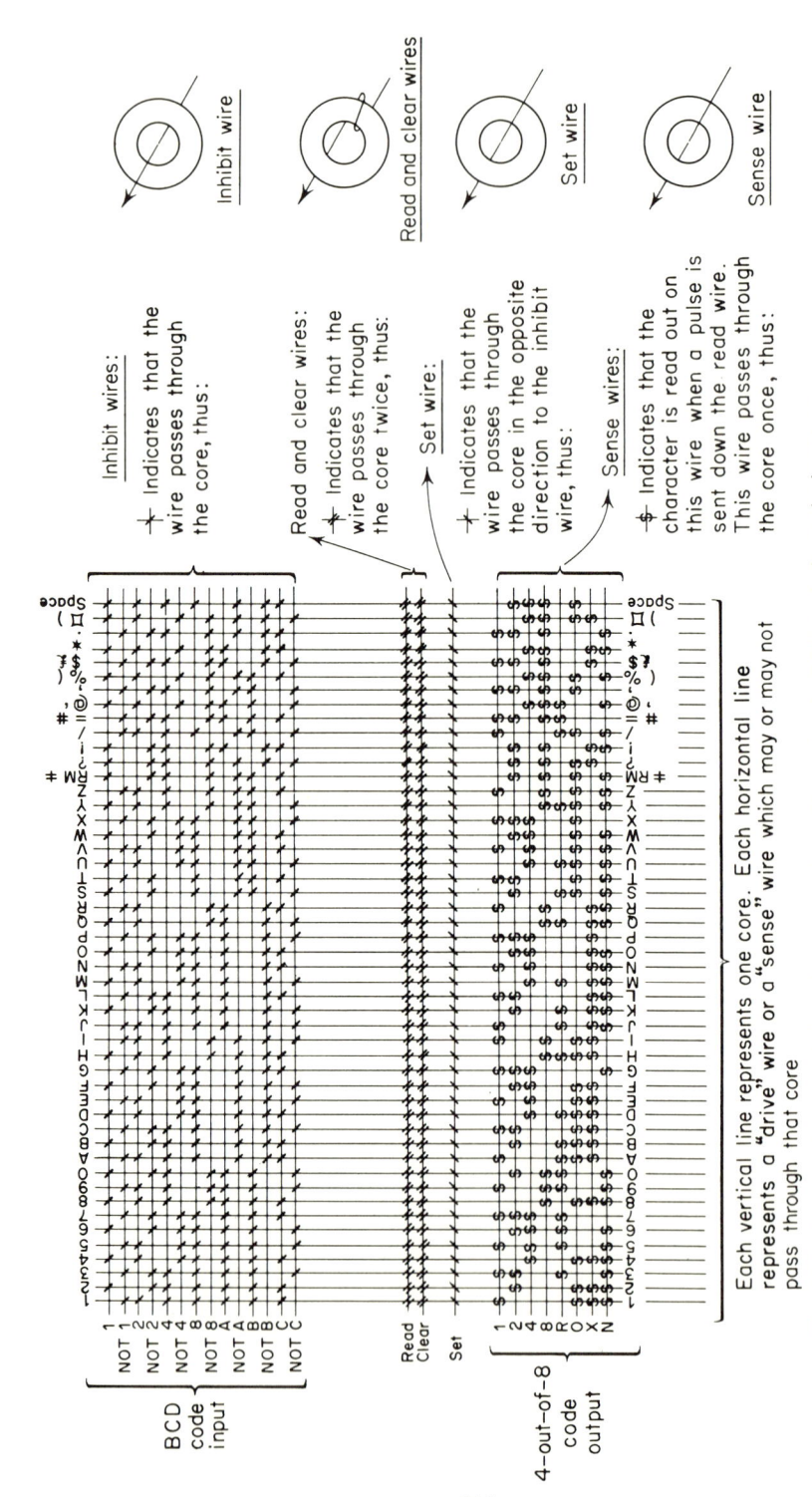

Fig. 6.14. A transmission code translator core plane. Characters in BCD code form the input on the inhibit wires. The sense wires read out these characters in 4-out-of-8 code, the code used for transmission.

POINT-TO-POINT LINE CONTROL 115

The core planes convert one character at a time. In Fig. 6.14 the BCD character is read in, the *inhibit* wires in the upper half of the diagram being used. If the BCD character contains a 1 bit, the 1 wire is energized. If it does not contain a 1 bit, the NOT 1 wire is energized. One of these two wires passes through each core. The same is true for all of the other bits in the BCD character.

Suppose that we are to attempt to translate an **E**: BCD character. An E is composed of 1, 4, A, B, and C bits. We therefore send a pulse down the inhibit wires labeled 1, NOT 2, 4, NOT 8, A, B, and C. Of all the cores shown in Fig. 6.14, only one will be *unaffected* by these pulses, and that is the E core. None of the above inhibit wires passes through the E core, but one or more of them passes through every other core.

To translate the character, three events take place, and these are shown in Fig. 6.15. First an impulse is sent down the *clear* wire. This resets all

Clear wire	⊓	First this wire resets all cores
Set wire	⊓	Second, this wire tries to set all cores but the
1, not 2, 4, not 8, A, B and C inhibit wires	⊓	inhibit wires negate its effect in all but the "E" core
Read wire	⊓	Third, at read time the "E" core is reset. As its magnetization
Output on the 1, 4, 0 and X sense wires	⋀	reverses direction, a pulse travels down the 1, 4, 0 and X sense wires and is amplified
Time	→	

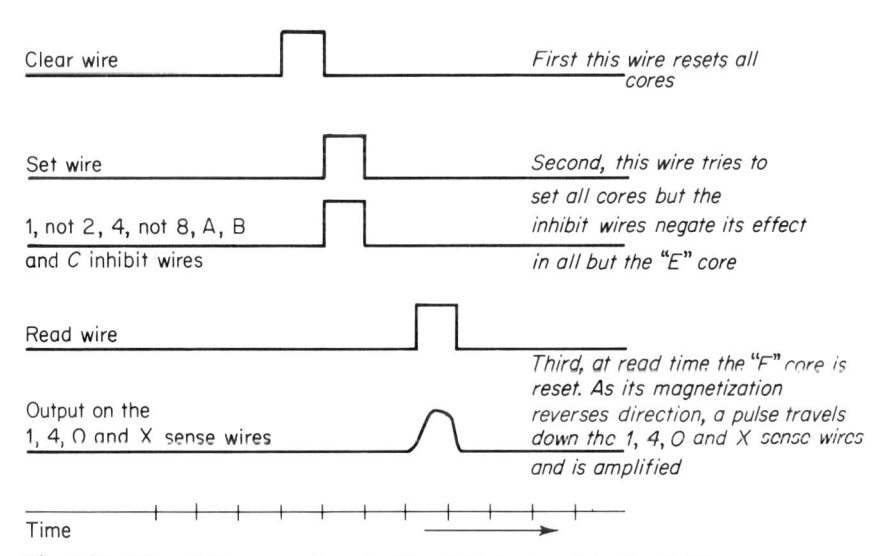

Fig. 6.15. Use of the core plane in Fig. 6.14 to translate the BCD character "E" into 4-out-of-8 code ready for transmission.

of the cores. Next, a pulse is sent down the *set* wire, but at the same time the above pulses are sent down the *inhibit* wires. The set wire attempts to set each core, but its effect is negated by the pulses on the inhibit wires for every core except the E core. As will be seen in Fig. 6.14, the inhibit wire passes through the core in the opposite direction to the set wire. Only the E core is set, because that does not have a pulse on any inhibit wire going through it. After this a pulse is sent down the *read* wire. This is strong enough to reset any core that is set. Only the E core *is* set, so that one alone switches its direction of magnetization. As it switches, it induces a pulse in any *sense* wire that passes through it. The 1, 4, 0, and x sense wires pass through the E core, so these wires alone are affected. The pulse generated on them is amplified, so we have produced the 4-out-of-8 character E.

The converse process happens when a 4-out-of-8 character is received on the communication line and must be translated to BCD. A similar mechanism can be used for recognizing special characters. These, instead of merely being code-converted, may cause a sense wire to be energized that gives rise to some special function.

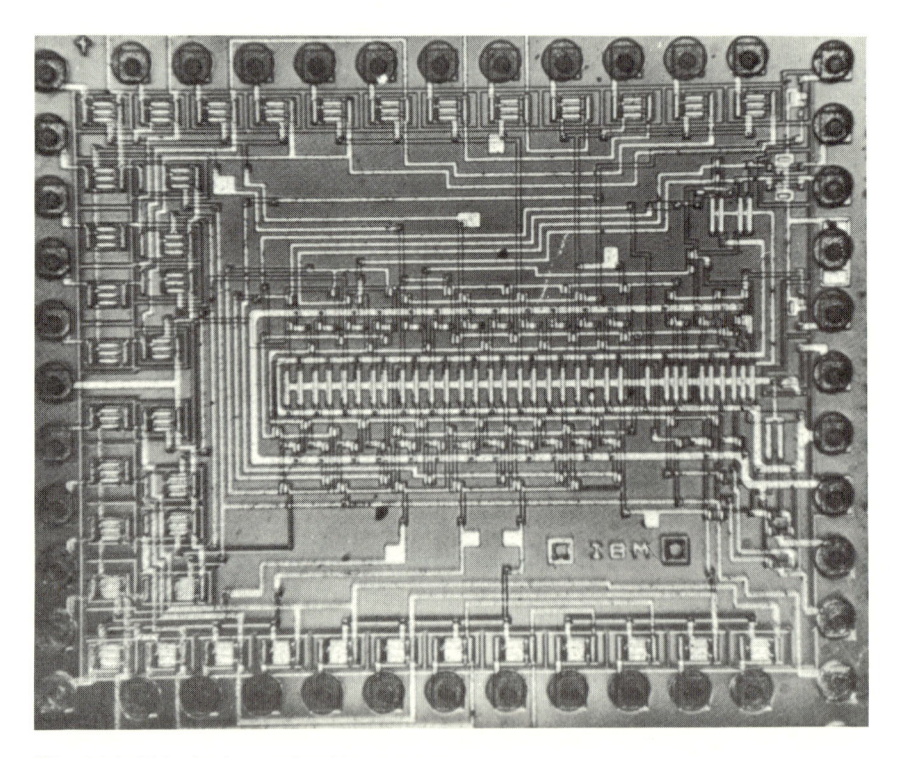

Fig. 6.16. This single metal oxide chip is little larger than one-tenth of a square inch, yet it contains circuitry which would have occupied a fairly large box of electronics in earlier years—a 32-channel multiplexer and a five-stage counter. Such "large-scale integration" circuitry can be mass-produced in great quantities by a process which in some ways resembles photography. If large enough quantities of such circuits can be sold, the cost of logic circuitry will drop to a fraction of today's cost. It will be economic to put much more logic in terminals and other tele-processing components.

LARGE-SCALE INTEGRATION

Where it is economic to do so such core mechanisms are being replaced by semiconductor electronics. *With large-scale integration* (LSI) circuitry the whole of the code translation mechanism of Fig. 6.14 could be placed on one tiny silicone chip. Figure 6.16 shows a typical LSI chip.

LSI circuitry offers two major advantages for teleprocessing equipment. First, it can be made to be highly reliable. This is important because the maintenance calls on a large network of terminals and could become prohibitively costly. Second, if made in large quantities, LSI circuitry can become inexpensive. It is very costly to set up the process for making an LSI chip, but once set up there is little extra cost in mass-producing it. LSI is therefore likely to be used for functions in which large quantities of the same complex logic circuit can be employed. There will be a major redesign of computers and other equipment to take advantage of LSI. The terminal area is one which lends itself to this. Many terminals will have their active logic circuitry, and buffers, on a single LSI chip.

The main disadvantages of LSI are the cost and difficulty of designing the chip and producing the masks needed to manufacture it, and the difficulty of modifying it once it is made. You cannot modify an LSI chip with a soldering iron as you could a transistor card. Where circuits are needed in smaller quantities, or are likely to need modifying, we will not use LSI. We may use a smaller level of integration or may use basic transistor cards and devices such as the core plane of Fig. 6.14.

Once the chips are under production, the cost per chip is not closely related to its logical complexity. A chip with 1000 transistors costs little more than one with 100. Therefore, if a terminal is going to have an LSI chip, it might as well have a complex one. On this basis we will be able to afford to put much more logic in our terminals than we do today. This is true for any telecommunication device that can be sold in quantities of several thousands. Some of the implications of this advantage will emerge in subsequent chapters. It means that the teleprocessing devices can be built to be more "intelligent" than they were before LSI. We can afford elaborate polling schemes, powerful error-detecting codes, error retransmission, logic circuitry to assist in editing and other functions, and multiplexing and concentration devices to conserve line cost. With this in mind, let us now explore the types of networks that can be built.

7 MEANS OF LOWERING COMMUNICATION NETWORK COSTS

As increasing numbers of terminals are used with one computer system so the need to devise means of lowering the overall cost of the network grows. Data transmission lines today work with various devices that were not found on early telegraph lines. The variety and scope of such devices are growing rapidly, and the systems analyst is faced with a confusing array of different machines and claims. It is intended that the remainder of this book should set these in perspective.

The purpose of much of the increased complexity is to reduce the overall network cost. The larger real-time communication networks of today would have been unthinkably expensive without the concentrators, multiplexors, or other devices that they use, and without their sometimes elaborate line control procedures. On such systems one can no longer simply connect each terminal by itself on a voice line, leased or dial-up, to the computer.

The techniques used depend upon the lengths of the communication lines and the numbers of terminals. If the lines are very long, they are expensive, so techniques for minimizing the cost of the lines needed dominate the network design. If the lines are short, their cost is of less concern, and the cost of the terminals and devices attached to the lines is of greater importance. Where the lines are *very* short, for example, all in one plant, one office block, or one campus, then their cost is of little significance in the design, and we often find their bandwidth being used quite wantonly. If a system has a large number of terminals, then the terminal cost becomes of major importance, and the network organization should use schemes that enable the terminal design to be as inexpensive as possible.

MEANS OF LOWERING COMMUNICATION NETWORK COSTS   119

1. SYSTEMS WITH VERY SHORT PRIVATE LINES

Let us start this discussion with the situation in which the lines are so short that we can almost ignore their cost. Bunches of wires, as in Fig. 7.1, travel

Fig. 7.1. Photograph of in-plant wiring.

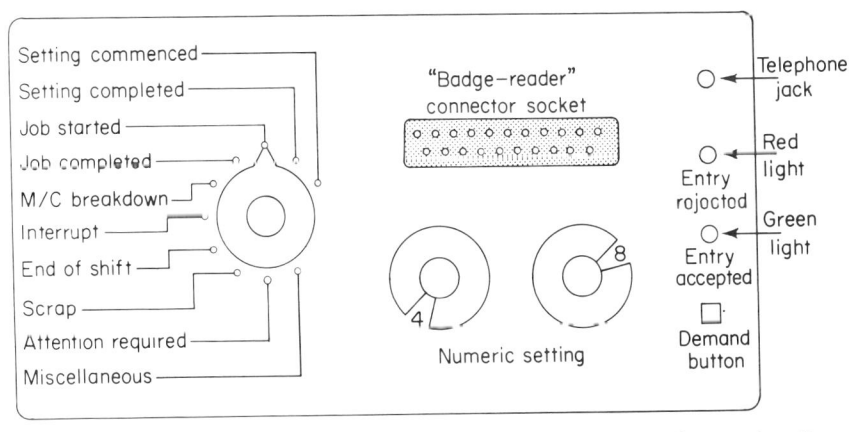

Fig. 7.2. A homemade terminal for data entry from a factory shop floor. The terminal is as inexpensive as possible, but the line requirements are high

120 MEANS OF LOWERING COMMUNICATION NETWORK COSTS

through a building to the terminals. We are prepared to lay many wires to one terminal if we can make the terminal cost sufficiently low.

The terminal might be a simple data entry device, for example, on the shop floor of a factory. Figure 7.2 shows such a terminal that is about as low in cost as possible. A number of companies have built their own homemade terminal out of inexpensive connectors, lamps, and rotary switches. A "badge-reader" in one case was fabricated from a cheap multiwire connector. The badges were appropriately labeled plugs with the pins wired across internally. Figure 7.3 shows a cheap 25-pin plug wired to represent up to ten digits.

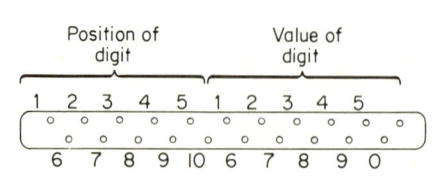

Let us suppose that such a device is used with the three ten-position rotary switches in Fig. 7.2. Suppose that no more than seven digits are coded into the badge-reader. We then have to read ten digits from the switches in the terminal. This would require ten wires into the terminal, one for each digit, and ten wires out for each of ten possible values of the digit.

Figure 7.4 shows the wiring from a scanning unit that is needed to read the badge and the switch settings of the terminal in Fig. 7.2. The scan control unit sends pulses one at a time down the ten wires to the terminal. The pulse is detected on one of the ten output wires, giving the value of that particular switch setting or badge digit. This scheme needs 20 wires to each terminal.

In fact, the terminal in Fig. 7.2 will need another five wires for the lights, demand button, and telephone jack. It is acceptable to lay down such multiwire cable over short distances.

Figure 7.5 shows wiring that could connect a large number of simple terminals to an IBM 1070 unit. In this case the IBM 1070 is remote from the computer and is connected to it over a conventional voice line, with modems. In order to attach as many terminals as possible to the IBM 1070

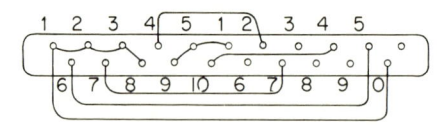

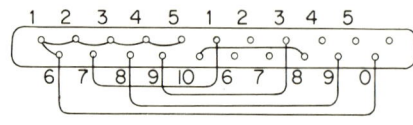

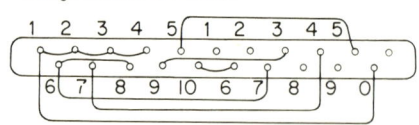

Fig. 7.3. A badge-reader in which the badge is made from an inexpensive 25-pin plug.

a specially engineered scanning device is used. This device sends the ten pulses down a multiwire cable to the terminal it is interrogating at that time. They return along a common group of ten wires to the IBM 1070. When a terminal raises the voltage on its "attention" line, it will be interrogated.

The total number of wires needed could be reduced by adding, somewhat, to the terminal cost. Ten wires are not needed to interrogate the ten-digit setting; four wires with binary coding could be used, but this would need a configuration of relays in the terminal to interrogate the individual switches and badge settings. Similarly, four wires with binary encoding could be used for the output, but this would require diodes in the terminals

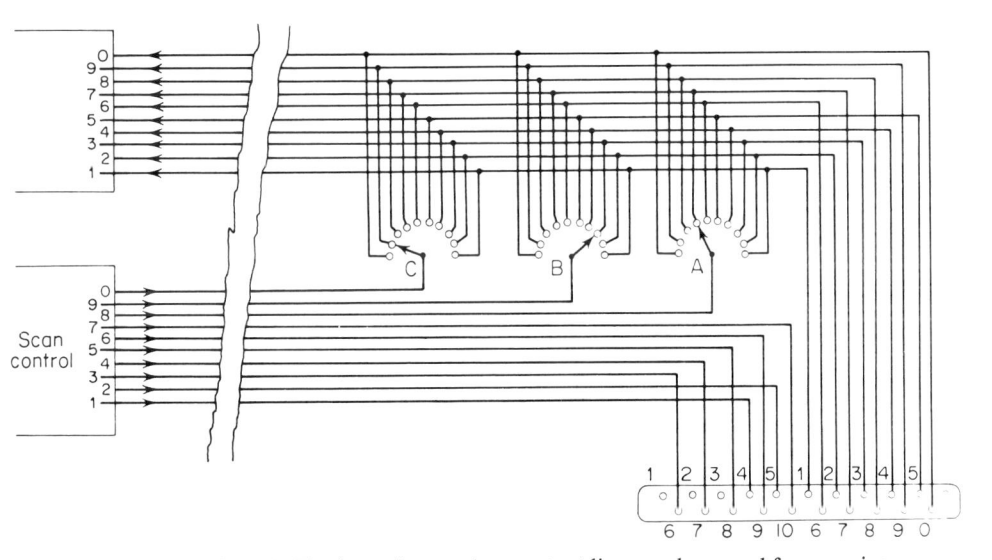

Fig. 7.4. Ten input lines and ten output lines are here used for scanning the rotary switches and badge-reader.

to prevent back-circuits. A still lower number of wires could be used if the terminal had a scanning mechanism inside it, to interrogate the digit positions in sequence. This would again increase the terminal cost. The terminal could, indeed, have only two wires going to it and could have a mechanism for sending a ten-digit word when interrogated, but the main objective of making extremely inexpensive terminals would have been lost. Where a factory or group of factories is to install many hundreds of terminals, this is important.

TERMINAL CONTROL UNITS

The terminals with their multiwire cable may be connected to an intermediate device which scans them and sends the data onwards to the computer on lines that are more efficiently used. Figure 7.5 is an example of this. Many other types of terminal control units with remote terminals are possible. On the terminal side of the control unit the cost of the wires, even though they may be several thousand feet in length, is subordinated entirely or partially to the cost of the terminal.

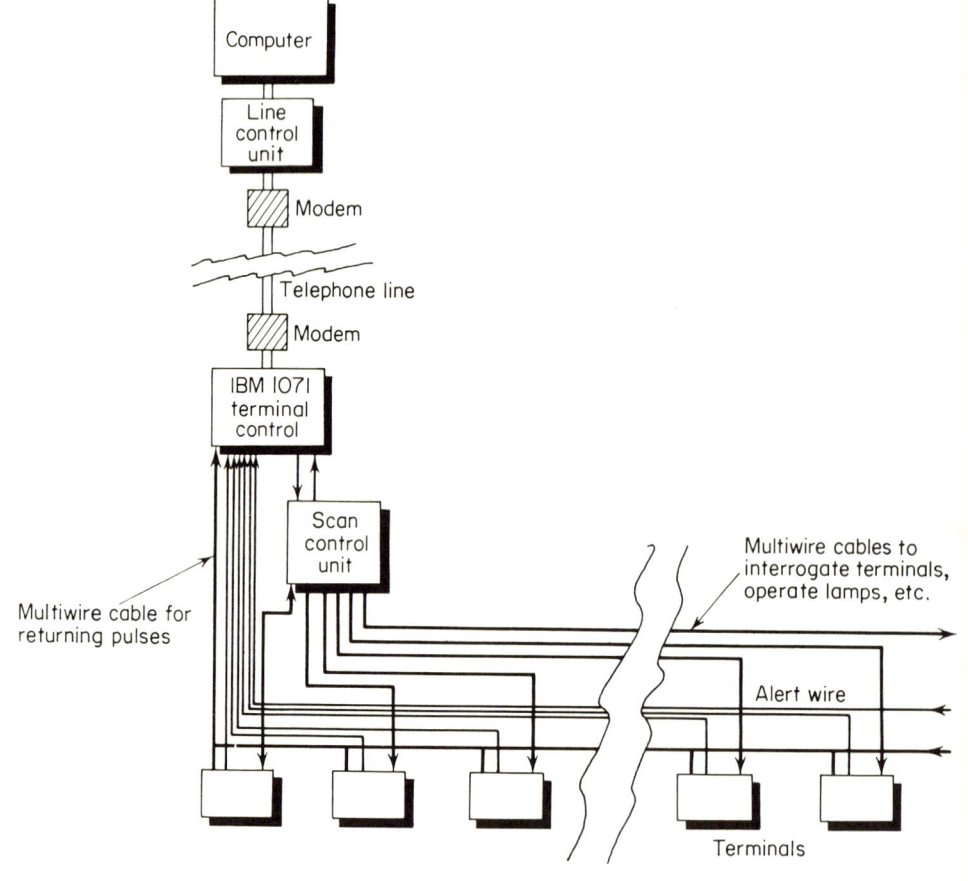

Fig. 7.5. A terminal control unit attached to a telephone line, with multiwire cables, as in Fig. 7.4, for reading information from many inexpensive terminals.

If it is short, the link from the control unit onto the computer may also be inefficiently used, though very much less so than the wiring to the terminals. An interesting example is the use of a loop wire carrying a high-speed pulse stream. This promising development in communications technology enables binary pulses to be sent around a simple loop consisting of a pair of wires at rates of 500,000 bits per second or higher. The wires have an inexpensive repeater every thousand feet or so, and this enables the line to be very long if necessary. It will always be a private line and cannot link directly to common carrier lines except at prohibitive cost, so we will find this pulse-carrying loop traveling around private premises and not normally beyond them. Because of the very high bit rate, the loop can wander around many control units, each with many terminals attached to it. It will carry data to, and take data from, all of the control units to which it is attached. The data characters from different devices will be mixed together on the loop and then sorted out into separate transactions at the computer that controls the loop.

The total information rate to and from factory data collection terminals on which such a system is used is very low compared with the 500,000 bits per second which the loop can carry (see diagram on pp. 124–125). We therefore find this bit stream used at a level of efficiency staggeringly low by the conventional standards of data processing. The line control procedures are designed with a very great overhead because it does not matter. The efficiency with which the bit stream is used compares with that with which the multiwire cables are used in Figs. 7.4 or 7.5.

2. SHORT PUBLIC LINES

We cannot be as wanton in our use of public lines or lines leased between locations from the common carrier. We would not employ multiwire connections such as that in Fig. 7.4 or the high-speed pulse stream discussed above. Nevertheless, on short lines the terminal cost still dominates the design. Short public lines still are used at efficiencies far below that of which they are capable.

It is indeed intriguing to reflect upon how inefficiently local telephone lines are used. The two wires from the telephone in my apartment travel to the local central office in a lead-shielded cable, like that in Fig. 7.6, which carries hundreds of other such pairs. This pair of wires is for my exclusive use as long as I pay my telephone bill. Nobody else will share it. This seems inefficient, because such a pair of wires has a bandwidth capable of carrying 24 or more telephone conversations at the same time. When such a pair of wires is used to link switching centers, it is common to find it carrying a "channel group" of 12 telephone conversations. Furthermore, my telephone is in use no more than an hour in total in an average week—0.6 percent utilization. One could say that the line carries $\frac{1}{24} \times 0.6$ percent = 0.025 percent of the voice transmission it is capable of carrying. Rarely in engineering can one find such an expensive facility as a nation's local telephone lines used so inefficiently. This, of course, is not true of the

USE OF A HIGH-SPEED PULSE-CARRYING LOOP:

A simple pair of wires can be made to carry a very high bit rate if repeaters are used at sufficiently frequent intervals to regenerate the bits. This is the principle of the Bell system T 1 carrier, using pulse code modulation of 1,500,000 bits per second. It is also used on some in-plant systems to give a high bit rate on loops of wire around a factory or laboratory. The IBM 2790 system uses a 500,000 bit per second pulse stream on loops of No. 20 AWG twisted wire pair with repeaters every 1,000 feet:

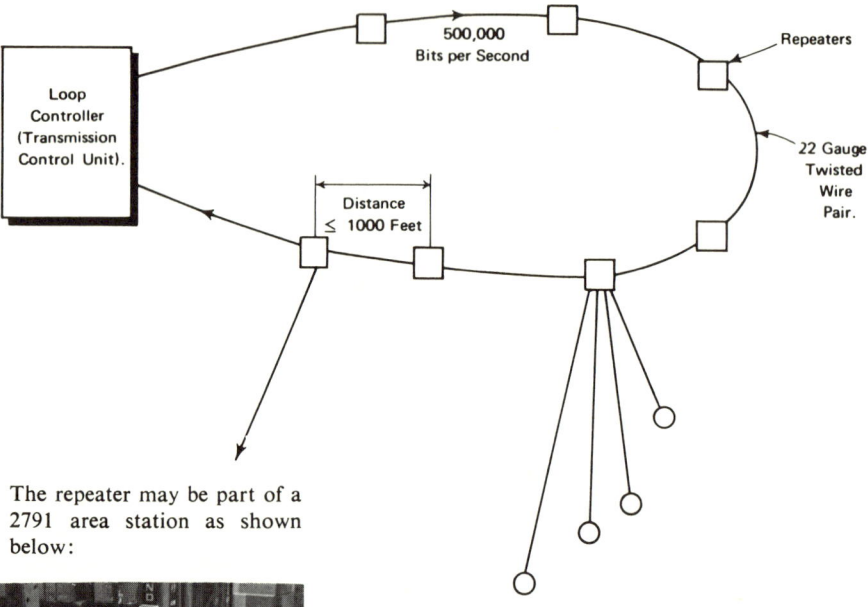

The repeater may be part of a 2791 area station as shown below:

Various devices may be attached to the area station on other wire pairs, for example, this 2796 data entry unit:

Synchronous frames, of 45 bits each, flow on the high-speed loop.

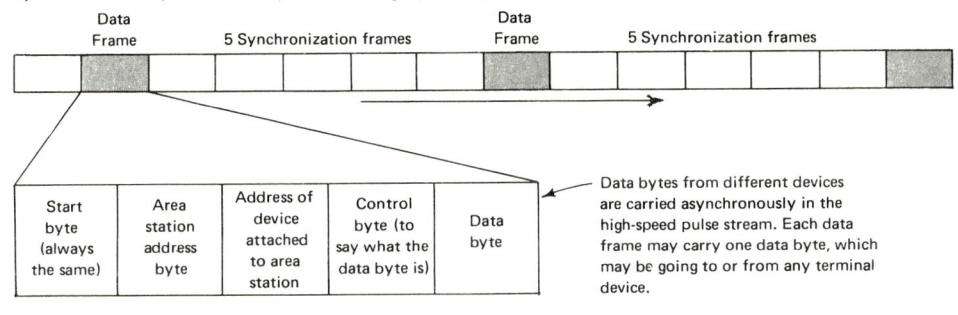

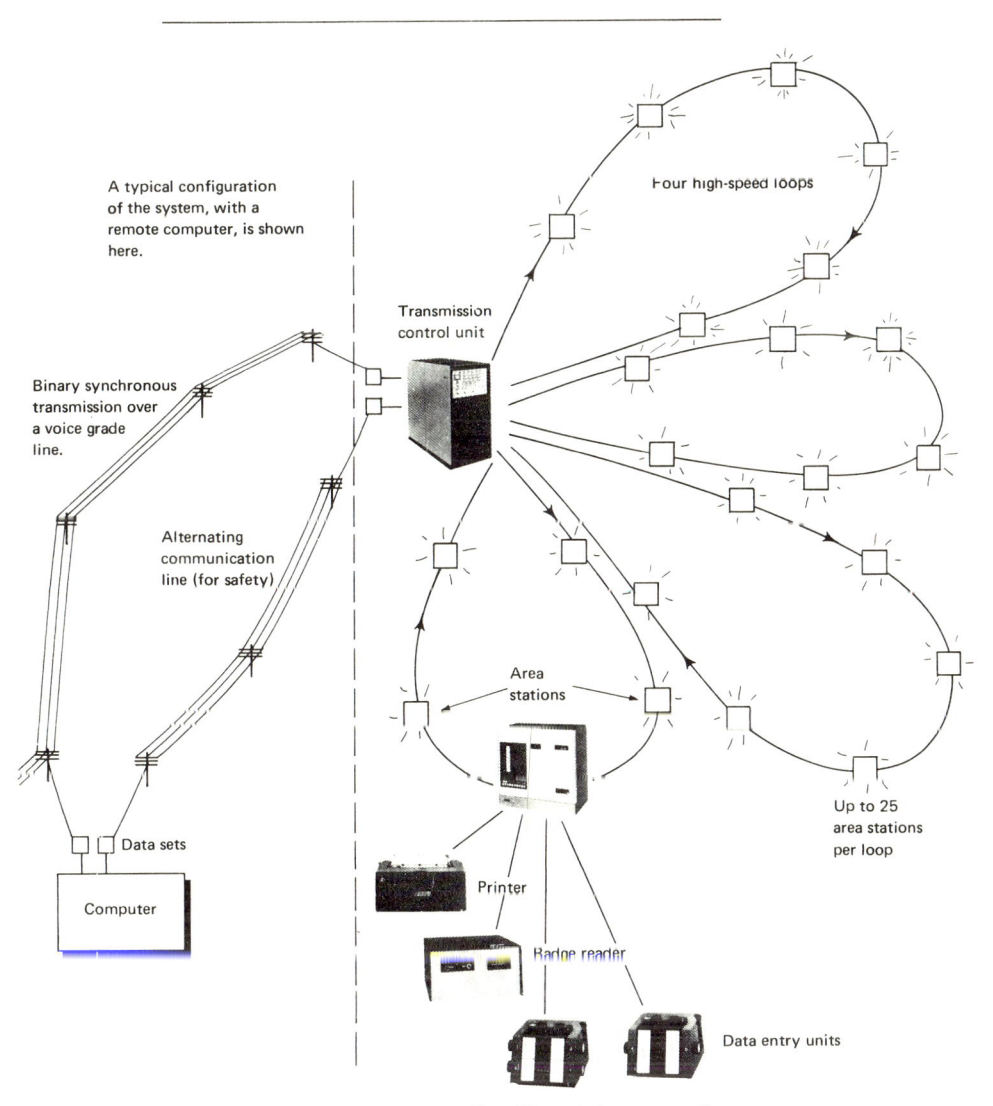

| Data Frame | 5 Synchronization frames | Data Frame | 5 Synchronization frames |

| Start byte (always the same) | Area station address byte | Address of device attached to area station | Control byte (to say what the data byte is) | Data byte |

Data bytes from different devices are carried asynchronously in the high-speed pulse stream. Each data frame may carry one data byte, which may be going to or from any terminal device.

A typical configuration of the system, with a remote computer, is shown here.

Binary synchronous transmission over a voice grade line.

Alternating communication line (for safety)

Data sets

Computer

Transmission control unit

Four high-speed loops

Area stations

Printer

Badge reader

Data entry units

Up to 25 area stations per loop

Up to 32 terminals per area station

long distance links. On these great ingenuity has been expended in making full use of the bandwidth.

When one is transmitting over a public or leased line it is normal to use a modem (data set) as shown in Fig. 1.5 to convert the square-edged pulses from the data machine to a form better tailored to the characteristics of the line. However, transmission *can* take place over pairs of wires—for example, those in Fig. 7.6—without a modem. Two-level telegraph signals have always been sent directly, without modulation, until relatively recent years, when it has become economical to send them over telephone lines with amplifiers that do not handle DC signals.

A signal from a data machine can be sent over a pair of wires in the DC form of Fig. 1.3, provided that:

1. The line does not have AC amplifiers.
2. It is not too long. (The longer the line, the more distorted the square-edged pulses become.)
3. The signal is not at too high a bit rate. (The faster the signaling rate, the more difficult it is to recognize the distorted pulses.)

Fig. 7.6. Many hundred wire pairs may be grouped together in a lead-sheathed cable like this. Such cables are laid under the streets of cities, and take pairs of wires from subscribers' telephones to their local central office (exchange). *Photo by author.*

In practical terms, on-off or double-current DC pulses can be sent without difficulty over today's wire pair lines for a distance of up to three miles or so, at speeds of not more than about 300 bits per second. That is, they can be sent without modems. This fact is useful because low-cost computer terminals with a printer generally operate at less than 300 bits per second, and systems often have areas less than three miles across in which many such terminals are needed. Inside such areas the terminals can be interlinked without data sets, possibly through an exchange. The links can be public lines if they are local "loops" which, as is normally the case with the line from a telephone to local central office, do not have amplifiers. Once this has been said, however, it should be noted that, at the time of writing, it is common to find data sets being sold for such systems.

Consequently, a large number of data sets are often found at the computer center, as is illustrated in Fig. 1.6, even when the terminals are within three miles and operate at less than 300 characters per second.

As was discussed in Chapter 3, acoustical coupling is often used on short lines and has the advantage that the terminal can be connected to any telephone. The device needed for acoustical coupling is low in cost. Another low-speed means of connecting a terminal to a voice line is to use a device directly wired to the line which codes the data as multiple frequencies in the audible range, as discussed on p. 40. Both of these means of transmitting data use the line at a bit rate far below that of which it is capable. In fact, to place one typewriter speed terminal on a voice line is to make very inefficient use of the line. However, it is acceptable, as with the inefficient use of domestic telephone lines, because of the convenience of obtaining a dial-up connection. The same dial-up connection could transmit at a higher character rate to a terminal fast enough to handle it, such as a visual display unit.

3. LONG LINES

When longer lines are used, the techniques change. Now the lines are more costly, and it is desirable to use them efficiently. If the lines are more than 100 miles or so long, the line cost dominates the network design.

The first step to efficient line utilization is good modem design. Throughout the 1960's much ingenuity was expended on designing modems that would enable higher speeds to be used, especially on voice lines. In the early 1960's speeds of 600 bits per second were used on dial-up lines and of 1200 bits per second on leased lines. By the end of the 1960's, speeds of 1200 bits per second were easily achieved on dial-up, and 4800 bits per second on leased, lines. As was mentioned before, modems existed, though they were not in general use, that could achieve 3600 bits per second or more on dial-up, and 9600 on leased, lines. This was a vast improvement upon the 14.8 characters per second used on many of today's voice lines.

Modems alone, however, do not solve our problem, except in the case of fast point-to-point operation—for example, magnetic tape-to-tape transmission or the sending of listings, invoices, etc., in quantity. On many commercial systems we wish to connect manual terminals to the long lines, or other devices which cannot transmit at the full line speed.

Suppose that terminals are to be situated at nine locations shown in Fig. 7.7 and connected to the computer many miles away. If the terminals have a high usage, it is cheaper to have a leased line to them rather than a dial-up line, as was discussed on p. 6. One approach would be to have a

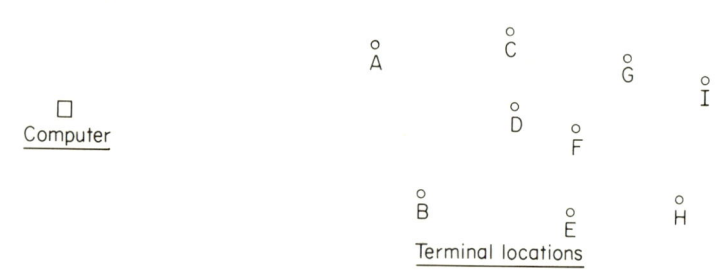

Figure 7.7

leased line to each of them, as shown in Fig. 7.8. Such lines, however, are likely to be inefficiently utilized for the following reasons:

1. Most of the terminals will not be transmitting or receiving constantly. A high-speed printer may only be used in occasional bursts. A real-time terminal may have relatively short messages and responses, interspersed with much lengthier periods of silence while the operator "thinks," uses the response, or talks to his client. We would like to organize the line in such a way that when one terminal is inactive another is using the line.
2. The line, equipped with suitable modems, has a certain maximum transmission speed. Most terminals are unlikely to operate at this speed. We would like to interleave the characters or messages to and from different terminals to make the most of available line speed.
3. The messages we send to the terminals may contain repetitive information or information that could be coded less redundantly. By building some form of intelligence into the network away from the computer, we could lessen this repetitive or redundant transmission.

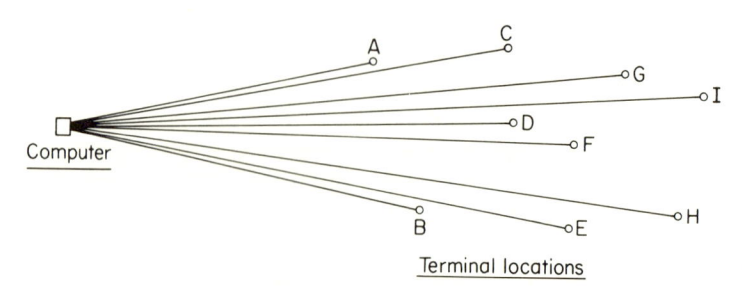

Figure 7.8

TECHNIQUES FOR MINIMIZING NETWORK COST

A variety of devices and techniques are available for achieving these ends. Their objective is to minimize network cost. Without such methods a large network would have prohibitively expensive line costs. Much of the remainder of this book is concerned with this problem. Let us now summarize some of the methods available. These will be explained in more detail in the following chapters. Combinations of the techniques below may be used in practice.

1. *Use of a Private Exchange*

The total line mileage can be cut greatly by using line switching. In Fig. 7.9 a switching mechanism is placed at location D. Messages from the computer to a terminal go through this switch. The messages are preceded

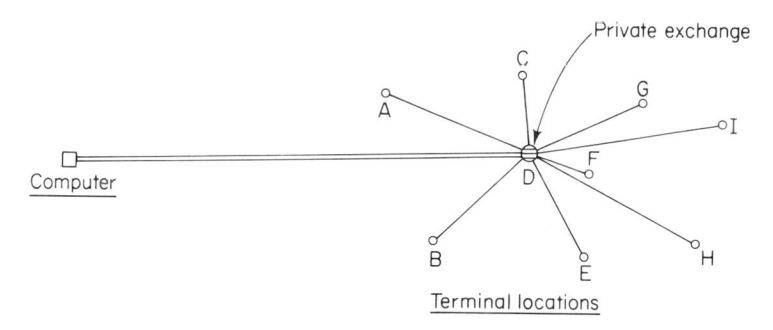

Figure 7.9

by addressing characters, which cause the line from the computer to be physically connected to the line to the appropriate terminal. Similarly, when the terminal sends a message to the computer, the appropriate connection must be established. There may be more than one line from the exchange to the computer, and the switching mechanism must be able to hunt for a free one.

The switch mechanism may be a privately owned switching device; it may be a private branch exchange at location D leased from the common carrier; it may be a mechanism installed in the local central office (telephone exchange) at D. Private exchanges will be discussed in Chapter 8.

The disadvantage of using an exchange in this way is that a terminal operator may be unable to obtain a line to the computer when he wants to because they may all be busy. Figure 7.9 shows two lines from the exchange to the computer. In practice there may be more terminals and more lines. The designer must calculate the probability of the terminal operator not being able to make a connection and being kept waiting for a long time.

130 MEANS OF LOWERING COMMUNICATION NETWORK COSTS

2. *Multidrop Lines*

The total line mileage can be cut further than in Fig. 7.9 by using a multidrop line. "Multidrop" (or "multipoint") means that several terminals are connected to the same line. The terminals can be in different locations with the line taking the shortest path between them, as in Fig. 7.10. It will be seen that the total line mileage in Fig. 7.10 is substantially lower than in Fig. 7.9.

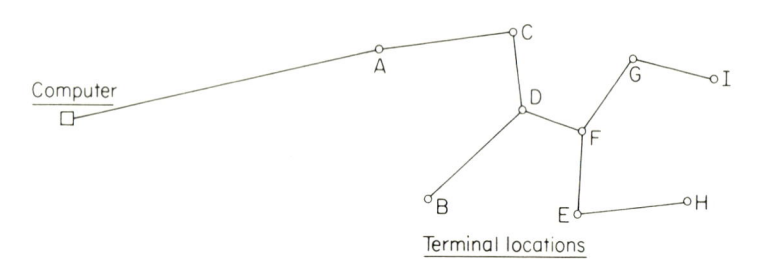

Figure 7.10

Two terminals on a multidrop line cannot transmit or receive at the same time. A discipline must be established on the line whereby the devices wait their turns to transmit. If a terminal user wishes to transmit, he must wait until the line is free, and on most systems must wait until the computer sends a signal asking that terminal if it has anything. The terminal cost will be somewhat higher because it needs circuits for recognizing its address and carrying out the line control procedures.

The terminal user may, thus, be kept waiting on a multidrop line just as on a system using an exchange. However, he may not be kept waiting so long. A multidrop line is occupied only for the duration of one message. If a real-time terminal sends a transaction to which a response is needed, the line on most systems is released after the transaction is sent and is not held up waiting for the response. (This is usually the case, but not always.) After the sending of the transaction and before its response, another terminal could use the line, either transmitting or receiving. On the other hand, if the exchange in Fig. 7.9 is used, the connection from one terminal to the computer may remain unbroken for the duration of a "conversation." Indeed, one terminal user often has the capability of hanging on to his connection. He may do so deliberately for fear of losing his link to the computer. However, using an exchange and not multidrop lines, the user then never has to wait for the line to become free, so he may receive somewhat faster response times.

If a user can hold his connection for the duration of a "conversation" with the computer, the long line from the exchange to the computer will be used inefficiently, because it will be idle during the time when the computer is preparing the response and when the user is reading it and thinking about his next input.

It would be possible to design an exchange (a line configuration such as that in Fig. 7.9 being used) in such a way that it switched lines quickly at the end of each message in accordance with the demands for connections. This would then not keep a user waiting for a lengthy period. Such an exchange would be likely to be electronically operated, rather than using electromechanical step-by-step switches as in most PBX's.

In Fig. 7.9 two lines are used from the terminal area to the computer, whereas only one is used in Fig. 7.10. Because of the way the users hold their connection, this might be the case in practice. In fact, more than two lines might be needed to give adequate service. The one line in Fig. 7.10, on the other hand, might be overloaded, so two lines would be multi-dropped to the nine terminal locations shown. This is illustrated in Fig. 7.11.

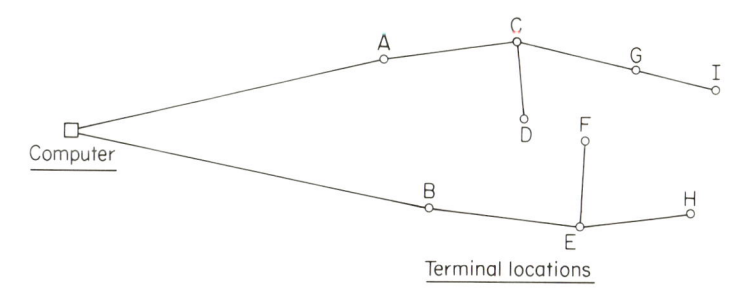

Figure 7.11

The line configuration here is arranged to give a balanced loading and again take the shortest path. The total line mileage is still shorter than in Fig. 7.9.

Multidrop lines will be discussed in detail in Chapter 9.

3. *Multiplexors*

Suppose that the terminals in the above illustrations are typewriter-speed devices that can operate on a 150 bit-per-second line. A conditioned voice line to the computer can carry 4800 bits per second. Such lines are commonly used today at 2400 bits per second with a not-too-expensive modem. If we could divide up this capacity we could use the 2400 bits per second as, say, ten entirely independent 150 bit-per-second channels. This gives an allowance for the inefficiency of the dividing-up process. [Here it is $(150 \times 10)/2400 = 62.5$ percent efficient.] Dividing up one line into several independent channels is called "multiplexing." Use of a multiplexor with

132   MEANS OF LOWERING COMMUNICATION NETWORK COSTS

our nine terminal locations is shown in Fig. 7.12. A *full-duplex* line is used from the multiplexing unit to the computer so that all the terminals can be either transmitting or receiving at the same time.

The way the configuration of Fig. 7.12 works can be thought of as being equivalent to Fig. 7.8, with separate lines to each terminal. The fact

[Figure 7.12: Computer connected via 2400 bit per second full duplex voice line to a Multiplexing unit at terminal location D, which connects to terminal locations A, B, C, E, F, G, H, I via 150 bit per second sub-voice grade half-duplex lines.]

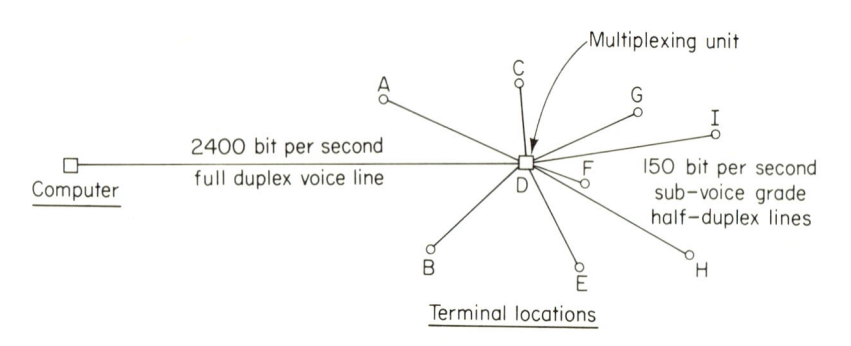

Figure 7.12

that the channels are grouped together by multiplexing does not affect the system performance or the programing. This is one of the attractive features of using a multiplexor. The response time or line availability is in no way degraded as it would be with a multidrop line or a configuration using an exchange. There are no addressing or polling messages needed; no headers for operating an exchange or making a terminal recognize that a message is addressed to it. The programer, indeed, need not know that multiplexing is being used. The multiplexing unit is transparent even to the line control units.

If 150 bit-per-second lines had been used in Fig. 7.10, however, the overall line cost of the scheme in Fig. 7.12 would have been higher.

The types of multiplexors are discussed in Chapter 11.

4. Buffers

Most transmitting and receiving devices operate at their own speed—a speed that is usually different from the optimum speed of the line. A severe example of this is a keyboard operator. A girl at a teletype machine presses the keys at her own rate, and the characters are transmitted at this rate regardless of the capacity of the line. An engineer at a time-sharing-system terminal may type with one finger. Worse, he pauses for lengthy periods to think about what he is going to type next. If the terminal has no buffer, he is tying up the communication line. On the other hand, if it has a buffer, his hesitant typing could be filling up the buffer without consideration of the line; the line could be transmitting other data. When he requests that the message be sent to the computer, the contents of the buffer would then be sent at the maximum speed of the line.

Communication line buffers are clearly of value when the line has many different devices contending for transmission, because it lessens the overall time that each device will occupy the line. Terminal buffers would be of value in Figs. 7.10 and 7.11. In Figs. 7.8 and 7.12 they would be less so because here there are no other devices contending for the line. Even in this case, however, the higher-speed transmission they make possible may lessen core and time usage in the computer.

The buffer is generally expensive on a terminal. Again, we are increasing terminal cost to improve line efficiency. The cost must be balanced against the saving, and the longer the lines, the more likely it is that terminal buffers save money.

As well as saving money, terminal buffers can have a major effect on the time an operator is kept waiting at a terminal. Suppose a typical operator keys data into a terminal at 1.5 characters per second on a multi-drop line that transmits at 15 characters per second. Suppose that an average length for a message keyed in is 75 characters and that you are waiting for another terminal to transmit. If there is no buffering you will wait 50 seconds. *With* buffering you will wait 5 seconds. If a voice line is used at a speed of 150 characters per second, and buffering is used, you will wait only half a second.

A cheap form of buffer is the use of paper tape. The operator punches her message into tape, and places the tape in a reader to be read when the computer requests transmission. The contents of the tape are transmitted over a telegraph line at its maximum speed.

5. *Shared Buffers*

Where any form of electronic storage is used, terminal buffering is expensive. One method that has been used to lower the cost is to share buffers between several terminals.

The unit holding the shared buffers may allocate a fixed area of buffer space to each terminal, or it may allocate the space dynamically on the basis of demand. When the device handles many terminals and randomly generated transactions of widely varying length, it is more economical for it to use dynamic buffer allocation.

6. *Terminal Control Units*

Where a unit is built to buffer the transactions from several terminals, there are other functions that may be built into the same device for controlling the terminals or improving line control. Some control units have a substantial amount of logic in them. These are discussed in Chapter 10.

The terminals are usually close to the control unit, connected with multiwire cabling. However, they may be far away, attached by private in-plant wiring, or possibly by telecommunication links.

7. *Concentrators*

Synchronous transmission, as we discussed on p. 50, is more efficient than start-stop transmission. To make the best use of a long voice line, or wideband line, we would like to transmit synchronously at its maximum speed. Many terminals are start-stop devices. Start-stop machines are simpler and less expensive than synchronous machines. Therefore, we find line configurations in which start-stop lines go to a concentration point, as in Fig. 7.14, and from there onwards the data are carried by synchronous transmission.

Similarly, a concentrator may link slow lines to a fast line—voice grade or wideband. It may link lines that are inefficiently or intermittently used to a line on which the transmission is organized as efficiently as possible.

The concentrator, as described here, is doing more than the multiplexor above, because it restructures the message or bit pattern in use. The multiplexor merely groups several transmissions on low bandwidth lines into one of higher bandwidth, without changing the structure or composition of any of them.

Common usage of the terms "concentrator" and "multiplexor" is unfortunately not always as clearcut as this. "Multiplexor" is sometimes used to mean what we here call a "concentrator," and vice versa. In this book, however, the word "multiplexor" is restricted to devices that combine signals or bit streams but do not change their structure.

Concentrators linking lines of different speeds carry out a buffering operation. They must, therefore, hold incoming messages on the low-speed lines until they can be transmitted onward on the high-speed line. Similarly, outgoing messages are stored while they are transmitted on the low-speed lines. Such devices are sometimes, therefore, called "hold-and-forward" concentrators. They do not store the messages semipermanently on secondary files such as disk units. A message switching system that we will discuss shortly does do this, and this is sometimes referred to as a "*store*-and-forward" system.

The functions of "hold-and-forward" concentrators are described in Chapter 12.

8. *Concentrators with Multidrop Lines*

Above we discussed improving the efficiency with which the lines are used by building features into the terminals which increase their cost. Now as we continue down this list of increasingly expensive items we can add complications to the concentrators. The intent, as before, is to reduce the overall network cost. The network may have far more terminal locations than in the above figures. Some networks have found it economical to couple two hundred or more terminals to one long line.

First, the concentrators can be multidropped on one line as were the terminals in Figs. 7.10 or 7.12. They must then have the logic necessary for responding to polling messages, recognizing messages addressed to them, and other line control functions discussed in subsequent chapters.

Second, the low-speed lines downstream of the concentrators may have several multidropped terminals on them. The logic must then exist to discipline the low-speed lines.

Third, in some large networks there is more than one level of concentration. Common control units such as that in Fig. 7.13 may feed voice lines that link into concentrators on wideband lines.

9. *Line Computers*

The complexity of the concentrator is now becoming sufficiently great for a stored-program computer to be used as the concentrator. This has been the case on many systems. The flexibility of this approach has permitted such a machine to vary its function. A line computer at regional centers has been used for concentration during the day when the system's real-time terminals are active, and then as a bulk transmission machine in the evening when the terminals close down, for example, printing listings and bills transmitted from the main computer.

10. *Line Computers with Files*

In may applications in which users carry on a "conversation" at a terminal, much verbiage is displayed on the screen in order to make the system easy to use. The next step in improving the efficiency with which the long lines are used is to reduce what is actually transmitted. If a stored-program machine is used either as a concentrator or terminal control unit, this can be used for generating terminal responses and for compacting the messages that are transmitted. Examples will be given in Chapter 13 of canned responses' being stored on secondary storage in a line computer so that many messages transmitted on the long lines are merely brief codes that cause the line computer to generate terminal responses.

11. *Line Computers with Application Programs*

Giving further responsibility to the down-line computers, these machines may have application-dependent programs for conversing with the terminal user. In many systems only a few of the transactions in a lengthy man-machine conversation need either the power of the central computer or its data base.

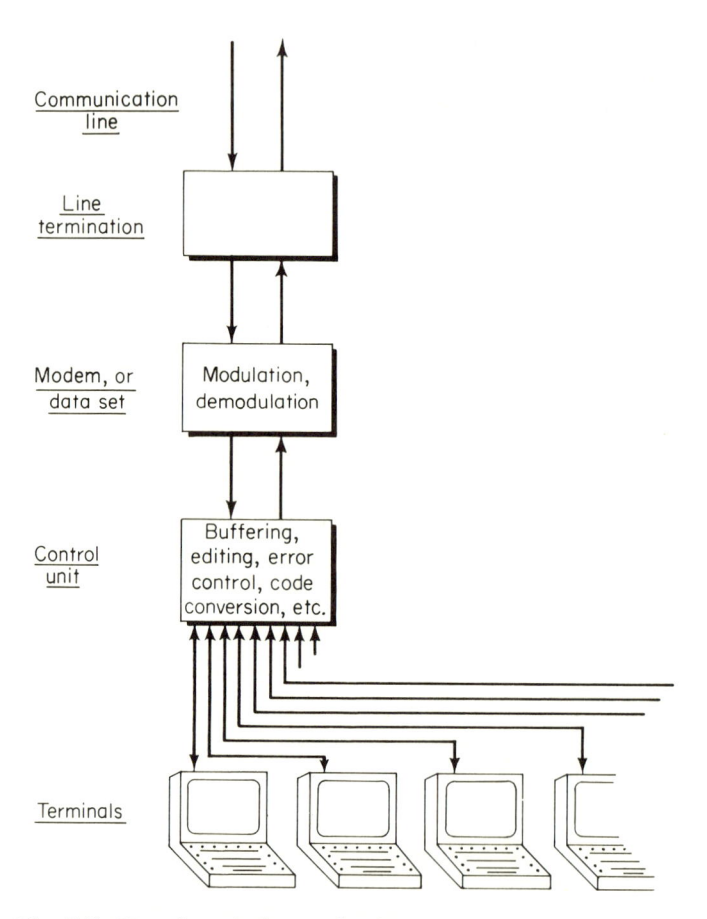

Fig. 7.13. Use of terminal control unit.

The philosophy on many large systems is becoming: when the lines are long, distribute the "intelligence" throughout the network so that the quantity of data transmitted is minimized. This is discussed in the last chapter.

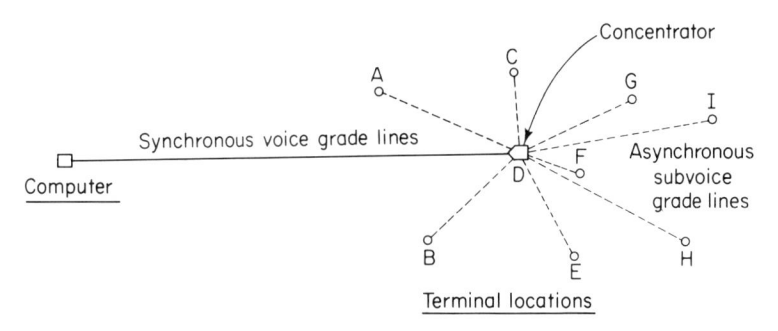

MEANS OF LOWERING COMMUNICATION NETWORK COSTS   137

Figure 7.14

12. *Message-Switching Systems*

For years message-switching systems have been used to relay cables between many locations. Now they are being used for computer data also.

Locations wishing to communicate with each other could send data directly by teleprinter links. However, where there are many widely separated locations, the cost of communications can be reduced by using some form of message-switching system. Consider the points illustrated in Fig. 7.15. These represent terminals located in eight different cities.

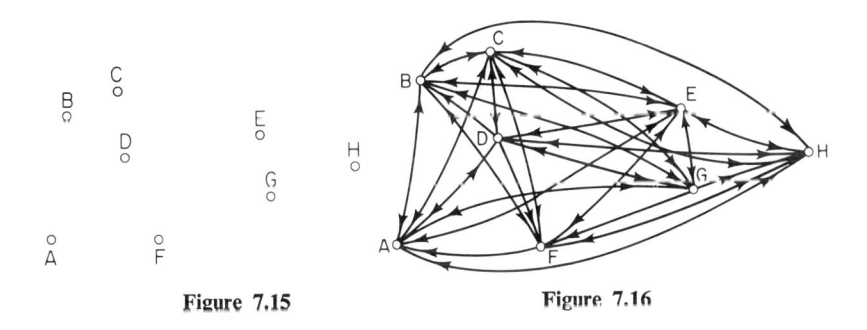

Figure 7.15   **Figure 7.16**

One solution to providing telecommunication links between these would be to have direct communications between every pair of points, as in Fig. 7.16. An alternative, which uses far fewer communication lines, is to connect all the points to a switching device, as shown in Fig. 7.17.

The larger the number of locations, the greater the saving in line costs when a switching center is used. A further reduction in line costs is achieved by using *multidrop* lines, in which one line connects several terminals (Fig. 7.18). This could be done if the traffic volume was not too high.

138 MEANS OF LOWERING COMMUNICATION NETWORK COSTS

There is a fundamental difference between line switching and message switching. Figures 7.15 through 7.18 could apply to either. With line switching a physical connection is made between the circuit paths as in a telephone exchange. With message switching this is not done; instead, the message is stored and then relayed onward to its destination or destinations. With message switching the terminals are not interconnected and enabled

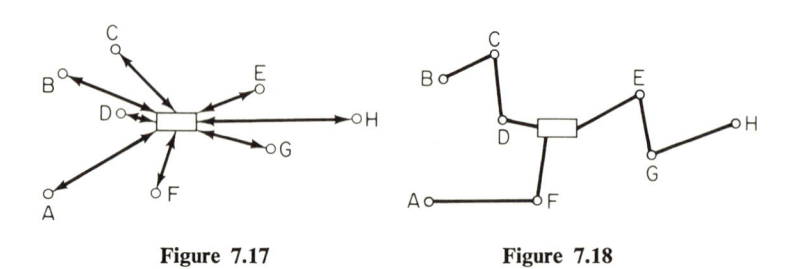

Figure 7.17

Figure 7.18

to respond to each other directly, in real-time, as with two telephones. Message switching, however, can give a better line utilization than line switching.

Today, if the traffic volume is high enough to warrant it, computers are used for message switching. A message-switching system sometimes becomes an important part of a computer network. Message switching is discussed in Chapter 14.

8 PRIVATE EXCHANGES

Most transmission devices are not in operation for the whole day, or even the majority of it; therefore, they do not require a line to be permanently connected to them. Communication cost can be saved by having some form of line switching.

The switching may be done by a private exchange, or it may be done by the switched public network. As we discussed on p. 6, private lines are less expensive than public lines if the connection is needed for more than a certain time per day. Consequently, the switching that is best on such systems involves private lines and a private exchange. Some systems use a combination of private lines, private exchanges, and the public network. Others make their connection solely by dialing on the public network, either by voice, Telex, or TWX facilities.

It should be noted that on many real-time systems the terminal is needed to be in operation most of the day. In this case the line is not normally switched but is permanently connected. In an airline reservation system, for example, the agents' sets are on leased lines with no switching. The same is normally true with bank teller terminals, production control stations, and many other terminals on commercial real-time systems.

In some cases the reason for switching has nothing to do with line costs but with the computing facility itself. A laboratory, for example, may have 100 terminals but a computer with only 30 "ports." Because of a restriction in the computer center, no more than 30 terminals can be connected at once. The 100 terminals are linked to an exchange that permits up to 30 connections to the computer. If another user tries to dial the system when all 30 connections are in use, he will receive a "busy" signal and will have to wait. It is necessary to determine the probability of this happening and design the system to give an acceptable grade of service.

140 PRIVATE EXCHANGES

Several types of exchanges are possible. It may be a manual switchboard with a girl who jack-plugs the appropriate connections along with telephone connections. At its simplest it may be a manually operated multiway switch or jack-panel. On the other hand, it could be an elaborate automatic private branch exchange leased from the telephone company.

Such a system may have the ability to "hunt" for a free line to the computer when one particular number is dialed. Again, the exchange may be a small inexpensive automatic device not leased from the telephone company, using step-by-step or cross-bar switches.

Figure 8.1 shows a computer with an off-line data transmission facility. Data are transmitted to and from a tape transmission unit on half-duplex leased lines. The unit can be connected either to a line to another such tape unit, or to an on-line computer at a different location, or

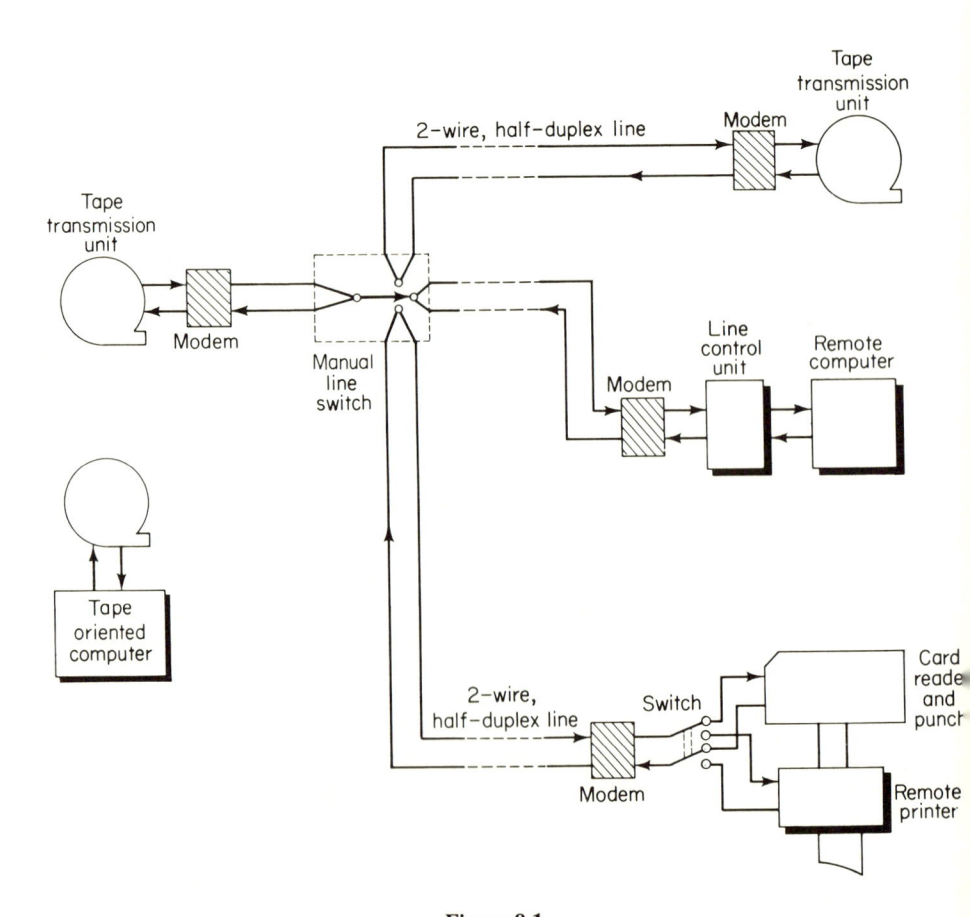

Figure 8.1

to a card reader, card punch, or printer. The connections between these locations are made with a manual switch.

Similarly, in Fig. 8.2 three computers can transmit to a fourth on full-duplex lines. The connection between them is switched manually with a switch at the location of the large computer.

Figure 8.3 shows connections made via the switched public network. A terminal can dial up different computers in different locations on public lines. It may also dial up and transmit to other terminals.

In Fig. 8.4, all of the equipment shown is in one location. This might be a laboratory, a factory, corporation offices, a college campus, and so on. The interconnections are private leased lines that go via a private branch exchange. All of this equipment and the data sets shown are installed by the common carrier. The private branch exchange is the same one used to handle telephone calls. One telephone can dial another extension in the same building or an outside number, and similarly one data device can dial another device in the same building or may possibly dial another number. Most of the data connections in the building are from a terminal to the computer. There are several lines from the private branch exchange to the computer, and this number of terminals can be using the computer simultaneously. Often on such systems there are more terminals than lines into the computer, as in the case mentioned above of the laboratory with 100 terminals but only 30 "ports" into the computer.

The lines into the computer may each be given a telephone number or extension number. The terminal user does not want to have to dial all of these numbers to find a free line, so the exchange has an automatic device that scans the lines into the computer, hunting for a free one. The data sets shown in the computer room each have a telephone dial and handset. The computer room personnel could use this to dial a terminal, or its operator, though in practice this is rarely used and seems a waste of hardware.

Figure 8.5 shows a similar arrangement, but the location in question is connected by tie lines to a distant location. The terminal user can now either dial the computer in his own building or dial the tie-line number of the distant computer. The tie line links two private branch exchanges. The arrangement of data sets for the computer at the distant location is here the same as that for the computer at the terminal location.

It is possible to use a less expensive data set, or no data set at all, if the lines are wire pairs with no multiplexing covering a distance of not more than a few miles and carrying data at a speed of no more than about 300 bits per second. These restrictive criteria would apply to many systems installed today, and so much of the cost of the data sets illustrated in Fig. 8.4 could be saved. The terminal user without a data set would not be able to dial directly to a computer outside his localized area, and this is one reason

Figure 8.2

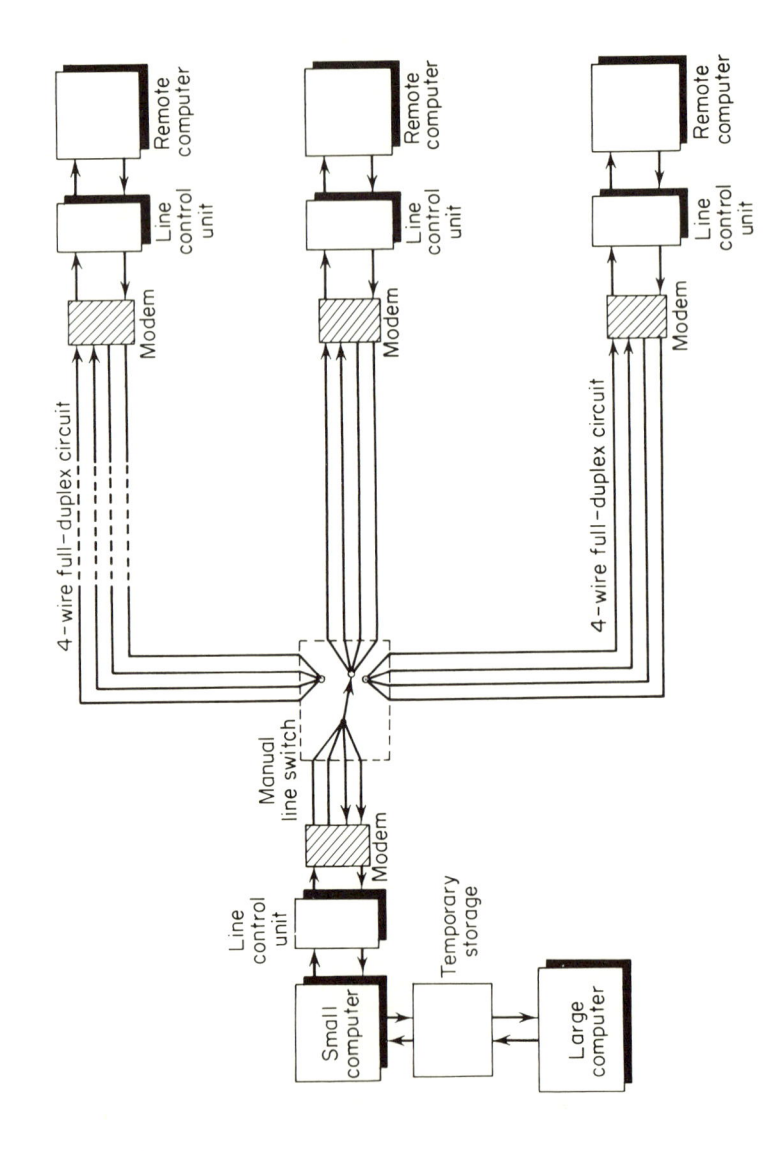

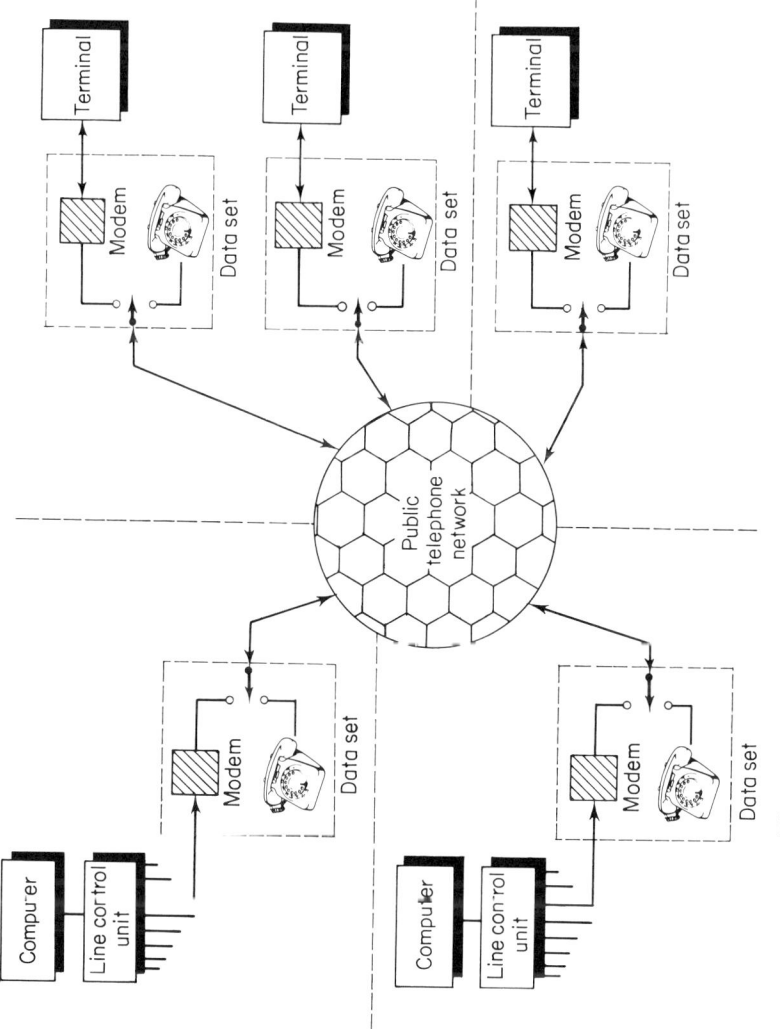

Fig. 8.3. Terminals that can dial different computers on public lines.

144 PRIVATE EXCHANGES

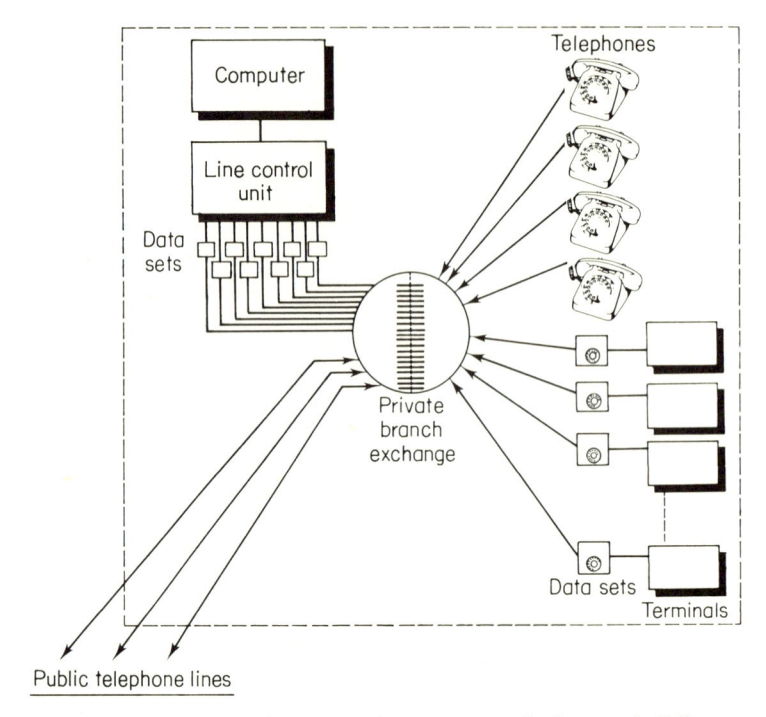

Fig. 8.4. Dial-up terminals connected to a computer in the same building via a PBX.

why the relatively expensive data sets such as those shown in Fig. 8.4 are generally used today. Data sets for outside calls could, however, be installed at the exchange location. If the user dials a computer in his own locality, he does not use modems. If he dials a distant computer, the modulation is done at his private branch exchange. Some experimental systems of this type have been installed, but at the time of writing are not generally marketed by the common carriers (who sell the data sets in Fig. 8.4).

In Fig. 8.6 the terminals can dial three types of computer location, one connected to their own exchange, one to tie-lines from their exchange, and the other anywhere on the public network. The terminal user might have three types of number he can dial:

1. For a computer in his own locality a local extension number, e.g., 7215.
2. For a computer on an exchange connected by tie-line:

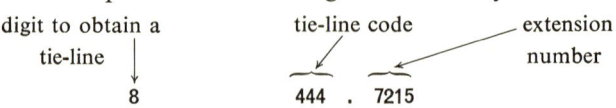

    digit to obtain a tie-line: 8
    tie-line code: 444
    extension number: 7215

    8 . 444 . 7215

3. For a computer dialable via a public line:

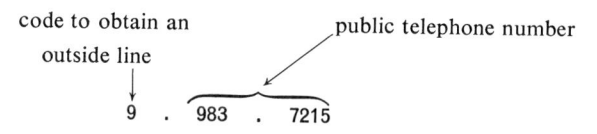

```
    code to obtain an            public telephone number
      outside line
           ↓                    ↙
           9   .   983   .   7215
```

Again the same data set is used for each of these connections. It makes little or no difference to the user which of the three computers his is connected to. His local computer may be a special-purpose service programed to run, for example, only BASIC or APL. If he wants a different service, he will dial elsewhere. If he cannot obtain a line to his local computer, he may dial another machine with the same facility, though this may not be satisfactory as the local machine contains his own data or programs.

In Figs. 8.4, 8.5, and 8.6 the terminals are shown in one location. In Fig. 8.7 they are in many locations and all need to contact one computer. This may be a computer dedicated to some commercial application, with files of commercial data. The terminals, for example, may be bank teller terminals in different branches, terminals of a management information system, order entry system, and so on. The distant terminals dial the computer location either on a tie line or on the public network. Figure 8.7 shows a location with a switchboard operator in addition to a private branch exchange. The connections should be arranged so that the switchboard operator does not interfere with the data transmission.

Where many locations have to be interconnected, as was discussed on p. 129, it is generally cheaper to interconnect them via an exchange. Figure 8.8 shows 23 such locations. The top part of the figure shows them interconnected via one exchange. The line cost is lower, however, if two exchanges are used and lower still with three exchanges, as shown in the rest of Fig. 8.8. More than three exchanges are not normally employed, and often only one is used.

Figure 8.9 shows a high-capacity data transmission network in use internally in the IBM Corporation. After initial success in connecting a programing center in New York to one in Poughkeepsie, 80 miles away, by means of a leased wideband link which transmitted from magnetic tape to magnetic tape, at 15,000 characters per second, it became desirable to interconnect many such locations. A private leased wideband exchange (an AT&T Type 758B PBX) enabled dial-up connections between IBM locations throughout the United States with a transmission speed of 40,800 bits per second. Within a period of only five years after the initial point to point link had been set up, the network had grown to a total circuit distance of 14,000 miles, with the wideband terminals in use an average of 80 hours per month. Shortly after this the network was extended

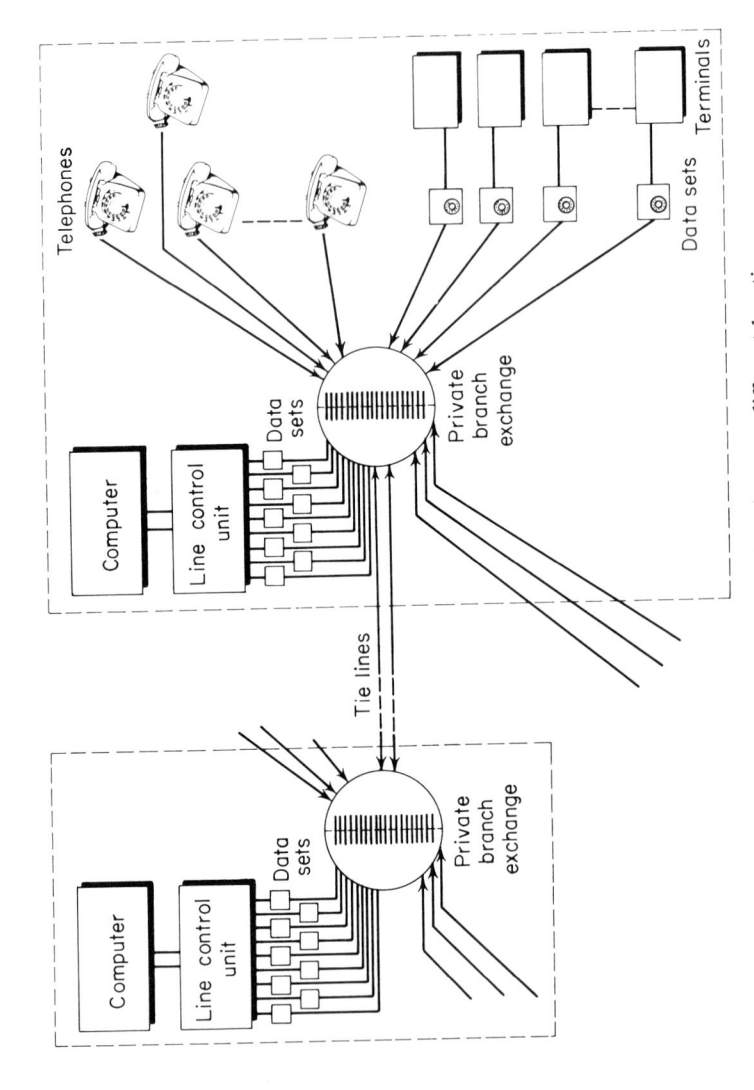

Fig. 8.5. Tie-line connections between different locations.

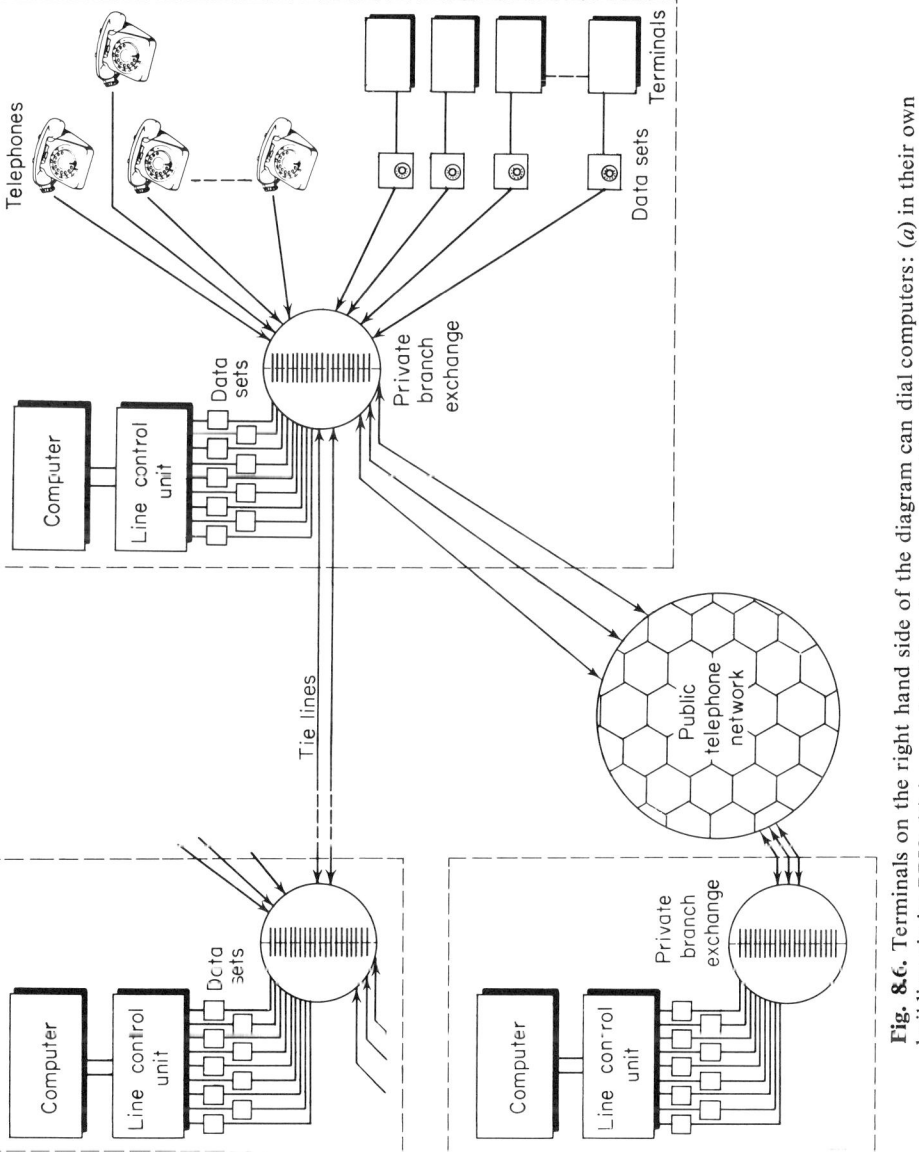

Fig. 8.6. Terminals on the right hand side of the diagram can dial computers: (*a*) in their own building via its **PBX**; (*b*) in associated locations connected by tie-line; (*c*) any location via the public network.

148 PRIVATE EXCHANGES

to include satellite channels to London and Paris. Other worldwide locations may be linked up in this way.

In firms with a lower volume of data to transmit, a similar network with voice grade lines is used, with a private exchange. Often such lines are used alternately for voice or data.

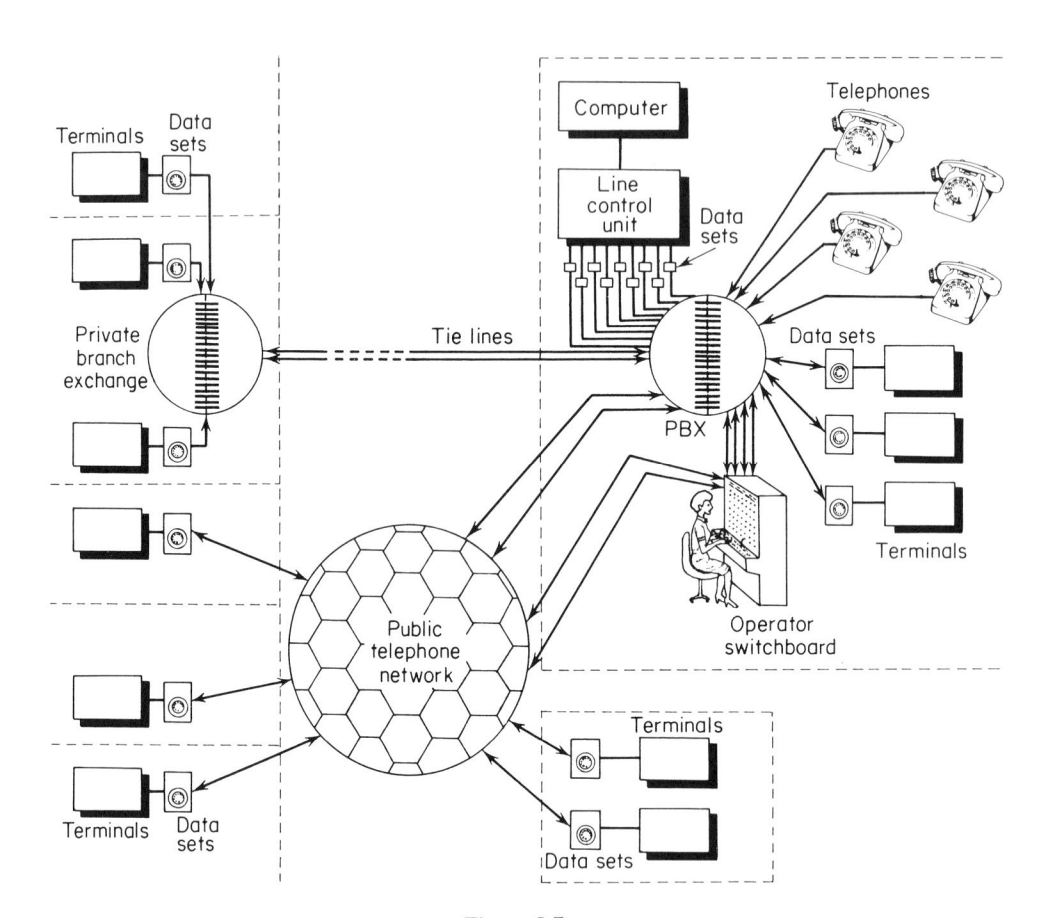

Figure 8.7

Some common carriers offer tariffs that enable a *group* of lines to be leased at a lower cost than the equivalent number of separate lines. Advantage can sometimes be taken of such tariffs in configuring the minimum cost network. These tariffs have been subject to change recently and are likely to change again. At the time of writing, the AT&T Series 5000 channels are such a tariff. Channel Type 5700 gives up to 60 voice grade channels or equivalent at $25 per airline mile per month, and channel Type

PRIVATE EXCHANGES 149

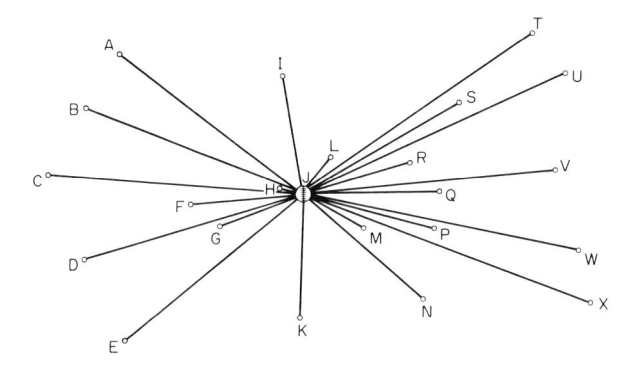

23 locations interconnected using leased lines and a private exchange

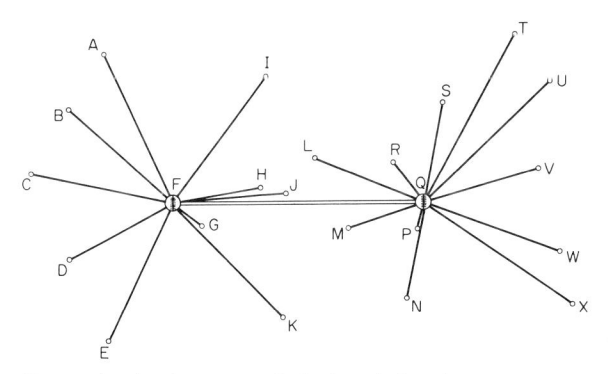

The same locations interconnected using two private exchanges gives a lower line cost

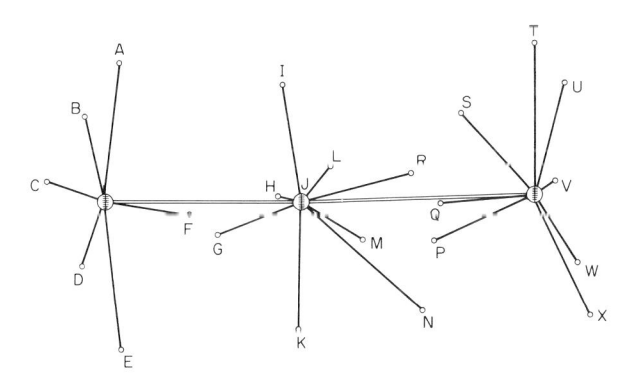

With three private exchanges the line cost is still lower

Figure 8.8

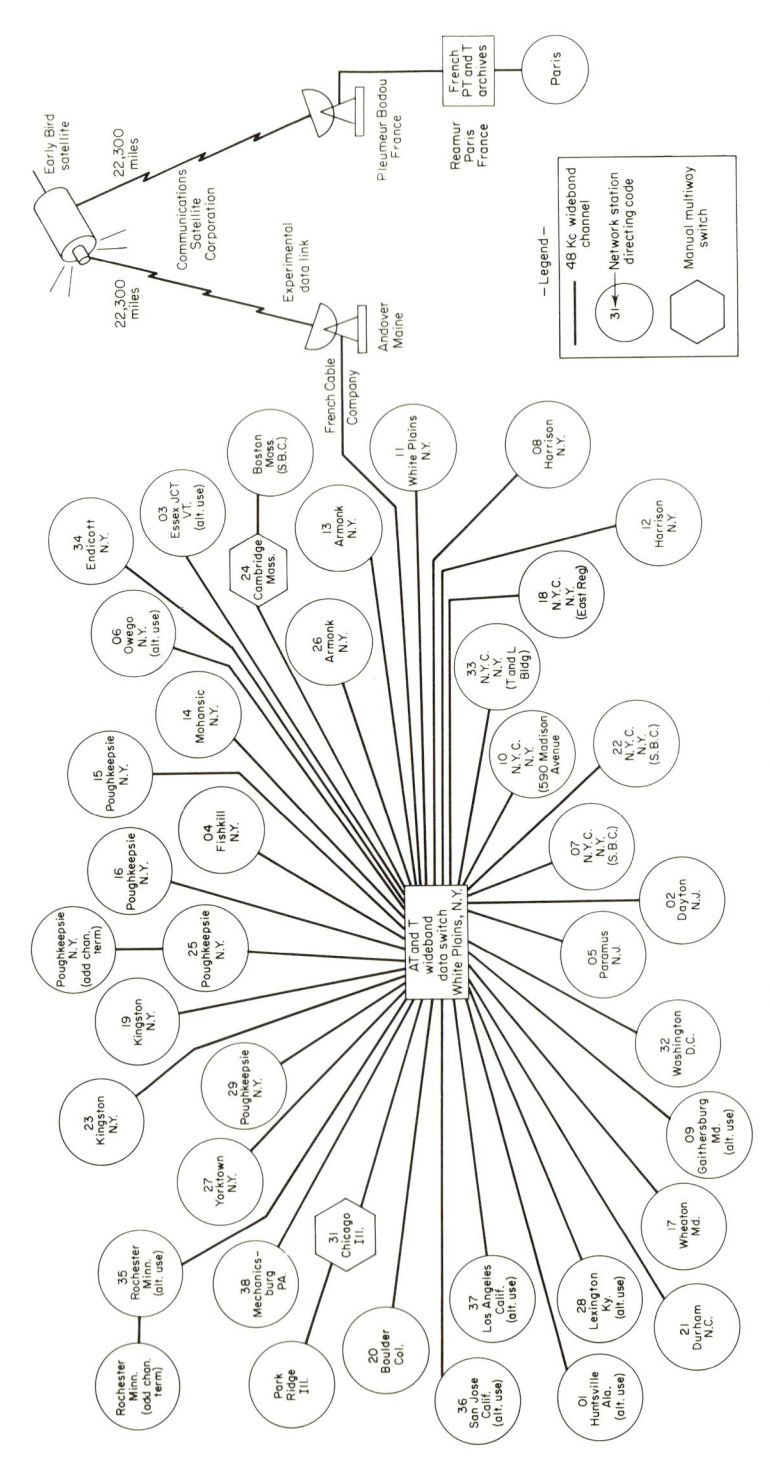

Fig. 8.9. IBM switched wideband data network (48-kHz service).

5800 gives up to 240 voice grade channels or equivalent at $45 per airline mile per month. A solitary voice channel in the United States costs $3 per mile for the first 25 miles, $2.1 for the next 75, $1.5 for the next 150, $1.05 for the next 250, and beyond that $0.75. These prices may change again perhaps even before this book is published, but we quote them here to illustrate that where many voice channels can be gathered together the cost drops dramatically, especially for the shorter distance links.

Channel Types 5700 and 5800 are sometimes referred to as Telpak C and Telpak D channels. Telpak A and Telpak B tariffs used to offer reduced rates for 12 and 24 voice channels or equivalent, but are now not available.

Figure 8.10 shows 13 locations which are to be interconnected and which require five voice-grade links between any two locations. At the top of the diagram they are linked by using exchanges at two of the locations and lines giving the minimum overall line mileage. With two exchanges this would be the lowest cost network, if there were no tariffs offering low-cost grouping. When channel Type 5700 (Telpak C) is used, the configuration in the center of Fig. 8.10 is of lower cost. The link from I to K is less than 25 miles, so with the cost figures above it is cheaper to route the 10 voice-grade channels over a Type 5700 link. This is not so with the link from B to D, which is more than 25 miles. The link from I to M is more than 25 miles but less than 100, and a Type 5700 link is cheaper than the 15 separate voice channels, even though some of these would have a shorter path.

The configuration at the bottom of Fig. 8.10 is another Type 5700 routing which might be less expensive than that above it, depending upon what the exact distances and terminal charges are.

152 PRIVATE EXCHANGES

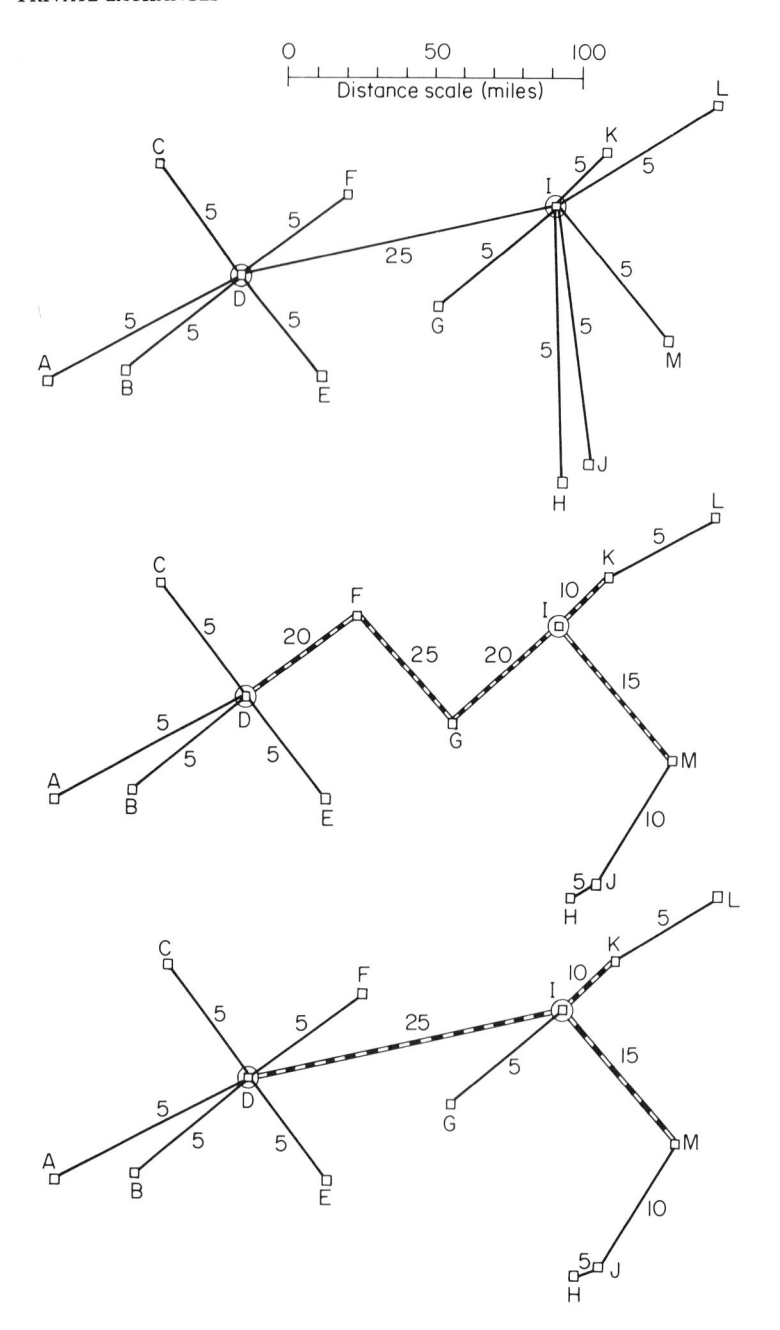

Figure 8.10

9 POLLING AND MULTIPOINT LINE CONTROL

When the lines on a data transmission system are expensive and the terminals are in use only a fraction of the time, it is usual to have several terminals on one communication line. The terminals may be all in one location, or the line may wander in a zig-zag fashion between different locations. By "in use a small fraction of the time," we could mean that a terminal occupies the line from only a few seconds in each minute, or we could mean that it spends much of the day unused. If the latter is the case, we must be sure that when it comes into use it does not prevent other users' obtaining the response time they need. Sometimes response time is not an important consideration. A terminal may be loaded with cards or paper tape, and the controlling machine polls it and reads its data when it is ready. Similarly, it transmits a response when convenient. The user is not sitting at the terminal waiting for an answer. In real-time systems response time is important. Often the user is carrying on some form of "conversation" with the distant computer, and he wants to obtain a response sufficiently quickly for there to be no interruption in his train of thought. Many systems require a response in three seconds or so. If the delay is longer than this, the user becomes impatient.

In Chapter 11 we will discuss multiplexing. This also permits many terminals to transmit over one line, but it is fundamentally different from the operations discussed in this chapter in that the terminals can transmit *at the same instant*. When multiplexing is used the line is behaving as though it were, in effect, several separate lines, or separate channels for data all independent of one another. In this chapter we are concerned with a line on which only one terminal can transmit at one instant. If at any time two terminals both want to transmit, one must wait until the other has finished.

The line, or one data channel of a multiplexed line, might have terminals attached, looking, for example, like the layout in Fig. 9.1. This diagram shows six terminals, A, B, C, D, E, and F, which must transmit to or receive data from a computer. The line can have any number of forks in it. Occasionally, when high reliability is important it may double back to the computer so that if a failure occurs at a point on the line the terminals on the far side of the break will still be linked to the computer. The terminals may be in the same location or different locations. Here they are shown as being in different locations attached to a line that follows the shortest path between them. Sometimes they are thousands of miles apart. Often the terminal user carries on a conversation with a distant computer, basically unaware that there are other terminals on the same

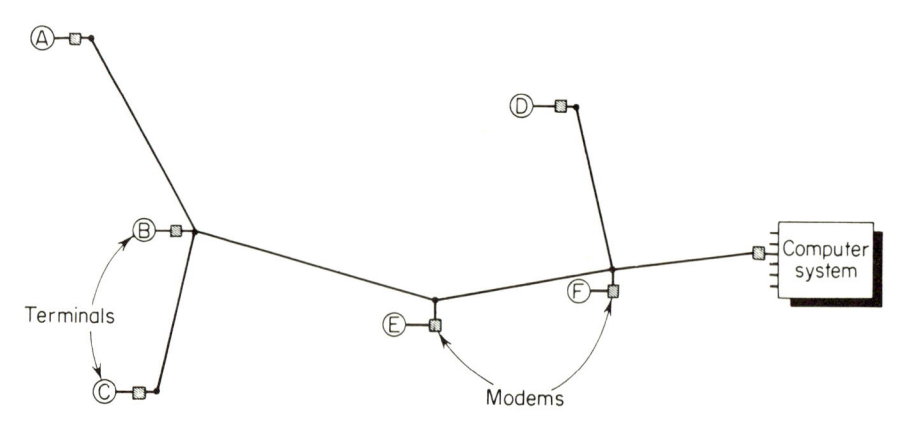

Figure 9.1

line. The only effect that the other terminals have on him is that they sometimes lengthen his response time or cause him to have to wait a while before he can begin transmitting.

Figure 9.1 could represent a wide variety of different types of system. The line could be a low-speed line with, for example, teletype machines on it. It could be a voice-grade line, or it could be a wideband link. The terminals could be operating in real-time, or they could be nonreal-time. They may transmit batches of data for remote entry into a batch-processing system. The line may be full duplex or half duplex. If it is full duplex an output to one terminal may be traveling at the same time as the input from another.

The line in any of these systems is generally referred to as a *multidrop line*, meaning that it has more than one terminal. When a real-time response is needed, we have to be careful in the design not to overload the line. Suppose that the six terminals in Fig. 9.1 are part of a real-time system. All six operators may be carrying on a conversation with the computer at the same time. All of them want a fast response. We must take into consideration the extent to which the response to one operator will be delayed by the transmissions to and from the others.

BUFFERED TERMINALS

Suppose that the terminals in question operate on a low-speed half-duplex line at 15 characters per second, that the average number of characters in a message to the computer is 30, and that the number in its response is 60. Suppose that the operator sends one such message on the average every 100 seconds. The remainder of the time is "think" time or time when he is discussing the transaction with a client. The total transmission time per message is $(30 + 60)/15 = 6$ seconds. If six terminals each transmit for a total of six seconds every 100 seconds, the line is clearly not overloaded. This assumes, however, that the terminal can transmit to the computer at its full speed. If transmission involves reading paper tape or sending the contents of a buffer, this is true. If, however, it is asynchronous transmission from a keyboard on an unbuffered terminal, then the characters are sent as the operator presses the keys, and this input could tie up the line for a lengthy period.

If the terminal is unbuffered, we have to take into consideration the typing speed of the operator. This can vary over a surprisingly wide range. A figure often used by systems analysts for the speed of a touch typist operating a terminal is three characters per second. The transmission of 30 input characters, then, ties up the line for about ten seconds rather than the two seconds if the characters are transmitted from a buffer. It is often, however, much worse than this. The terminal operator may be a one-finger typist who is lucky to average one character per second. Worse still, he may stop and "think" in the middle of a transmission, reading documentation or composing what he is to type next. If the terminal is unbuffered, he will exclude all other users from the line during his pauses—probably without knowing it.

Schemes have been devised for locking the keyboard during a terminal operator's pauses. If no character is received for more than a given time, say five seconds, the computer suspends transmission from that device to see whether there is anything to transmit for any other terminal. The keyboard is locked until the computer returns its attention to that terminal. In some cases, however, this system reckoned without the ingenuity of the terminal operator. The news went around the operators that during a period of pause they should idly move the SHIFT key up and down, and this would prevent their keyboard's locking. Thus a steady stream of shift-change characters were sent to the computer. This did not, in effect, alter the message content, but the computer never had five seconds without receiving a character and so did not lock out the operator. The other terminal users on that line, however, really had a long wait.

Suppose that our operator types at three characters per second with no pauses. The total transmission time for the input and output will be $\frac{30}{3} + \frac{60}{15} = 14$ seconds. If this timing applies to each of the six terminals, there will be a total of 84 seconds line occupancy every 100 seconds. The line will be tied up 84 percent of the time, and in practice, as we will see later, there are some times which we should add to this for line control operations. A use of queuing theory[1] indicates that it is inadvisable to load a communication line more than about 70 percent; otherwise, high queues build up and the response time is degraded.

If our operator typed at one character per second, the total transmission time for one terminal would be $\frac{30}{1} + \frac{60}{15} = 34$ seconds. With six such terminals the line would clearly be overloaded.

In this case it would be unwise to recommend that the line have six terminals, even if the operators were touch typists, unless buffered terminals were used. With a buffer it does not matter how fast the operator types or whether he has lengthy pauses. He can fill the terminal buffer slowly; its contents are always transmitted at the line speed of 15 characters per second.

RESPONSE TIME

Without buffering we might, perhaps, have three of these terminals on one line. The lack of buffering will, however, mean that a terminal will sometimes give a poor response time. A one-character-per-second operator will tie up the line for 30 seconds, and more if he stops to deliberate. Another operator may be held up for this time before he can start to transmit, or, if it is a half-duplex line, the reply to him may be held up by this amount. If he shares a line with *two* one-character-per-second colleagues there is a small probability that the delay will be twice this long.

Occasionally, on the other hand, he will receive a very fast response because his colleagues are not transmitting at that moment. The system will therefore appear to him to be somewhat erratic in its performance—sometimes keeping him waiting half a minute before he can transmit, sometimes delaying half a minute before responding, but sometimes reacting very quickly. The situation would be worse if, instead of talking about a 30-character input, we were talking about a much longer one.

Some multidrop terminals are designed so that many can be put on one line. Clearly, we cannot take advantage of this without careful examination of response times. Buffered terminals should always be used on such lines

---

[1] *Design of Real-Time Computer Systems*, by James Martin, Prentice-Hall, Inc., Englewood Cliffs, N.J., 1967.

when the response time is critical. When fast response time does not matter, as in batch operations, a large number of inexpensive unbuffered terminals such as paper tape readers and printers may be connected to one line. Sometimes, as discussed in Chapter 7, an operator effectively creates his own buffer by punching his transaction into paper tape or cards, and then transmitting those.

MODEMS The modems on a multidrop line can be exactly the same as a point-to-point line. Each line termination will need a modem, as shown in Fig. 9.1. The total number of modems will therefore be less than with the same number of terminals on point-to-point lines. A line like that in Fig. 9.1 with N terminals needs $N + 1$ modems, rather than $2N$.

SELECTIVE CALLING As the terminals are connected to the same line, no two terminals may transmit at once. We may think of a half-duplex line, no matter what its physical nature, as though it were a pair of copper wires that connect the computer to the different terminals on it. We may think of a full-duplex line as though it were two pairs of wires, one for input and one for output. If we send an output pulse down the line, it will reach all of the terminals. If one terminal sends an input pulse, this will travel down the line, not only reaching the computer, but also reaching all of the other terminals on the line. It is clear, then, that if two terminals transmitted at once, the pulses from them would be mixed up. We need a line discipline that prevents this.

Equally important, each terminal must have a means of recognizing which signals are meant for *it*; otherwise, it will react to the messages to and from the other terminals on the line. If a message is sent from the computer to a terminal, it may contain an address of that terminal near the front of the message. The terminal whose address this is, recognizes it, and takes appropriate action. All of the other terminals on the line ignore it. On the majority of systems it is not the message with the data or text to be transmitted which contains this address, but rather a preliminary control signal that is sent down the line to say, in effect: "Terminal number X, I have a message for you. Are you ready to receive it?" The terminal replies "yes" or "no" by sending back a control character, or characters. If it replies "yes," the computer sends the data and all but that terminal ignore it.

The control characters can cause the line to become, in effect, a point-to-point connection for a limited period of time. All of the other terminals ignore what is transmitted until a further signal is sent which ends the private interconnection of these two points. The control character, or group of characters, which releases the interconnection is called an "end-of-transmission" or "end-of-message" signal. Once it is sent, all of the terminals on the line reset and listen for the next signal to be sent. Most commonly the "end-of-transmission" is sent by the computer, but it could also be sent by a terminal. During that period in which our multipoint line has become, in effect, a point-to-point connection, any of the point-to-point line control procedures discussed in Chapter 6 could apply.

Systems for permitting the computer to send a message to one terminal on a multiterminal line are referred to as *selective calling*. In addition to having the ability to address one terminal, some systems also have a "broadcast" code that can cause a message to be accepted by all terminals.

In order to recognize its address and carry out the various line control functions needed, the terminal must have a unit with some logic capability. In telegraphy this is referred to as a "stunt box." In many computer manufacturer's terminals the requisite logic is in the terminal control unit. These logic circuits make the terminal somewhat more expensive than otherwise. A terminal without them cannot be used on a multidrop line.

CONTENTION AND POLLING

For transmission *to* the computer, several terminals may wish to transmit at the same time. Only one can do so, and the others must wait their turn. To control this, the computer, when it is ready, will signal the terminal to transmit. This may automatically cause transmission from paper tape or from a buffer, or it may display a light telling the operator to proceed. There are two methods in common usage for organizing this: "contention" and "polling."

In a "contention network" the terminal makes a request to transmit. If the channel in question is free, transmission goes ahead. If it is not free, the terminal must wait. A queue of "contention requests" is built up, and this is scanned either in a prearranged sequence or in the sequence in which the requests were made.

Far more common than contention is a *polled* line. In this form of line organization the computer, or master station, asks the terminals one by one whether they have anything to transmit.

A *polling message* is sent down the line to a terminal saying, "Terminal X, have you anything to transmit? If so, go ahead." If terminal X has nothing to send, a negative reply will be received, and the next polling message will be sent, "Terminal Y, have you anything to transmit? If so, go ahead."

Normally the computer organizes the polling. The computer may have in core a *polling list* telling the programs the sequence in which to poll the terminals. The polling list and its use determine the priorities with which terminals are scanned. Certain important terminals may have their address more than once on the polling list so that they are polled twice as frequently as the others. Certain terminals may always be polled before the others.

Once polling has established an interconnection between a terminal and the computer, the transmission, again, can proceed much as the point-to-point transmission described in Chapter 6, with appropriate error control. On some line control systems the computer can respond with data before it breaks the interconnection. On others this is not possible. Again, an end-of-transmission signal breaks the interconnection, and causes all the terminals on the line to reset and await the next transmission.

LINE CONTROL

A multidrop line may be in any of various types of status. Three of these are *receive status, transmit status,* and *control status.* As the lines are scanned by the transmission control mechanism, their status is examined, and this informs the program or hardware whether to take, receive, transmit, or control actions. All addressing, polling, and answering of polling messages take place in control mode.

Various characters or sequences of characters must have special meanings during the transmission. In some systems, certain characters are reserved as control characters. A variety of these are seen in the ASCII and other codes illustrated in Chapter 2. (See Figs. 2.5, 2.7, 2.9, and 2.10.) In five-bit or six-bit telegraphy codes, however, no characters were designated for these purposes (Figs. 2.3 and 2.4), so sequences of characters are used. For example, on many systems the sequence *Figures shift character —H—Letters shift character* means "end of transmission."

Some illustrations will clarify this:

First, let us examine the characters that flow on a typical multidrop telegraph line. Each terminal on the line has an address of two alphabetic characters. This address must be used when one is selectively calling the terminal. A *call-directing code,* CDC, is a sequence of characters which is used for addressing. On most systems this consists of the two characters of the address followed by a letters-shift character. On some the letters-shift character is not used.

Several terminals can be addressed at once by sending several consecutive call-directing codes. Each terminal, if it is ready to receive, sends a letter V as an answer. The computer or master control must end the addressing sequence with a designated group of characters before it sends the text. An *end-of-address* sequence commonly used is *carriage return character—line feed character.* This must be followed by a letters-shift character on some systems, but not on others.

The sequence of operations for sending data to the telegraph machines may then be as shown in Table 9.1.

Table 9.1

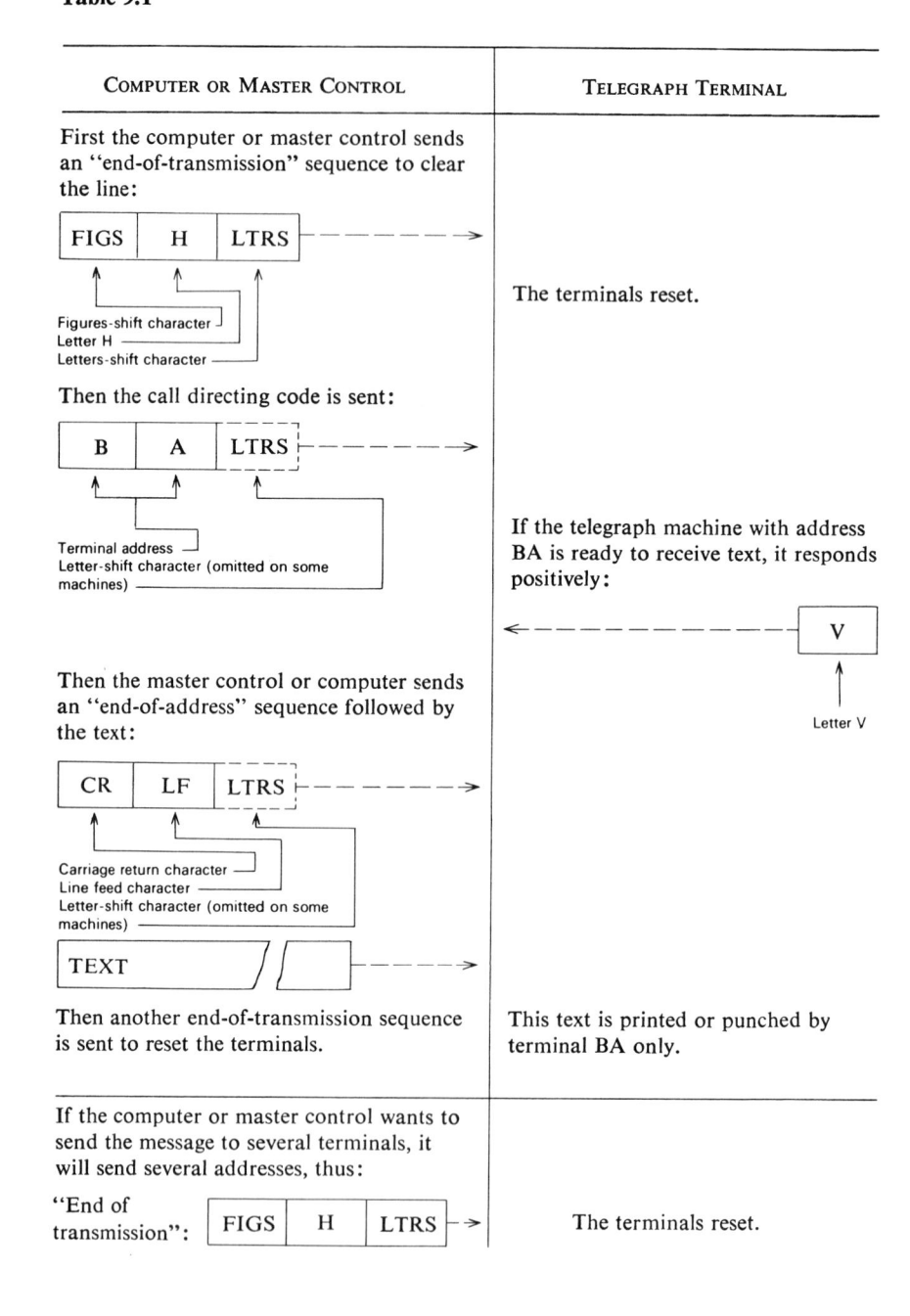

Computer or Master Control	Telegraph Terminal
First the computer or master control sends an "end-of-transmission" sequence to clear the line: | FIGS | H | LTRS | - - - - → Figures-shift character ┘ Letter H Letters-shift character Then the call directing code is sent: | B | A | LTRS | - - - - → Terminal address ┘ Letter-shift character (omitted on some machines)	The terminals reset. If the telegraph machine with address BA is ready to receive text, it responds positively: ← - - - - - - - - | V | Letter V
Then the master control or computer sends an "end-of-address" sequence followed by the text: | CR | LF | LTRS | - - - - → Carriage return character ┘ Line feed character Letter-shift character (omitted on some machines) | TEXT / / | - - - - → Then another end-of-transmission sequence is sent to reset the terminals.	This text is printed or punched by terminal BA only.
If the computer or master control wants to send the message to several terminals, it will send several addresses, thus: "End of transmission": | FIGS | H | LTRS | →	The terminals reset.

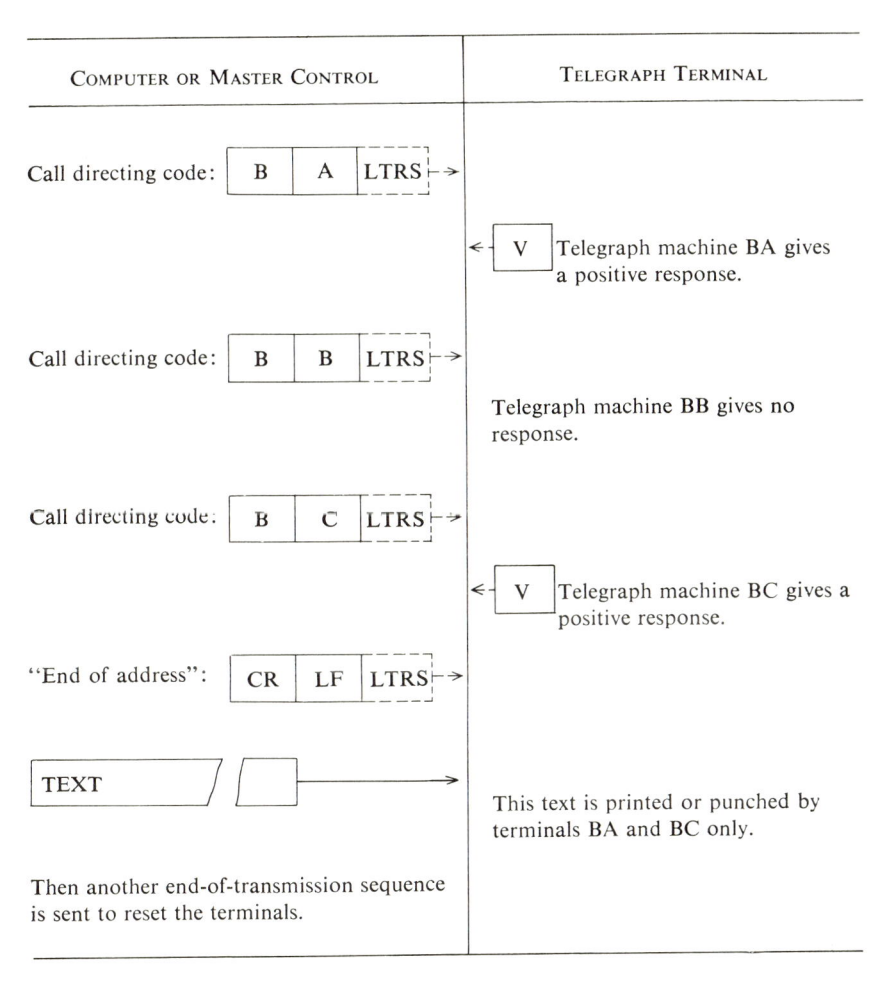

POLLING AND MULTIPOINT LINE CONTROL 161

Computer or Master Control	Telegraph Terminal
Call directing code: B A LTRS →	
	← V Telegraph machine BA gives a positive response.
Call directing code: B B LTRS →	
	Telegraph machine BB gives no response.
Call directing code: B C LTRS →	
	← V Telegraph machine BC gives a positive response.
"End of address": CR LF LTRS →	
TEXT →	This text is printed or punched by terminals BA and BC only.
Then another end-of-transmission sequence is sent to reset the terminals.	

When a terminal has data to send, it must wait until it is polled and then send it. During this sequence of operations a letter V answer back from the terminal means that the terminal has nothing to send. V is now a negative response, not a positive one. The sequence of operations is as shown in Table 9.2.

Line control for most of the computer (and other) manufacturers' terminals is similar in principle, though often different in detail. Transmission codes with seven or eight information bits per character usually have *one* character meaning end of transmission (EOT), and *one* character meaning "end of address" or "start of text" (STX). It is not necessary to use multiple characters for these controls (see Figs. 2.3, 2.5, and 2.7). Some codes with six information bits per character also have such characters (Fig. 2.8). This shortens the control sequences somewhat.

162 POLLING AND MULTIPOINT LINE CONTROL

Table 9.2

Computer or Master Control	Telegraph Terminal
First the computer or master control sends an "end-of-transmission" sequence to clear the line: | FIGS | H | LTRS | – – – – – →	
Then the terminal address is sent. | B | A | – – – – – – – →	The terminals reset.
(This is referred to as a "transmitter start code")	Telegraph machine BA has nothing to send and so responds negatively. ← – – – – – – – | V |
The computer or master control unit clears the line and tries the next terminal: "End of transmission": | FIGS | H | LTRS | →	The terminals reset.
Address of next terminal: | B | B | →	
	BB has something to send and so transmits it: ← –| TEXT / /
The end-of-transmission code is sent. | FIGS | H | LTRS | – – – – – →	
This could have been sent either by the terminal or the computer (or master control unit).	The terminals reset.

POLLING AND MULTIPOINT LINE CONTROL 163

Similarly, the address of the terminal can be one character. On the other hand, one terminal may consist of several devices, and an extra character is needed after the terminal address to say which device is being addressed.

Many transmission devices carry out checking of the data transmitted and then the receiving device sends a positive or negative response to say whether or not it has been received correctly.

Let us illustrate this by describing the characters that flow when IBM 1050 terminals are used on a multidrop line. 1050-type line control is used on a wide variety of terminals in addition to the IBM 1050. The complete list of IBM 1050 control characters is shown in Fig. 2.3. Those that appear in the example below are as follows:

BIT PATTERN

	S	B	A	8	4	2	1	P
Ⓒ : End of transmission	1/0	0	0	1	1	1	1	1/0
Ⓓ : End of address	1/0	0	0	1	0	1	1	1/0
Ⓑ : End of block	1/0	0	1	1	1	1	0	1/0
(This is used to terminate the longitudinal redundancy check accumulation.)								
Ⓨ : Positive acknowledgment	0	1	1	1	0	1	1	0
Ⓝ : Negative acknowledgment	0	1	0	0	0	0	0	0

Shift bit can be either a 1 or 0 (The other six bits are the same for either shift.)

Parity check bit

Suppose that the computer has some data that it wants to send to a terminal on a multidrop line. The address of the terminal is F. The sequence of operations is as shown in Table 9.3.

Table 9.3

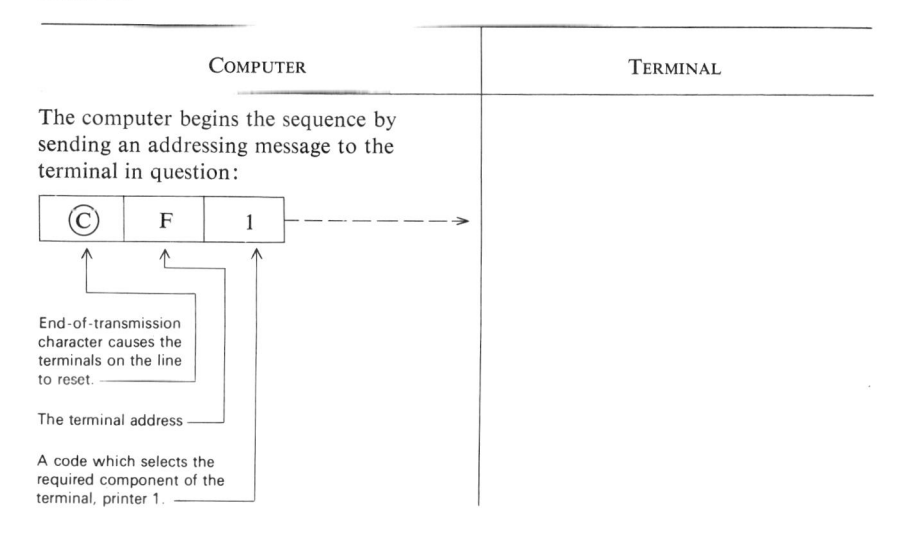

Computer	Terminal
The computer begins the sequence by sending an addressing message to the terminal in question: | Ⓒ | F | 1 | - - - - - → End-of-transmission character causes the terminals on the line to reset. The terminal address A code which selects the required component of the terminal, printer 1.	

164 POLLING AND MULTIPOINT LINE CONTROL

Table 9.3—cont.

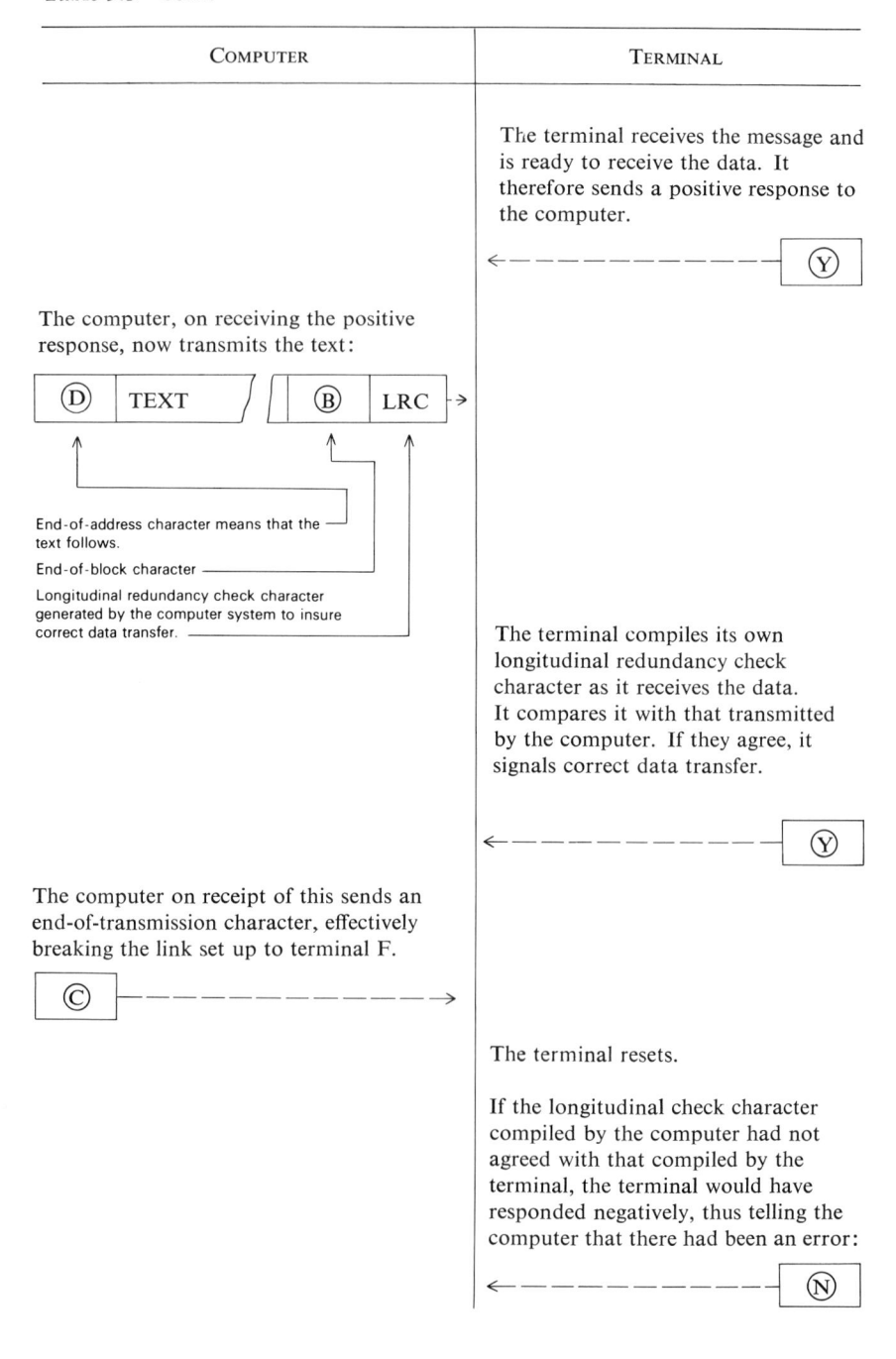

COMPUTER	TERMINAL
	The terminal receives the message and is ready to receive the data. It therefore sends a positive response to the computer. ←— — — — — — — —⏐ Ⓨ
The computer, on receiving the positive response, now transmits the text: ⎡ Ⓓ ⎸ TEXT ⎸ Ⓑ ⎸ LRC ⎤ → End-of-address character means that the text follows. End-of-block character Longitudinal redundancy check character generated by the computer system to insure correct data transfer.	The terminal compiles its own longitudinal redundancy check character as it receives the data. It compares it with that transmitted by the computer. If they agree, it signals correct data transfer. ←— — — — — — — —⏐ Ⓨ
The computer on receipt of this sends an end-of-transmission character, effectively breaking the link set up to terminal F. ⎸ Ⓒ ⎸— — — — — — — — →	The terminal resets. If the longitudinal check character compiled by the computer had not agreed with that compiled by the terminal, the terminal would have responded negatively, thus telling the computer that there had been an error: ←— — — — — — — —⏐ Ⓝ

COMPUTER	TERMINAL
The computer would then retransmit the message: | TEXT	/	ⓑ	LCR	→		
	If this time it is received correctly, the terminal responds positively. ← ——————	Ⓨ	
... and the computer signals "end of transmission": 	Ⓒ	————————→	The terminal resets.

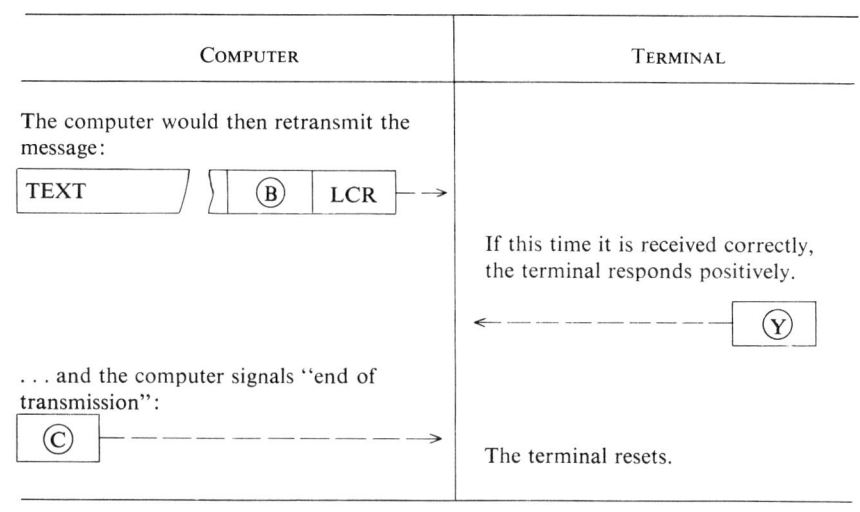

If it is the terminal that has something to send, it must wait until it is polled by the computer. The operator at a keyboard, for example, presses the "request" key, and waits. When the terminal is next polled, it will reply positively. The computer tells it to go ahead. A light comes on telling the operator to begin his keying. The sequence of operations is as shown in Table 9.4.

Table 9.4

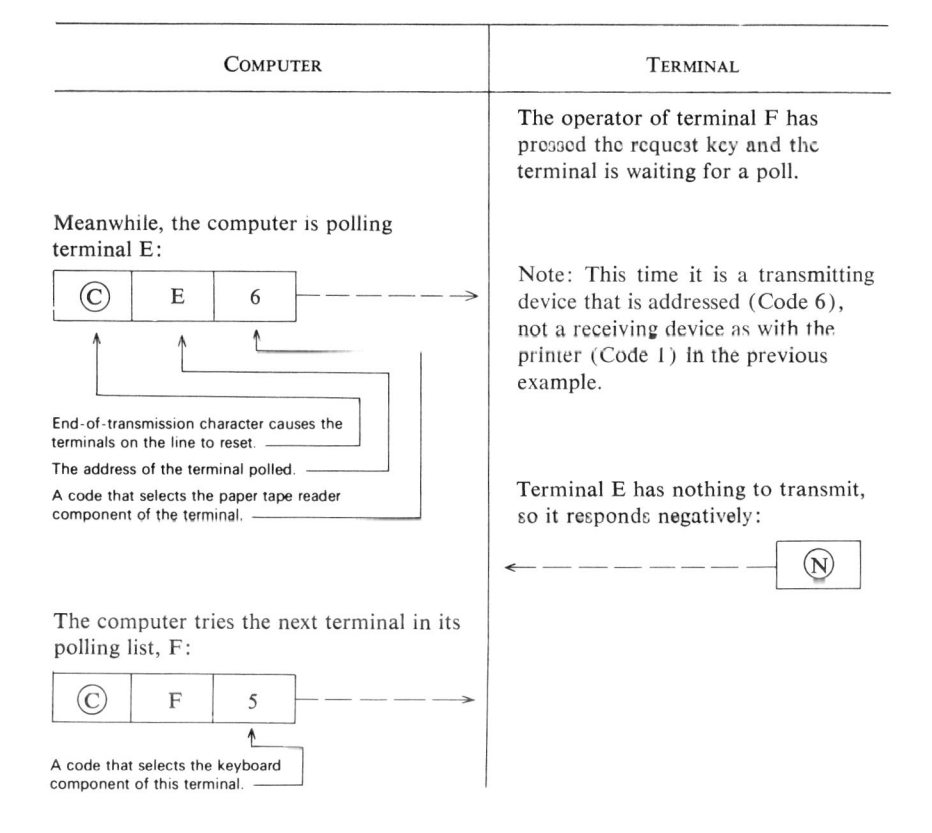

COMPUTER	TERMINAL
	The operator of terminal F has pressed the request key and the terminal is waiting for a poll.
Meanwhile, the computer is polling terminal E: | Ⓒ	E	6	————→ End-of-transmission character causes the terminals on the line to reset. The address of the terminal polled. A code that selects the paper tape reader component of the terminal.	Note: This time it is a transmitting device that is addressed (Code 6), not a receiving device as with the printer (Code 1) in the previous example. Terminal E has nothing to transmit, so it responds negatively: ← ——————	Ⓝ	
The computer tries the next terminal in its polling list, F: 	Ⓒ	F	5	————→ A code that selects the keyboard component of this terminal.	

Table 9.4—cont.

COMPUTER	TERMINAL
	The terminal responds positively by sending the data. In this case a light comes on telling the operator to start keying, and the keyboard unlocks. The operator keys in his transaction:
	← ⓓ | TEXT | // | ⓑ | LRC End-of-address character indicates that text follows. End-of-block character. Longitudinal redundancy check.
The computer's longitudinal redundancy check agrees and so it signals that the message was received correctly: ⓨ ┤ ─ ─ ─ ─ ─ ─ ─ ─ ─ ─ ─ ─ →	

TIME-OUT Sometimes a terminal may fail to reply entirely to a polling message (for example, if its power is switched off). To deal with this situation the computer or master control must be equipped with a time-out mechanism that enables it to go to another operation after waiting for a given period.

TWO TYPES OF POLLING The illustrations we have used above are examples of "roll-call" polling. Most systems today use this form of polling. With *roll-call polling* the computer works down a list of terminals and asks them one by one: "Have you anything to send?" Some terminals may be polled more than others to give them a better response time, or priority sequencing may be used; but still the computer (or master control) polls the terminals one at a time.

An alternative method of line control is called "hub go-ahead" or sometimes just "hub" polling. This is advantageous on long (and hence expensive) lines and on lines with fast turnaround time and many terminals. With *hub go-ahead* polling the computer addresses only the terminal at the end of the line, and the terminals pass the polling message down the line. The computer polls the farthest terminal, A (see Fig. 9.2), and says: "Have you anything to send?" If A does not, A sends the poll to its neighbor B. If B has nothing, it sends it to C. C sends it to D, and so on until it reaches the computer.

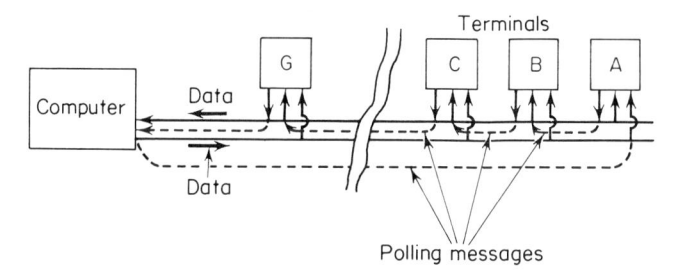

POLLING AND MULTIPOINT LINE CONTROL   167

Fig. 9.2. Hub go-ahead polling.

AN ILLUSTRATION OF HUB POLLING

During normal operations, hub polling at its simplest requires only two types of message: the polling message and the data message. If no data are being sent, the polling message will travel from terminal to terminal on the line.

Let us take our illustration this time from a line in which the main consideration is to obtain the maximum speed and response time—the maximum line efficiency. This is the situation in which hub polling is likely to be used. In this example the transmission is synchronous, as opposed to asynchronous in the above cases. Synchronization characters therefore have to be used both on the data messages and the polling messages. Six-bit characters are used with no parity check, but a cyclic redundancy check character is used.

The computer begins its line scan by sending a polling message to device A (Fig. 9.2):

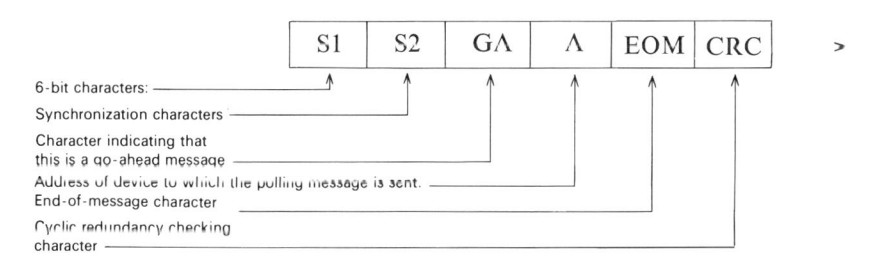

| S1 | S2 | GA | A | EOM | CRC |

6-bit characters:
Synchronization characters
Character indicating that this is a go-ahead message
Address of device to which the polling message is sent.
End-of-message character
Cyclic redundancy checking character

If device A has nothing to send, it sends a similar message on to device B:

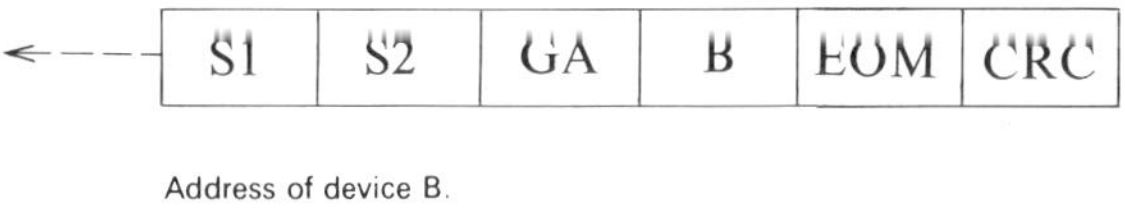

| S1 | S2 | GA | B | EOM | CRC |

Address of device B.

168 POLLING AND MULTIPOINT LINE CONTROL

If B has something to send, it sends it:

← — | S1 | S2 | B | TEXT // | EOM | CRC |

The computer receives this and then resumes polling at device C:

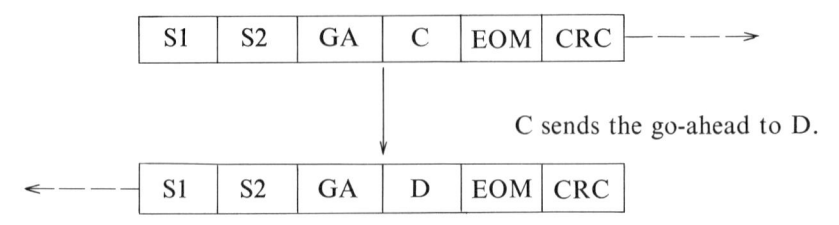

| S1 | S2 | GA | C | EOM | CRC | — — →

C sends the go-ahead to D.

← — — | S1 | S2 | GA | D | EOM | CRC |

... and so on until the last terminal on the line sends it back to the computer.

If the computer has an output message for a terminal it simply sends it, without prior calling:

| S1 | S2 | B | TEXT // | EOM | CRC | — — →

The absence of a positive or negative answerback means that the receiving device cannot notify the transmitting device if an error is detected. There is no automatic retransmission. Instead, the operator is notified, which on some systems is quite good enough. An alternative is to send an error message to the transmitting device if, and only if, the message received is incorrect or incomplete.

One more control message is needed on a hub go-ahead line. If a device fails, or is switched off, a vital link in the line control has gone. If the cleaning lady knocks out the plug of device B, then device B cannot send the go-ahead message on to device C. The computer can detect this because when it fails to receive a reply it can send polling messages down the line to detect which device is causing the trouble. Device A must then be made to by-pass the inoperative B, and send its go-ahead message on to device C.

To tell device A to make this change a "change-interchange-address" message—CIA message—is sent (nothing to do with the Central Intelligence Agency!). The format of this is as follows:

```
| S1 | S2 | CIA | A | C | EOM | CRC |
```

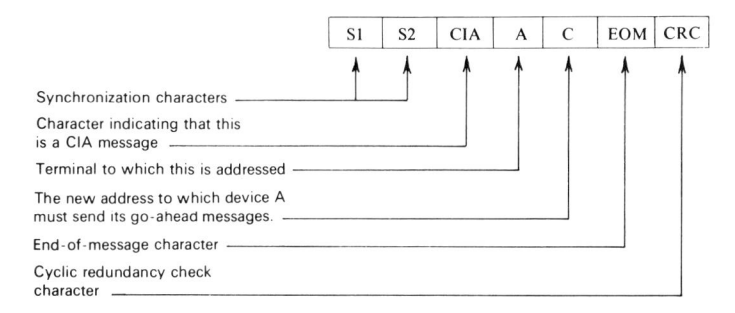

- Synchronization characters — S1, S2
- Character indicating that this is a CIA message — CIA
- Terminal to which this is addressed — A
- The new address to which device A must send its go-ahead messages. — C
- End-of-message character — EOM
- Cyclic redundancy check character — CRC

ADVANTAGES AND DISADVANTAGES OF HUB POLLING

1. The main disadvantage of hub polling is that more logic capability is needed at the terminals. They must have the ability to send the go-ahead messages on to the next terminal. They must have the ability to deal with a CIA message.

2. A second disadvantage is that an extra modem is needed at each station (or at least that half of a modem that *receives* data). As can be seen in Fig. 9.2, the modem can receive data from either the output line or the input line.

3. A disadvantage that is usually of much less importance is that (unless the scheme is more complex) terminals cannot be given preferential treatment when they have a greater load or need a faster response.

4. The main advantage lies in lessening the number of line turnarounds in a scan of the terminals. Because we are using fast transmission and hence perhaps quaternary phase modulation, we may take 20 character transmission times or more to turn around the line. With 10 terminals on the line the total of the turnaround times equals 400 character transmission times. Hub polling has therefore been used on systems with short response time requirements and long lines where it is desirable to put as many terminals as possible on a line.

5. The number of control characters flowing on the line in the above example is considerably less than would be the case for roll-call polling with the same synchronous transmission and error-checking. The line can, therefore, handle a higher throughput. This is especially significant if the messages are short, as on many real-time systems, and the response time requirement is high.

As in other aspects of line organization, more logic capability in the devices on the line can improve the overall line capability.

10 TERMINAL CONTROL UNITS

The nature of the terminal control units can have a major effect on the organization and cost of the communication line network. We can specify certain features that we would like to have on the terminal control units in order to minimize the network cost. Some terminals have these and some do not. Often we are asking for features that are expensive, so the desirability of the feature depends upon the cost balance between the terminal control units and the communication lines.

Let us first list the control unit features that affect the network organization:

1. *Ability to Operate on Multidrop Lines*

In order to put more than one device on a communication line, some means of handling polling or contention is needed, as discussed in Chapter 9. A control unit or stunt box must be able selectively to recognize messages sent to it. It must not print those that are intended for a different terminal. On a polled line it must recognize and interpret polling messages addressed to it, and must be able to respond appropriately. It must be able to generate or interpret all of the control messages used, for example, in the scheme shown on p. 160.

If hub go-ahead polling is used to improve the efficiency of long lines, the terminal control unit (or concentrator) must be able to pass the go-ahead message onward to the next unit. It must have the logic to bypass the next unit if so instructed because that unit has failed.

2. *Buffers*

Where more than one device is used on one communication line, buffering becomes important. It is desirable to transmit at the maximum speed of the line, or the maximum speed that the modems in use permit. Many terminal machines operate at a speed different from this; in particular, a typist at a keyboard generates characters at a speed often far different from the optimum line speed.

As was discussed on p. 155, if a keyboard terminal on a multidrop line has no buffering, as is the case on many systems installed, then the operator at that keyboard can hold up the other terminals by slowly typing a lengthy message, he can often hold up the line while he sits and thinks at the terminal. On a system that needs a fast response time to almost all of its terminals, a multidrop line with unbuffered devices is unlikely to give it unless all of the messages are kept very short. Even then an ill-trained or malevolent operator can degrade the line performance.

To avoid the cost of buffer storage, telegraph machines, as we mentioned previously, use paper tape as a buffer. The keyboard operator punches the transaction into tape and puts it in the tape reader. It is then transmitted under the control procedures at the full speed of the telegraph line. A similar technique is being adopted by some other terminal manufacturers, and media other than paper tape are being used. In some devices inexpensive magnetic tape as on a domestic tape recorder is the storage medium. This is reusable and can transmit at the full speed of a voice line. On some machines many transactions are collected on tape and then transmitted together on a dial-up line. Other machines use a magnetic disk or drum for collecting transactions so that many can be transmitted together relatively quickly, taking advantage of a dial-up tariff. Such schemes, however, do not solve the problem of obtaining a fast response to a real-time terminal. For this, on a multidrop line, buffering is needed.

For most display screen terminals it is necessary to store the information displayed on the screen in a control unit. The image must be constantly regenerated by a cathode ray beam sweeping across it, and the data for the regeneration are in the control unit storage. The transmission buffer is the same storage. The control unit must have the logic to put characters received in the correct position in this storage, and to transmit only that part of the data which the operator keyed in.

3. *Ability to Control More Than One Device*

Sometimes several transmitting or receiving machines are attached to one control unit. These may all constitute one terminal station with one operator, or they may be independent stations. Figure 10.1 shows one terminal station with one control unit, but several different devices.

Fig. 10.1. Control unit for visual display terminals: The IBM 2848. Functions: buffering; generation of characters on screens; translation to and from transmission line code; logic for multidrop line control and polling; logic for error checking, cursor positioning, etc.

4. *Shared Buffers*

Where several separate terminals share a control unit, one area of buffer storage may be shared between them. This may be allocated in a fixed manner with each terminal having its own slice, or it may be allocated dynamically with blocks of storage being used as and when needed by different terminals. The control unit in Fig. 10.1, which controls IBM 2260 visual display stations, uses magnetostrictive delay lines as a buffer allocated in a fixed manner. This storage can handle up to eight terminals with 960 characters displayable on the screen, or else 16 terminals with 480 characters displayable, or 24 terminals with 240 characters displayable. In addition, the control unit can handle a printer operating at 14.8 characters per second which also has a fixed area of buffer. The cables connecting the display stations to the control unit restrict their distance apart to 2000 feet.

With devices that do not require continual regeneration, as with this screen unit, the cable between the terminal and shared buffer could be designed to carry bits at a much lower rate, and hence could be a very long connection. It could, in fact, be another telecommunication link.

5. *Error Checking*

An important function of the control unit is to perform error checking, as discussed in Chapter 4.

6. *Retransmission*

When an error is detected, many systems automatically retransmit the message in question. Some merely give a warning to the terminal operator that an error has occurred. If the terminal control unit is required to retransmit a message when an error is detected, it must have some means of storing the message until it has received an acknowledgment of correct transmission. This means may be a buffer, or in some terminals it may involve backspacing a paper-tape reader to retransmit or rereading a punched card. Similarly, on reception the terminal control unit must have the means of requesting a retransmission from the sending machine.

7. *Control Functions*

The control unit may have to control a variety of mechanical actions on the terminal, such as feeding forms, skipping, tabbing, detecting when paper has run out and signaling the computer not to transmit, backspacing, and so on.

8. *Home Mode*

Where there are several devices on one control unit, the control unit may permit these devices to communicate with each other without using the communication line. This is sometimes referred to as "home mode" operation. Paper tape may be punched, cards listed, tape-to-card operations carried out, documents prepared that involve both typing and printing from tape, and so on. The multidevice terminal can become a small unit record data-processing installation in its own right.

9. *Editing Logic*

A terminal control unit may have the ability to do certain types of editing. The main purpose of this would be to lessen the total number of characters traveling on the communication line. Editing may be done by the operator before her message is transmitted from the control unit buffer, or it may be performed on messages coming from the computer before they are printed or displayed. In the former case the operator may want to correct mistakes before she transmits the data. The control unit may give her the ability to backspace and change what she entered, or to insert or delete words or items on a screen. In editing the message coming from the computer, it is often necessary to print or display a neatly formatted table, statement, set of data, or document. This could be done either by transmitting all of the blanks, heading, straight lines, and so on, that are used, or by transmitting only the essential data and doing the formatting at the control unit. The formatting logic is then somewhat equivalent to such logic in a punched-card accounting machine. The instructions for doing the formatting may reside temporarily in the storage of the control unit, as when one control panel is inserted into an accounting machine, and can be changed by transmission from the computer.

10. *Input Syntax and Format Control*

When data are manually entered into a keyboard, it is necessary to ensure that the computer interprets them correctly. When there is more than one field in a message, the computer must interpret the format and attach a correct meaning to each input character. A group of characters might be typed into a terminal in an airline reservation system, giving a flight number, class, date, code for the airport of departure, code for the airport of arrival, departure time, and a code indicating what action the computer is to take. The same group of characters might be typed in as one long string, with a quite different meaning. Some of the fields might be variable in length. It is necessary for the computer to sort out the format of messages of this type.

One method of ensuring correct formats is to use preprinted stationery or templates so as to indicate the positions and sizes of the fields. This method can be inconvenient where there are many different types of input. It is also undesirable where the fields vary considerably in length.

Another method is always to key in fields in a given sequence and to put a hyphen or other symbol between the fields. Again, the operator may enter the field label in a given code before entering the data. A field code or a message-type code can be useful when there are only a small number of fields in the message. However, with long messages, fields may be forgotten. If fields are separated only by hyphens, there is a danger that they may be entered in the wrong sequence.

The type of method that has proved most satisfactory is one in which the computer has carried on a "conversation" with the user, telling him what data to enter, step by step or field by field. This technique is particularly valuable when the data entry is complex, as when one is ordering complex machinery, or entering elaborate details of a customer order. It is desirable when the operator is likely to be not too well trained or to use the terminal only occasionally, unlike a dedicated operator who is highly trained and has plenty of time to practice. The conversation an airline booking agent carries on with a terminal screen is different from one suited for a computer salesman, for example, because the former spends almost all of her working life using the terminal and is highly trained, whereas the salesman uses it infrequently and may want to carry out more than one type of operation.

The disadvantage of a conversation directed by the computer is its effect on the communication line design. It requires many more characters transmitted, sometimes several times more. The prompting, syntax-checking, and completeness-checking for this type of data entry may be done by an "intelligent" control unit or concentrator. There is a strong argument in a far-flung and expensive data-transmission network for locating some of the "intelligence" at locations other than the control computer; we will take up this argument in later chapters.

POLLING

When we use multidropped control units with many terminals attached to them, the polling discipline becomes a little more complicated than the relatively straightforward polling typified by the illustration on p. 160. The polling message now addresses a control unit rather than a terminal, as that is what is attached to the line. The poll could be one of the two types: it could be a *specific poll*, which addresses a specific terminal (via the control unit), or it could be a *general poll*, which says to the control unit, in effect, "Have you anything to send from any of your terminals?" A specific poll must have two addresses in it—the address of the control unit and the address of the specific terminal attached to it.

A good illustration of this is the polling discipline used with the IBM 2848 control unit, which handles up to 24 IBM 2260 alphanumeric visual display terminals and can also have an IBM 1053 printer attached to it. The control units contain an area of core for each device they control. They may be attached in a multidrop fashion to a voice line. The line discipline is effective but not unduly complex. Figure 10.1 shows the IBM 2848.

Seven control characters are used in the line discipline, as follows:

Character:	Meaning:	Bit pattern:
STX	Start of text	0100000
ETX	End of text	1100000
CAN	Cancel	0001100
ACK	Acknowledge affirmatively	0110000
NAK	Acknowledge negatively	1010100
SOH	Start of heading	1000000
EOT	End of transmission	0010000

It will be seen that these are a small subset of the control characters provided by the seven-bit U.S. ASCII code (Fig. 2.8). When these characters, or data characters, are transmitted on a line using IBM 2848's they are sent as eight-bit characters, the eighth being a parity check.

The format of the basic polling or address messages that travel on such a line is as follows:

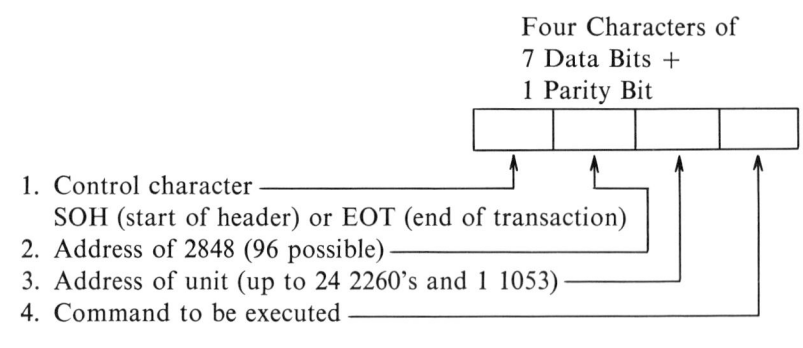

Four Characters of
7 Data Bits +
1 Parity Bit

1. Control character
   SOH (start of header) or EOT (end of transaction)
2. Address of 2848 (96 possible)
3. Address of unit (up to 24 2260's and 1 1053)
4. Command to be executed

Polling or addressing messages of this format are sent down the line to the 2848 by the computer. The fourth character, the command to be executed, can be one of the following.

Write commands

1. Write 2260, starting at the present cursor position.
2. Write 2260, starting at a given line address.
3. Erase screen and write 2260, starting at the top left-hand corner.
4. Write printer.

Poll commands
1. Specific poll to a given 2260.
2. Specific poll to the printer.
3. General poll.

Read entire buffer for a given terminal.

When the computer has information to write on the screen of a terminal, there are three ways to do it, using the three Write 2260 commands above. First, it can write starting at the present cursor position, wherever that might be on the screen. Second, it can start writing at the leftmost position of a given line. Third, it can erase the screen and start writing at the leftmost position of the top line.

The sequence of events will be as shown in Table 10.1.

Table 10.1

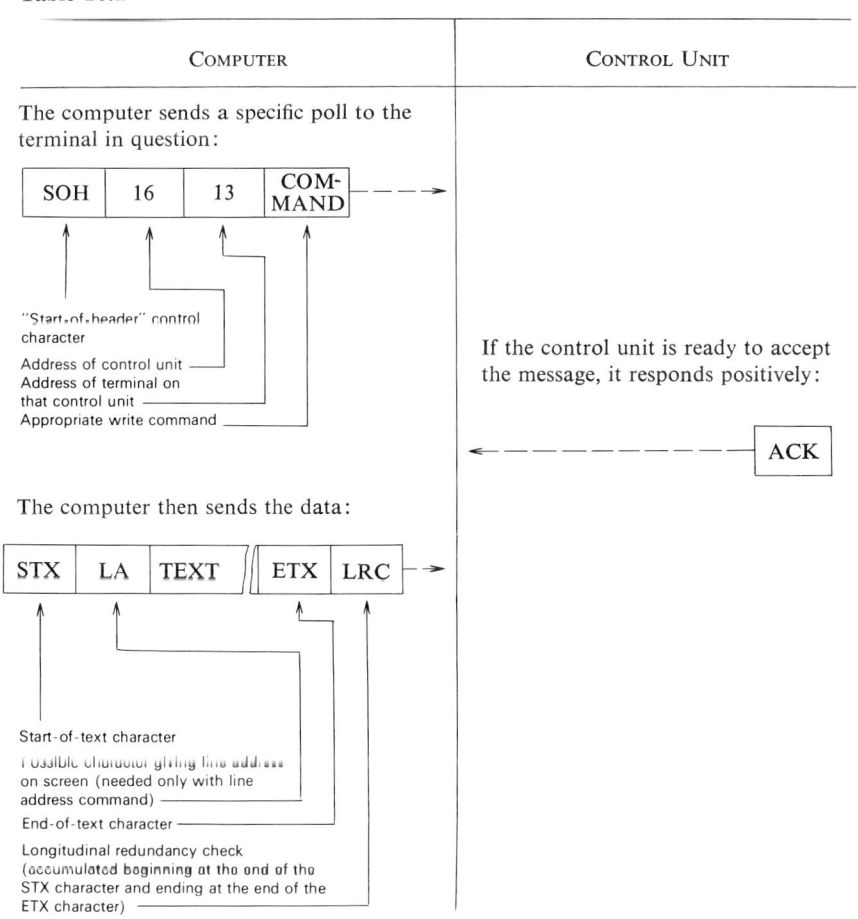

Computer	Control Unit
The computer sends a specific poll to the terminal in question: | SOH | 16 | 13 | COMMAND | "Start-of-header" control character Address of control unit Address of terminal on that control unit Appropriate write command	If the control unit is ready to accept the message, it responds positively: | ACK |
The computer then sends the data: | STX | LA | TEXT | ETX | LRC | Start-of-text character Possible character giving line address on screen (needed only with line address command) End-of-text character Longitudinal redundancy check (accumulated beginning at the end of the STX character and ending at the end of the ETX character)	

Computer	Control Unit
	The control unit checks the parity of each character received and also accumulates a longitudinal redundancy check character as the message is being received. If these are correct it responds: ←————————— ACK If a parity or longitudinal check disagreed, the control unit would respond: ←————————— NAK
and the computer would retransmit the data:	STX	LA	TEXT // ETX	LRC	→	
	←————————— ACK
On receiving the acknowledgment of correct receipt from the terminal the computer may terminate the operation by sending an "end-of-transmission" character:	EOT	————————→	This effectively breaks the link between the computer and control unit.
Or it may send more data:	STX	LA	TEXT // ETX	LRC	→	
	← ACK or NAK If the data had been lost or not completely received by the control unit (no ETX character), the control unit would have responded: ←————————— EOT
This effectively breaks the link between the computer and control unit.	The computer would attempt to reestablish the connection.

A similar sequence would be followed if the computer were attempting to write on the printer of this control unit. The control unit would respond ACK to the addressing message if the printer were ready to receive, NAK if it were "not ready" (that is, if it were out of paper, if its power were switched off, or if it were otherwise inoperative), and EOT if it were busy printing something else either from an earlier transmission or directly from a display terminal screen.

If the control unit were thus unable to accept a message for the printer, it would nevertheless remember that the computer had one, by storing a "printer request" indicator. It would respond later to a general poll by saying that the printer was now ready to receive the message for it.

When the terminal operator has something to transmit to the computer, he types it into the keyboard. It appears on the screen and is stored in the control unit buffer for that terminal. He can take his time over this because he is not tying up a piece of equipment other than his own buffer. Transmission and reception to and from the other terminals go on regardless. He can change what he has written with backspace and erase characters. When he is happy with it, he presses the ENTER key and then it is transmitted. The data transmitted begin with the START symbol on the screen, ▷, and end at the place where the ENTER key was pressed. The start symbol may have been put there by the computer, or may have been typed by the operator.

The polling line control sequence now takes place, as shown in Table 10.2.

Table 10.2

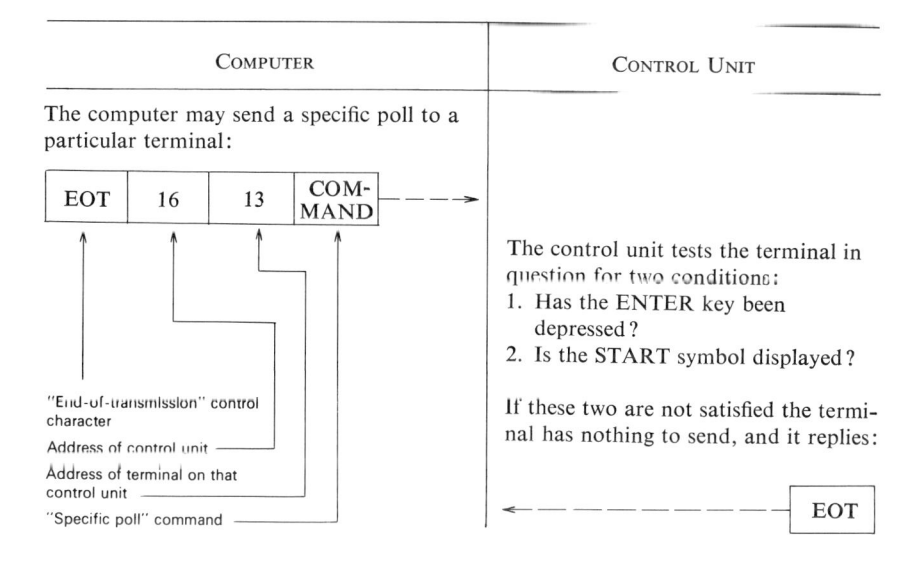

Computer	Control Unit
The computer may send a specific poll to a particular terminal: | EOT | 16 | 13 | COM-MAND | ----→ "End-of-transmission" control character Address of control unit Address of terminal on that control unit "Specific poll" command	
	The control unit tests the terminal in question for two conditions: 1. Has the ENTER key been depressed? 2. Is the START symbol displayed? If these two are not satisfied the terminal has nothing to send, and it replies: ←-------| EOT |

180   TERMINAL CONTROL UNITS

Computer	Control Unit
This effectively breaks the link between the computer and control unit. The computer will resume its polling, or other transmission.	The control unit might make NO RESPONSE. This could be caused by a defective communication line, illegal commands or addresses in the above message from the computer, or parity errors in the addressing sequence.

If the computer receives nothing after two seconds (a two-second timeout) it will try again:

| EOT | 16 | 13 | COM-MAND | ----> |

If the control unit has something to send, it sends it thus:

<-- | STX | 13 | TEXT | CAN | ETX | LRC |

Terminal address
Start of text
"Cancel" character. This is placed here if the control unit has detected an error while the message was being transmitted.
End of text
Longitudinal redundancy check

If the computer receives this correctly, it will respond positively:

| ACK | ---------------->

The control unit unlocks the terminal keyboard, which has been locked since the ENTER key was pressed, erases the START symbol from the screen, and replies:

<---------------- | EOT |

This effectively breaks the link between the computer and control unit, so the computer can take its next addressing action.

If the computer had not received the message correctly, it would have responded

| NAK | ---------------->

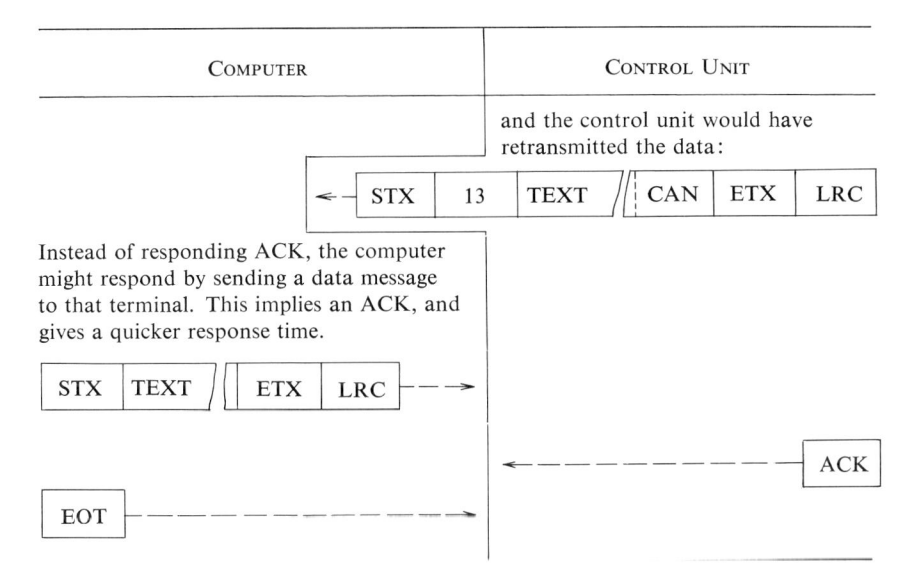

TERMINAL CONTROL UNITS   181

Computer	Control Unit
	and the control unit would have retransmitted the data:

```
          ← ─ STX | 13 | TEXT // CAN | ETX | LRC
```

Instead of responding ACK, the computer might respond by sending a data message to that terminal. This implies an ACK, and gives a quicker response time.

```
STX | TEXT // ETX | LRC ├──→
```

                                                   ←─────────────── ACK

```
EOT ├──────────────→
```

GENERAL POLL   Instead of sending a *specific poll* the computer could have sent a *general poll*, addressing all terminals on a particular control unit. The sequence of events would be as shown in Table 10.3. It will be seen that the general poll can result in far fewer control characters flowing on the line and fewer line turnarounds, and so a faster message response time.

Table 10.3

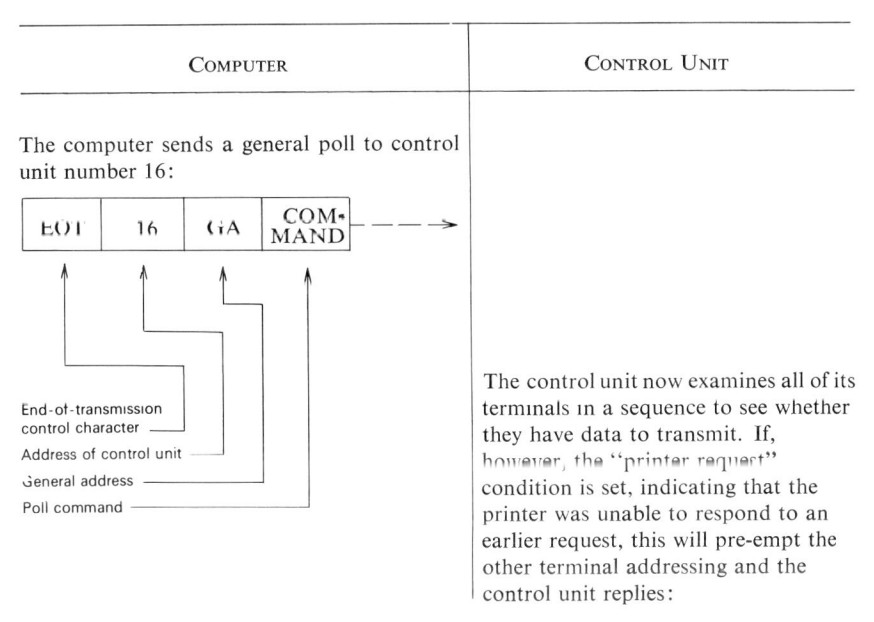

Computer	Control Unit
The computer sends a general poll to control unit number 16:	

```
EOT | 16 | GA | COM-
              MAND  ├ ─ ─ →
 │    │    │    │
 │    │    │    └── Poll command
 │    │    └────── General address
 │    └─────────── Address of control unit
 └──────────────── End-of-transmission control character
```

| | The control unit now examines all of its terminals in a sequence to see whether they have data to transmit. If, however, the "printer request" condition is set, indicating that the printer was unable to respond to an earlier request, this will pre-empt the other terminal addressing and the control unit replies: |

182 TERMINAL CONTROL UNITS

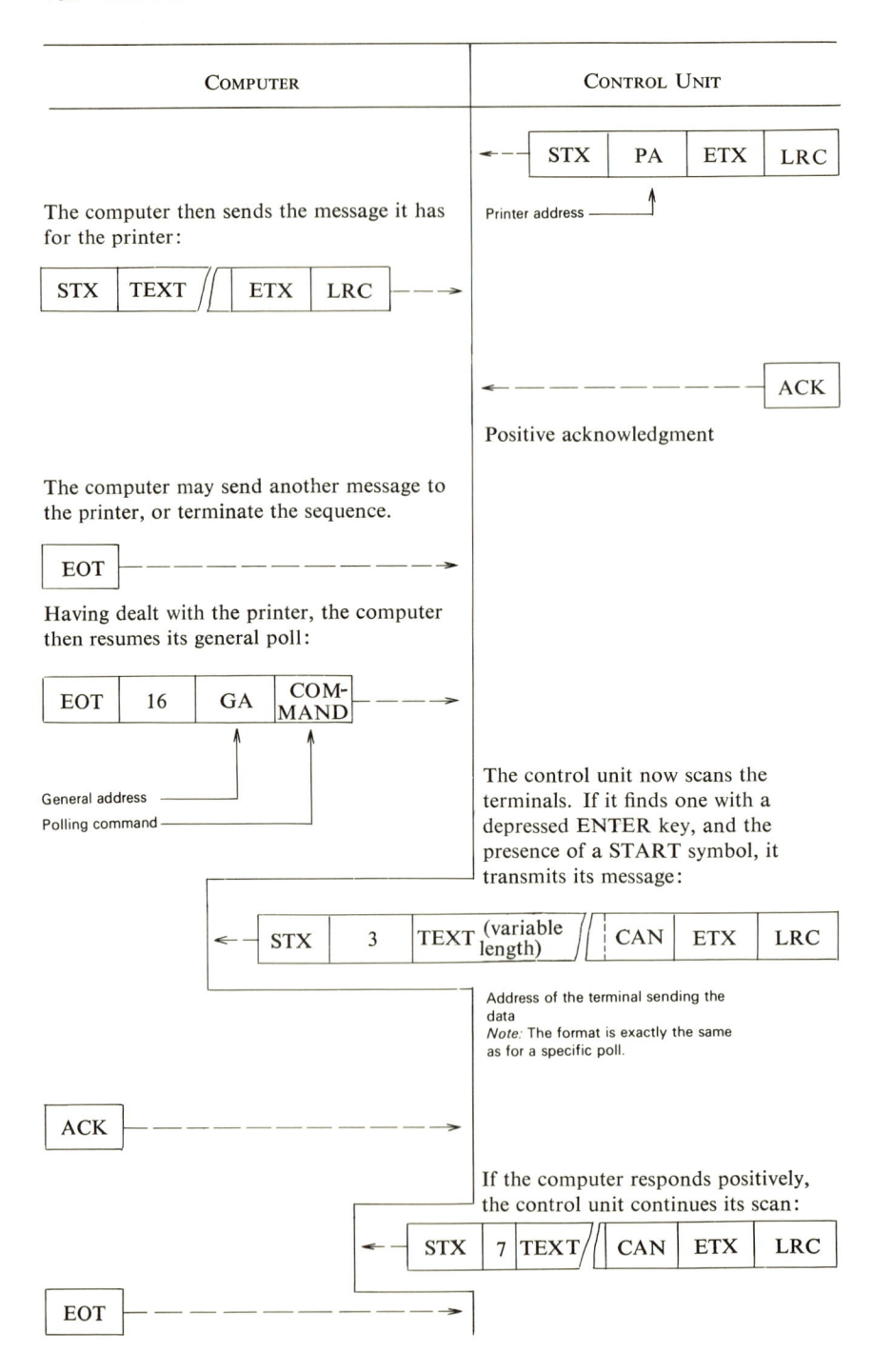

COMPUTER	CONTROL UNIT

CONTROL UNIT → COMPUTER:

| STX | PA | ETX | LRC |

Printer address ↑

The computer then sends the message it has for the printer:

| STX | TEXT | // | ETX | LRC | →

← ——————————— | ACK |

Positive acknowledgment

The computer may send another message to the printer, or terminate the sequence.

| EOT | ——————————— →

Having dealt with the printer, the computer then resumes its general poll:

| EOT | 16 | GA | COMMAND | ——— →

General address ↑
Polling command ↑

The control unit now scans the terminals. If it finds one with a depressed **ENTER** key, and the presence of a START symbol, it transmits its message:

← | STX | 3 | TEXT | (variable length) | // | CAN | ETX | LRC |

Address of the terminal sending the data
Note: The format is exactly the same as for a specific poll.

| ACK | ——————————— →

If the computer responds positively, the control unit continues its scan:

← | STX | 7 | TEXT | // | CAN | ETX | LRC |

| EOT | ——————————— →

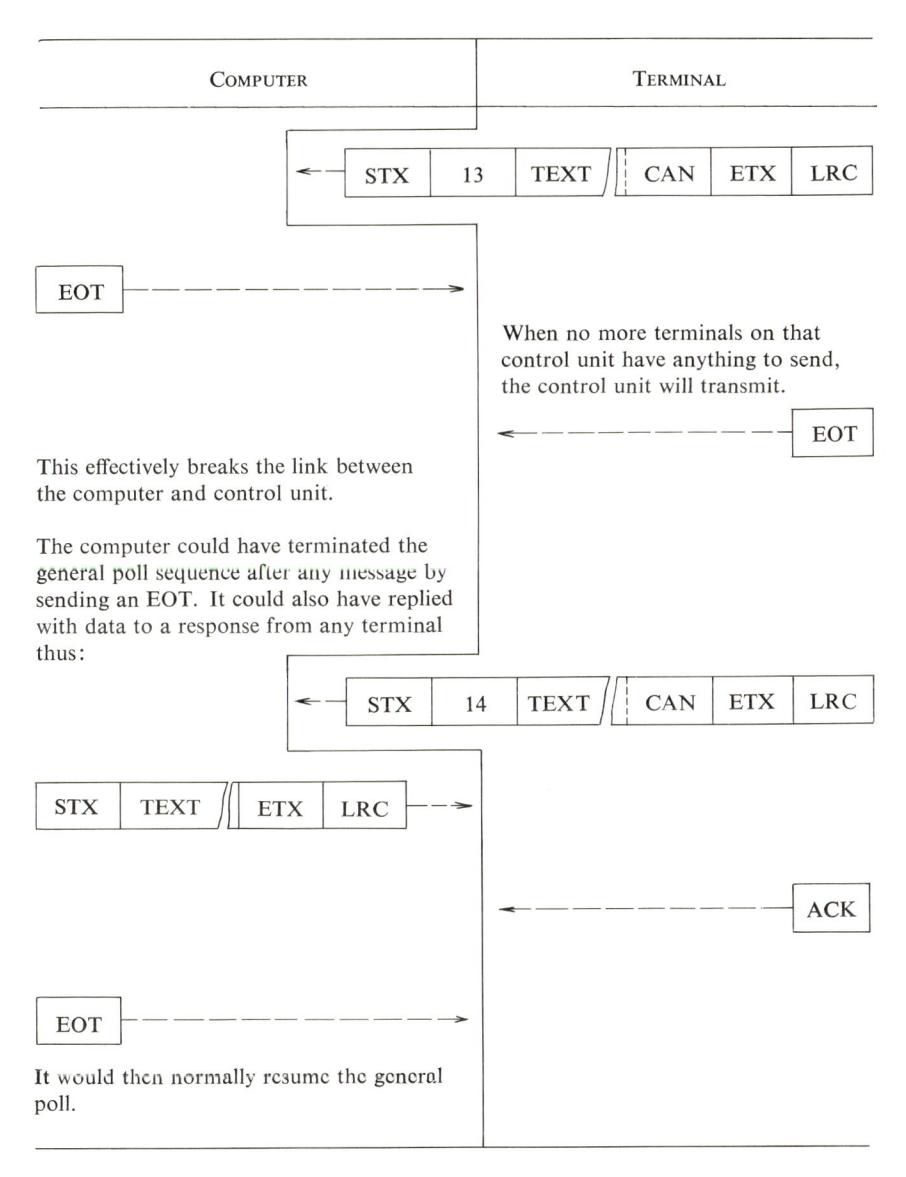

COMPUTER	TERMINAL

```
                    ← STX | 13 | TEXT // CAN | ETX | LRC

EOT ----------------→

                                When no more terminals on that
                                control unit have anything to send,
                                the control unit will transmit.
                                        ←-------------- EOT
```

This effectively breaks the link between the computer and control unit.

The computer could have terminated the general poll sequence after any message by sending an EOT. It could also have replied with data to a response from any terminal thus:

```
                    ← STX | 14 | TEXT // CAN | ETX | LRC

STX | TEXT // ETX | LRC ---→

                                        ←-------------- ACK

EOT ----------------→
```

It would then normally resume the general poll.

The above polls have been reading only that portion of the screen contents between the START symbol and the point at which the operator pressed the ENTER key. It is not normally necessary, and certainly not desirable, to transmit the entire screen contents. The computer can, however, send a command to read the entire terminal buffer if so desired. The control unit then no longer makes tests to see whether the ENTER key has been pressed and the START symbol is present; it transmits the entire buffer regardless.

11 REMOTE MULTIPLEXORS

Terminal buffers are used in order to transmit at the maximum speed of the line, even though the terminal does not operate at that speed. The disadvantage of terminal buffers is that they add considerably to the cost of the terminal. In order to bring terminal facilities to as many people as possible, we would like the terminal to be as inexpensive as possible. The computer industry has much to learn from the telephone companies. The Bell System subscriber set has a production cost of only about $14.00, but it provides simple access to immensely complex multibillion-dollar facilities.

Another way of bridging the gap between the speed of an inexpensive terminal and the maximum speed of the line used, is to employ, as do the telephone companies, *multiplexing*—in other words, send the signals from several terminals over the same communication line simultaneously. Devise a means of splitting up a line of a given capacity into several lines of proportionately lower capacity. Almost all physical transmission media have a capacity great enough to carry more than one voice channel. In other words, their bandwidths are considerably greater than the approximately 3400 cycles per second needed for transmitting the human voice. Each open-wire pair hanging from telegraph poles may carry about 12 telephone channels. The thicker cables with many twisted-wire pairs that hang from poles or lie under the ground carry many more telephone channels. Coaxial cable and microwave systems commonly carry bands of 600 to 1860 voice channels, or many more, and circular waveguide systems that could be constructed now if it were economically desirable will carry as many as

100,000 or 200,000 voice channels. At the lower end of the scale, each voice channel may be split into 12 or 24 telegraph channels.[1]

Where a facility is set up, such as a chain of microwave links, which has a high capacity, it is very desirable to make the maximum use of this capacity by making it carry as many channels as possible. The worldwide demand for communication facilities of all types is increasing at a tremendous rate, and economics and the need to conserve precious radio-frequency space demands that the common carriers devise means of increasing the capacity of their facilities. Building new communication links is expensive. It is often desirable, therefore, to construct a communication link with as wide a bandwidth as possible and then divide the bandwidth between as many users as possible.

The technique of carrying several channels over one telecommunication facility is referred to as *multiplexing*. In a multiplex system, two or more signals are combined so that they can be sent together as one signal. The original signals may be voice, telegraph, data, or other types of signals. The resulting combined signal is transmitted over a system with a suitably high bandwidth. When it is received it must be split up into the separate signals of which it is composed.

The word "multiplexing" is also used in other connections in data processing. For example, a "multiplexing channel" on a computer is one on which several devices can operate at the same time. Several printers, card readers, or paper tape punches may be operating simultaneously, and the bits that are sent to them, or received from them, are in some way intermixed as they travel along the single channel. Also, "multiplexor" is sometimes used as the name of a device that receives, transmits, and controls data on several communication links at the same time. The bits arrive or are transmitted at a rate that is slow compared with the machine's scanning speed and, therefore, the machine is capable of overlapping its handling of many links. The word "multiplexing" has also been used in connection with time-sharing systems to refer to the use of one computer handling in real time the communication with several terminal or console operators. The computer, working at much faster speed than its human operators, switches its attention rapidly from one to another, and one operator need not know that the computer is, in fact, interleaving the conversations of many such people.

The word "multiplexing," then, in general means the use of one facility to handle in parallel several different but similar operations, and in particular, in telecommunication language, means the use of one telecommunication link to handle several channels of voice or data. Multiplexing is

---

[1] *Telecommunications and the Computer*, by James Martin, Chapter 15. Prentice-Hall, Inc., Englewood Cliffs, New Jersey, 1969.

possible, and of economic value, because the operations that are multiplexed take place at a considerably slower speed than the optimum operating speed of the facility in question.

TWO METHODS OF MULTIPLEXING

The systems analyst is presented with a small list of line types that he can use in the system he designs. He knows their cost, the frequencies at which they operate, and, if he also purchases data sets, the speed at which these transmit data. There are two basic types of devices he might use to split up the line he elects to use. The first of these uses a technique referred to as *frequency-division multiplexing*; the second, *time-division multiplexing*. The former splits up the bandwidth into several smaller bandwidths at different frequencies. Each of these can then be used individually as though it were a separate line. The latter uses a high-speed bit stream to interleave the bits from several slower streams. With either method one line would be employed to transmit a number of different signals in parallel. One voice line, for example, nominally operating at 1200 bits per second might carry 14 channels using teletype machines operating at 75 bits per second. (75 × 14 = 1050—there is some overhead involved in combining the channels.)

FREQUENCY-DIVISION MULTIPLEXING

When frequency-division multiplexing is used the systems analyst specifies a "line adapter" or "data set" that provides him with several groups of channels over one line. It performs the functions of a normal modem, but in addition carries out the multiplexing function.

The data set, instead of using a single "carrier," as with the modems described on p. 17, uses several, one for each of the signals that are to be transmitted in parallel (sometimes one carrier is used for two signals). An example of this that has been in use for some years is IBM's "shared-line adapter," which is designed to send four independent 150 bit-per-second signals over a leased voice channel. IBM makes a variety of terminals that operate at up to this speed. The four subchannels operate over approximately the following frequency ranges:

Subchannel 1: 735–1075 hertz
Subchannel 2: 1145–1485 hertz
Subchannel 3: 1555–1895 hertz
Subchannel 4: 1965–2305 hertz

These frequencies fit easily into that part of the voice channel in which attenuation does not vary greatly with frequency. They avoid the Bell System signaling frequencies (see Fig. 1.2). It is interesting to note, however, that they would not avoid the signaling frequencies used in Britain and some other countries, so this machine would be permitted on Britain's public lines without redesign.

Frequency shift keying is used in each subchannel, a 1 bit or stop signal being sent as a low frequency and a 0 bit or start signal being sent as a high frequency. Each subchannel is filtered to prevent stray frequencies' interfering with the other channels. The filters' characteristics and frequencies used are shown in Fig. 11.1. It will be seen that a substantial "guard band" between the derived channels is used to prevent interference.

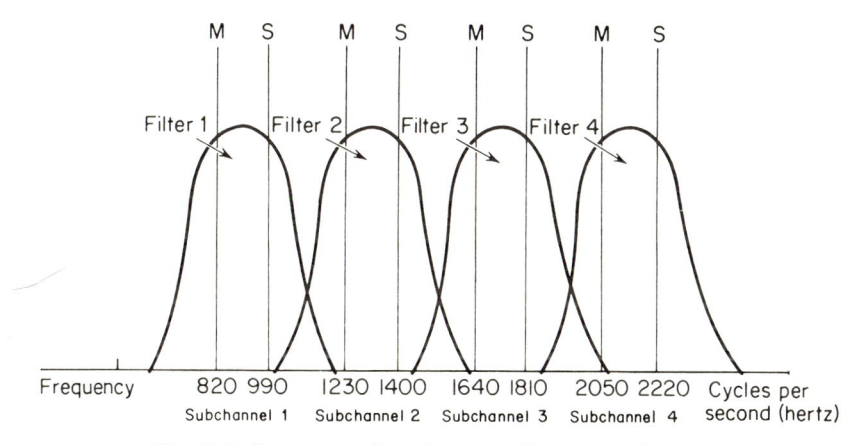

Fig. 11.1. Frequency allocations and filter ranges for IBM shared-line adapter.

If a conditioned voice line is employed, a somewhat wider frequency range can be used for multiplexing. The Rixon Frequency Division Multiplex Modems derive eight channels of 150 bits per second from a conditioned voice line. Again, frequency shift keying is used with the following channel assignments:

	\multicolumn{2}{c}{Frequency (hertz)}	
	Mark	Space
Subchannel 1	425	595
2	765	935
3	1105	1275
4	1945	1615
5	1785	1955
6	2125	2295
7	2465	2635
8	2805	2975

188 REMOTE MULTIPLEXORS

With more expensive circuit design and a more elaborate form of modulation, more channels than this can be packed into the frequency range available. The Collins "Kineplex" TE-202 data set uses four-phase modulation of 20 separate carriers, spaced 100 hertz apart in the voice band from 700 to 2700 hertz. Each carrier is modulated with a "dibit" stream and so carries two data signals. The data rate of the signals used is 75 bits per second. The data set can hence transmit 40 teletype signals over a leased conditioned voice line. A synchronizing signal at 2900 cycles is also sent. The line must not use Bell Systems signaling, as that takes place within the above frequency range (2400 or 2600 hertz).

With frequency division multiplexing the termination of the derived subchannels need not be at one location. The terminals on the multiplexed line can be scattered geographically as shown in Fig. 11.2.

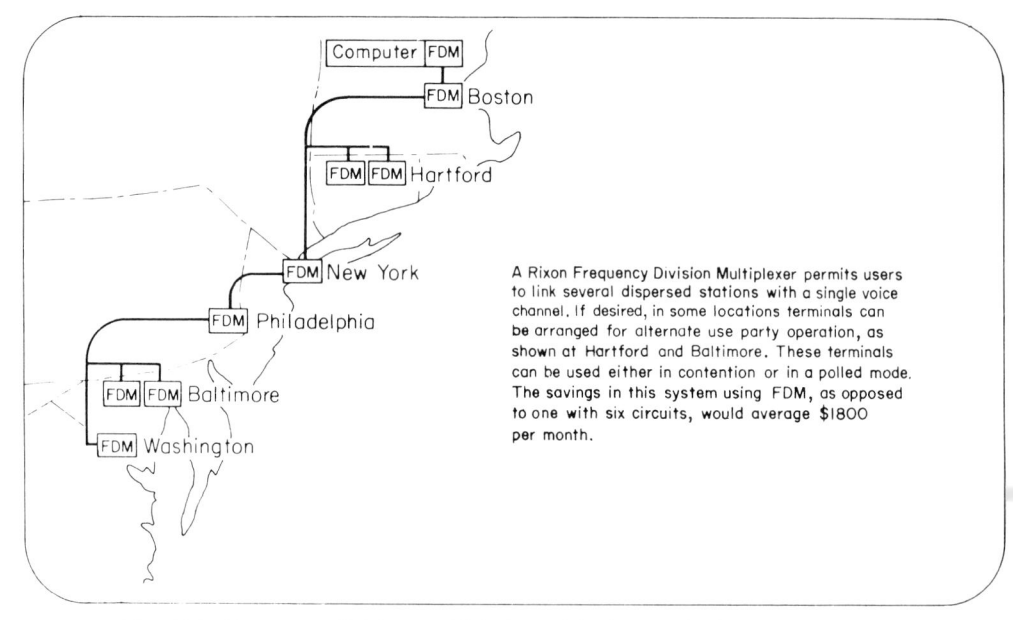

A Rixon Frequency Division Multiplexer permits users to link several dispersed stations with a single voice channel. If desired, in some locations terminals can be arranged for alternate use party operation, as shown at Hartford and Baltimore. These terminals can be used either in contention or in a polled mode. The savings in this system using FDM, as opposed to one with six circuits, would average $1800 per month.

Fig. 11.2. Frequency-division multiplexing with the terminals on one line geographically separated. *Courtesy Rixon Electronics, Inc.*

TIME-DIVISION MULTIPLEXING

The alternative to frequency-division multiplexing in any communication system is *time-division multiplexing*. Here the time available is divided up into small slots and each of these is occupied by a piece of one of the signals to be sent. The multiplexing apparatus scans the input signals in a round-robin fashion. Only one signal occupies the channel at one instant. It is thus quite different from frequency multiplexing, in which all of the signals are sent at the same time, but each occupies a different frequency band.

Time-division multiplexing may be thought of as being like the action of a commutator. Consider the commutator sketched in Fig. 11.3. The

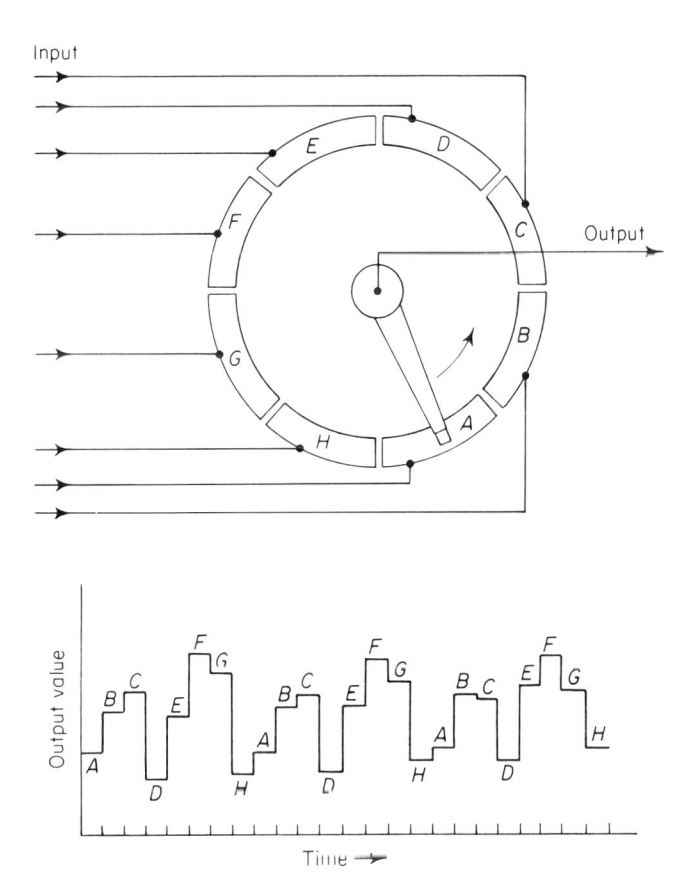

Fig. 11.3. A commutator illustrates a simple form of time-division multiplexing.

mechanically driven arm of this device might be used to sample the output of eight instruments. Providing the values of the voltages from the instruments are not varying too rapidly compared with the rotation time of the arm, the individual inputs can be reconstructed from the composite signal.

Such a device is used in telemetering. To separate the signals when they are received, a commutator similar to that illustrated might be used, but with the input and output reversed. The receiving commutator must be exactly synchronized with the transmitting commutator. The time multiplexing devices we meet in telecommunications today are normally electronic and of much higher speed, but in principle are similar to the commutator.

The systems analyst can obtain a relatively simple device that takes several low-speed bit streams, as for example from telegraph machines, and combines them into one high-speed bit stream. Figure 11.4 shows a typical

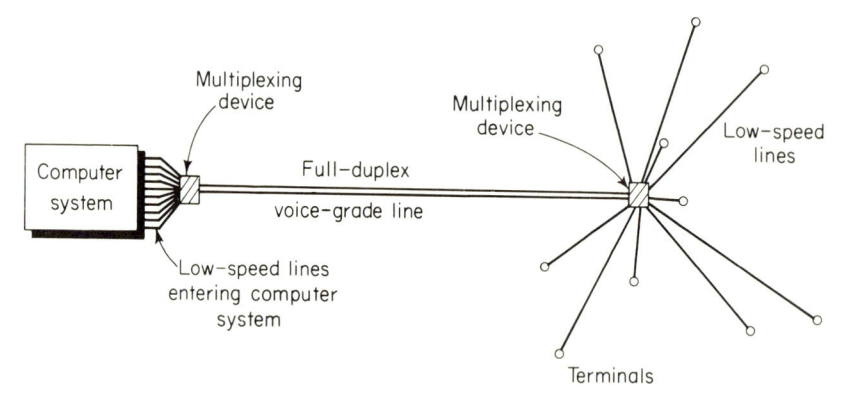

Fig. 11.4. Bit streams from low-speed terminals (for example, teletype machines or other keyboard machines) are multiplexed together to travel on one voice-grade line. This costs less than having a low-speed line from each terminal location to the computer.

arrangement. A group of relatively inexpensive terminals (no buffering, no multidrop logic, asynchronous operation) are each attached to a low-speed line. The low-speed signals are multiplexed together to travel over a leased voice line to the computer center. The low-speed lines are half duplex, but the high-speed line is full duplex, so some of the terminals can be transmitting at the same time as others are receiving. Such a scheme is used where its overall cost is lower than that of taking low-speed lines directly from the terminals to the computer center.

In the diagram the voice line is shown entering a multiplexing device at the computer center, which is identical to the remote one. Low-speed lines then enter the computer system as though they were linked directly to the terminals. An alternative would be to take the high-speed lines directly into the computer or its line control unit, and this will then assemble and disassemble the high-speed bit stream.

MULTIPLEXING WITH MULTI-DROP LINES

Any characters traveling on the low-speed lines can be multiplexed into the high-speed bit stream in this way, including control or polling characters. The low-speed lines can therefore take on any configuration that they would have had if the multiplexing device had not been present. In particular, such configurations are often found with multidrop lines. Figure 11.5 illustrates this. It requires a complex calculation to work out the lowest cost configuration of lines and location of multiplexing unit for a network such as that in Fig. 11.5.

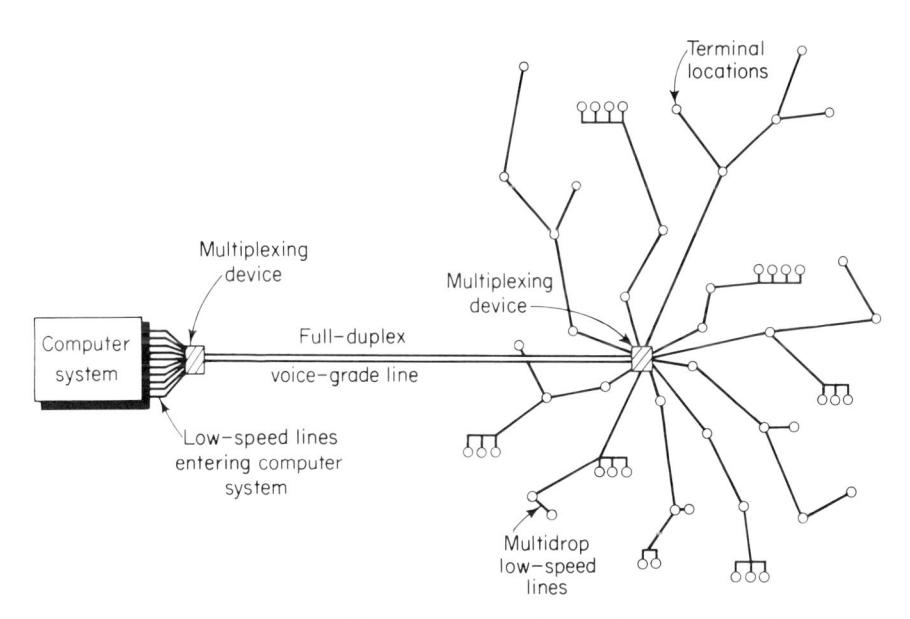

Fig. 11.5. Multiplexing on a voice line used to connect multidrop low-speed lines to a computer center.

A typical remote multiplexor of this type is the IBM 2712 (shown in Fig. 11.6). Let us use this to illustrate the interleaving of the low-speed bit streams.

The 2712 is attached to a voice line as in Fig. 11.5. The multiplexing/demultiplexing at the computer is done by a "2712 adapter" on the transmission control unit. Model 1 of this machine serves up to ten low-speed lines operating at speeds of 134.5 bits per second (the speed of a variety of types of IBM terminals). Model 2 serves up to 14 lines operating with 74.2-baud telegraph terminals.

The low-speed lines are sequentially scanned by the remote multiplexor. During each scan cycle a single bit of data is taken from each low-speed line. These bits, together with a synchronization bit, are formed into a word that is transmitted on the voice line. This word therefore contains 11 bits for model 1, and 15 bits for model 2. Bits on the 74.2-baud telegraph line arrive once every 13.47 milliseconds; hence one 15-bit word must be transmitted down the voice line to the computer, and a second one *from* the computer, every 13.47 milliseconds. An unconditioned voice line is adequate for this (Bell System line type 3003). Data sets are needed, as normal, on the high-speed line and also on the low-speed lines for terminals not at the same location as the multiplexor.

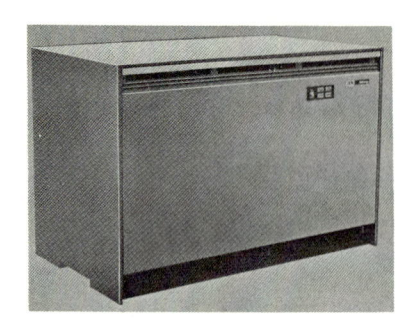

Fig. 11.6. IBM 2712 Remote Multiplexor.

Figure 11.7 illustrates the bit multiplexing process. It shows the high-speed bit stream format traveling down the voice line from the computer. In each scan cycle of 13.47 milliseconds one synchronization bit is sent, followed by one bit for each of the 14 low-speed lines. The multiplexor distributes these to the lines as shown. When there is no character being sent and the line is idle a mark condition (stop bit) is transmitted. The converse process applies with data going to the computer. The multiplexor scans the lines, taking one bit from each and forming a high-speed word similar to that in Fig. 11.7.

Because of the differences in character time on the low-speed lines, and the difference in speed between the low- and high-speed lines, buffering is needed for the bits transferred. Fifteen words of core storage are used, one for each low-speed line and one for the synchronization bits on the high-speed line. These are used for buffering and control.

Before transmission begins, synchronization must be established between the two multiplexing devices. This is done by sending a synchronization pattern. Once the two devices are synchronized, line scanning begins. From then on the synchronization bits in each high-speed word carry an alternating bit pattern, and this is used for maintaining synchronization. If a multiplexor finds that a synchronization bit is not the opposite of the previous bit, then it knows that synchronization has slipped and the machines must be resynchronized. The transmitting machine changes the phase of each even-numbered data bit transmitted on the high-speed line. For example, the character 0000111 would be sent as 0101101. A continuous mark condition representing no data on the line, 1111111, would be sent as 1010101. This is done to provide the maximum number of bit transition points and hence enable better clock synchronization.

Error checking is done by using the data's own checking techniques, whatever these might be, as though the data streams had not been multiplexed.

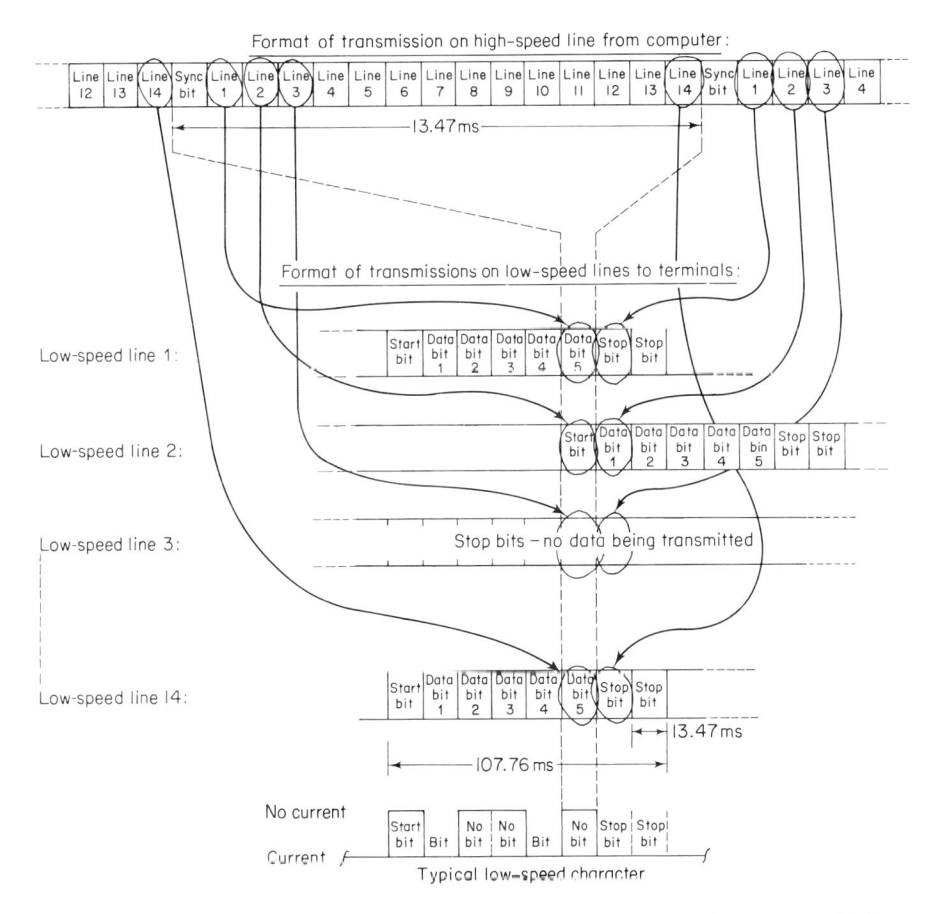

Fig. 11.7. Bit demultiplexing in which the stream of bits on the high-speed line is distributed to 14 low-speed lines.

ADVANTAGES The straightforward multiplexing methods discussed in this chapter have a number of advantages over the other means of improving line utilization:

1. They are relatively inexpensive—certainly more so than the concentration techniques discussed in the next chapter.

2. They do not affect the programing in any way. The multiplexors are "transparent" to the programs which send and receive data and control signals on the low-speed lines as though those lines were still separate and nonmultiplexed.
3. Being simple, the multiplexing devices generally have a high reliability.
4. If it is desired to send a long continuous stream of data on some of the lines, as in a remote batch operation, the multiplexors can handle this without interruption of the stream or interference with other users.
5. They cause no significant increase in response time.
6. The low-speed lines are connected to the computer center at all times. The terminal user will not have to wait to obtain a line as he often does with dial-up public lines, or systems using a private exchange.

Time-division multiplexing is generally more efficient than frequency-division multiplexing in that more subchannels can be derived. Frequency-division multiplexing has the advantage, however, that the derived subchannels need not all terminate together as in Fig. 11.4. The termination points can be geographically separated as in Fig. 11.2.

12 HOLD-AND-FORWARD CONCENTRATORS

This chapter, like the last, discusses devices for taking data from several lines and sending them, more economically, down one higher-speed line.

As mentioned in Chapter 7, the terminology in common usage is not quite as clear-cut as these two chapter titles might suggest. The devices discussed in this chapter are occasionally also referred to as "multiplexors," although they are doing more than a simple multiplexing function. In this book we have reserved the word "multiplexor" for a device that is doing a multiplexing function without storing the data transmitted or changing its content or format.

The concentrators discussed in this chapter store the data and retransmit them. They often change their format, and some such devices change their content. These "hold-and-forward" concentrators are more expensive than the devices discussed in the previous chapter. Because of their increased complexity they cause more problems. The justification for using them rather than simple multiplexors should be that they can further reduce the cost of the network.

Consider Fig. 11.3 again. What can we do to improve upon this? First, in the configuration shown, when a low-speed line is idle it is nevertheless occupying signal space on the high-speed line. This might represent a very severe wastage, because the users sitting at each of the terminals do not transmit or receive data continuously. In fact, the user is likely to spend a large proportion of his time "thinking," talking to a client, preparing his next transaction, or carrying out other activities that do not occupy the communication line. In many applications he will spend most of his time doing this with only an occasional quick burst of transmission. The line sits idle most of the time. It would be better if we could occupy signal space on the high-speed line only when there was something to transmit.

In this case we might, indeed, have high-speed lines on the terminal side of the concentrator capable of operating high-speed terminals such as displays, instead of only low-speed terminals. The transmissions from such terminals would be stored and relayed to the computer as soon as the ongoing line was free.

Second, the high-speed line in Fig. 11.3 is transmitting a collection of start-stop signals. This can be inefficient. As we commented before, if the operator pauses for a second between pressing keys on the terminal, the line will be occupied throughout this pause. It would be better to have buffering and synchronous transmission.

Third, we may be sending characters on the high-speed line which need not be sent. These include all of the characters we discussed in Chapter 10 when saying that an "intelligent" terminal control unit could save transmission time. Indeed, most of the functions we discussed in the chapter on control units for lessening communication line cost could reside instead in the concentrator. Many applications need the generation of lengthy English language responses at the terminal. These could be stored at the concentrator instead of being transmitted, and, if the concentrator can be programed, many elements of the man-machine conversation that is used might take place at the concentrator rather than involving transmission to the computer.

BASIC FUNCTIONS OF A CONCENTRATOR

We can divide the functions of a remote concentrator into two categories: those that are dependent on the application (such as the generation of terminal responses, the handling of the man-machine conversation, and checking of control totals and so on), and those that are not. The latter category, concerned solely with the organization of bits flowing on the lines, can be built into the machine logic and used in a variety of differing applications. Application-dependent functions are likely to require a stored-program concentrator that can be programed specifically for the application in question.

In the next chapter we will discuss application-dependent concentration techniques needing a stored-program computer. In this chapter we will consider machines that are application-independent—not concerned with the data content of the messages transmitted. The latter may be small special-purchase machines, although sometimes stored-program computers are used. The functions concerned with the organization of bits flowing on the lines are likely to be as follows:

1. *Buffering*

The transactions coming in on the low-speed lines must be stored so that we can carry out the necessary manipulations on them. Whereas the bit multiplexing technique discussed in the previous chapter required the buffering of bits on each incoming line, here we need to store complete messages, or at least blocks of many characters. As the characters arrive on the incoming lines, they are stored in the memory of the machine until the complete transaction is assembled. Conversely, outgoing transactions are transmitted from the storage.

2. *Allocation of Storage and Control of Queues*

The allocation of storage may become fairly complex. At one time many terminals may be transmitting; at another time, few. The transactions may differ widely in length. If this is the case, then some form of dynamic allocation of memory is needed rather than the fixed allocation of an area of core per terminal. Queues of transactions build up waiting for transmission on the high-speed line. This queuing is particularly significant if the concentrator transmits on the high-speed line only in occasional bursts. This would be the case when the high-speed line is multi-dropped, serving several concentrators as in Fig. 12.1. The messages are queued until it is that concentrator's turn to transmit. It then sends all it has. Conversely, it periodically receives a burst of messages which it must distribute on the low-speed lines. The storage must be organized to hold the varying numbers of items sent and received. The items in the queue must be chained together so that they can all be sent in a stream when their time comes.

The mechanism for allocating storage to incoming messages must have some means of knowing which portions of storage are now free and can be allocated. When a message is transmitted and is known to have been received correctly, the portions of storage it occupied will be returned to the list of available storage.

3. *Receipt of Messages on the Low-Speed Lines*

The concentrator has circuits continuously sampling each of the low-speed lines, waiting for a character to arrive and then storing the data bits of that character. This mechanism must be able to handle random and unrelated inputs from all terminals simultaneously. Some concentrators have the ability to handle simultaneous transmission from different types of terminals operating with different codes and at different speeds. This complicates the receiving logic, as the concentrator must know which line is which.

Typically the concentrator scans the lines at electronic speeds, sampling them sequentially. Suppose that this circuit has a cycle time of 20 microseconds and can handle up to 50 lines; each line will be sampled once per millisecond. They may carry 150 baud transmission, in which case the duration of a bit will be 6.67 milliseconds and so it will be sampled six or seven times. If the samples differ, the majority vote is taken. As most noise spikes last less than 3 milliseconds this gives a measure of protection from noise. It can also give a high protection from bit timing distortion.

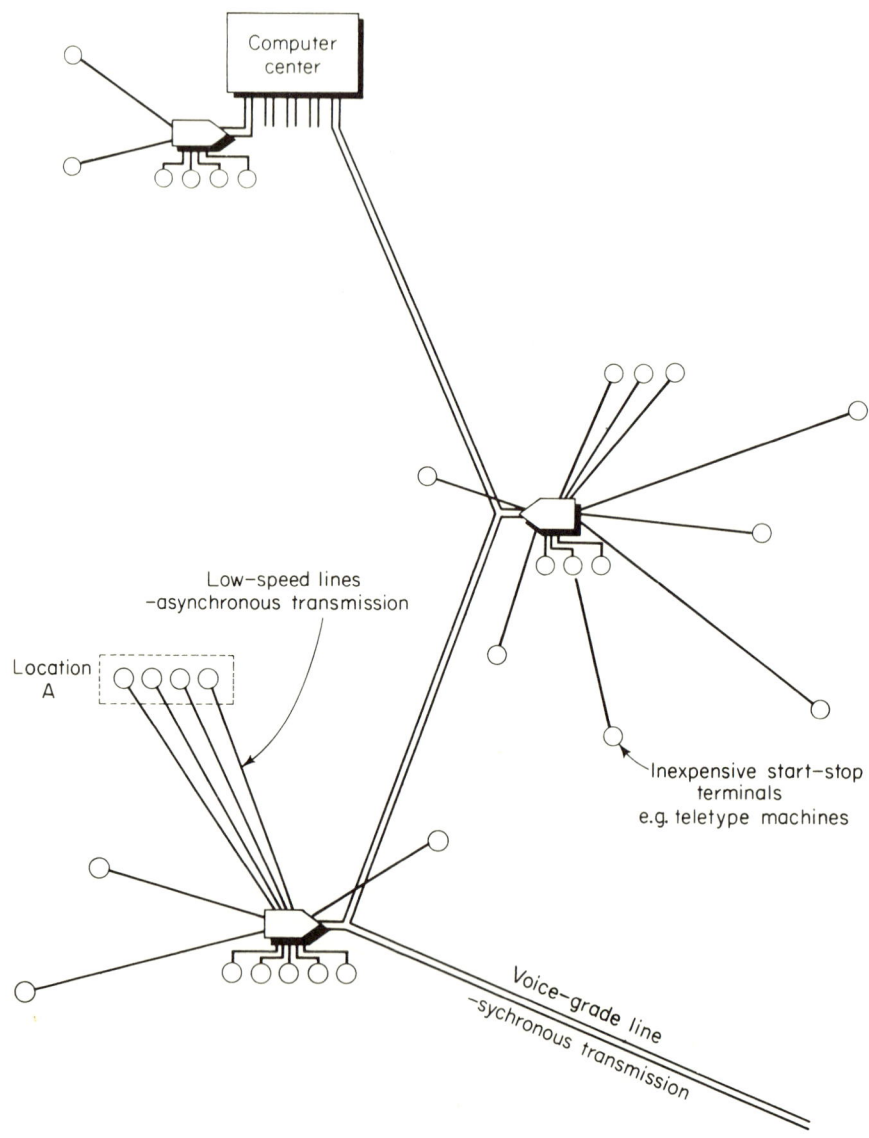

Fig. 12.1. Hold-and-forward remote concentrators connected to a multidrop voice line.

4. Code Translation

The code used by the central computer is often different from that transmitted from the terminals. In some cases the code transmitted on the high-speed line is different from that on the low-speed lines. The high-speed line may carry ASCII code (Fig. 2.8), for example, but the terminals not. On the other hand, the concentrator may, in the interest of simplicity, do no code translations of this type.

The concentrator does enough character modification to convert the low-speed lines' start-stop characters into characters designed for the high-speed synchronous transmission. This may entail solely the removal of the start, stop, and (if any) parity bits. Where telegraph signals enter the concentrator, it must be able to recognize a telegraphic sequence of characters meaning "end of message." This may be a three- or four-character sequence, typically **NNNN**. The synchronous transmission normally has its own end-of-message indicator (p. 57). Other telegraph sequences may also be translated or deleted, for example, "start of message," **ZCZC**. If Baudot code is used, the "letters shift" and "figures shift" must cause a switch to the appropriate alphabetic or numeric translation.

The converse translation will take place on output.

5. Assembly of Messages for High-Speed Transmission

The characters as they are received are stored, after code conversion sometimes, in an area from which they will later be transferred on the high-speed line. The computer must know which terminal the transaction came from, so the address of the terminal will be stored preceding the data characters. If the message is long and has been split up because of the needs of storage allocation, a further character will be needed to indicate which segment of the message this is. When receipt of the transaction is complete, an end-of-message character will be stored.

6. Transmission of Messages on the High-Speed Line

The messages are transmitted from the concentrator storage to the computer in synchronous blocks, much as from any other buffered synchronous device. The concentrator sends first a synchronization pattern as in Fig. 3.9 or a start-of-message character, then a character giving the concentrator address (if there is more than one concentrator per line), and then the message, or a block of messages, each one having its end-of-message indicator. If more than one message is sent in a continuous stream, an end-of-transmission character will follow. As the transmission is proceeding, a longitudinal checking character or group of characters is compiled. This will be added to the end of the message.

7. Error Checking

The characters on the high-speed lines may or may not have parity checks. A message or block of messages transmitted synchronously also normally contains longitudinal error-checking patterns or characters, as discussed in Chapter 4. As the concentrator transmits it compiles an error-checking character or group of characters. Those will be added to the end of the message, and checked by the computer. The converse occurs on transmitting from computer to concentrator. It is usually desirable to have a means of retransmitting if an error is found, although some systems merely show a warning light at the terminal when an error is detected.

On the low-speed lines start-stop transmission is usual, so the only checking is likely to be parity checks on characters.

8. *Polling on the High-Speed Lines*

It is often desirable to have more than one concentrator on the high-speed line, as shown in Fig. 12.1. In this case there must be a polling function in the concentrator logic. The concentrator must be able to recognize its own address on messages sent down the line, and must be able to respond appropriately to polling and control signals sent to it.

9. *Polling on the Low-Speed Lines*

On large networks it may be economical to have multidrop lines on the terminal side of the concentrator. Some locations on a real-time or time-sharing system have many terminals. Where possible, the concentrators are located at these places to minimize low-speed line cost. It may not be economical to position a concentrator at all such locations, however. Location A in Fig. 12.1 has four terminals. Four low-speed lines connect these to the nearest concentrator. This is less expensive than putting a concentrator at location A. It would have been better, however, to connect a multidrop line to all the terminals at location A. (Alternatively a multiplexor or small private exchange could have been used in connecting location A to the concentrator.)

If we put multiple terminals on the concentrator's low-speed lines, we must have some way of controlling them. Normally they would have to be controlled by a *polling* discipline. This means that we now have two levels of polling—polling the concentrators on the high-speed line, and polling the terminals on the low-speed lines. The polling of the low-speed lines could be done in one of two places—in the main computer or in the concentrator. If it is done by the computer, high-speed-line messages to the concentrator would have low-speed-line polling messages embedded in them. As we will see in an illustration shortly, this grossly increases the complexity of the line control. The alternative is to have the concentrator itself do the polling. This substantially increases the quantity of logic needed in the concentrator. The concentrator must have a polling list and the logic for addressing the terminals on this list one by one, asking whether they have anything to transmit and interpreting their responses. The polling list must be easily modifiable as in a stored-program computer, because the terminals actually connected will be subject to constant change. The concentrator must be able to cope with line errors, terminal malfunctions such as printers running out of paper, and terminal failures.

EXAMPLE As an illustration of a remote concentrator we will describe briefly the working of the IBM 1006. This machine is referred to as a "terminal interchange." It has been used extensively for some years on airline reservation systems, in which several hundred to more than a thousand low-speed terminals are connected to a central computer. Figure 12.2 shows a 1006.

A conditioned full-duplex voice line from the computer, operating at 2000 or 2400 bits per second, has several IBM 1006 machines attached to it.

Fig. 12.2. IBM 1006 Terminal Interchange. Link between low-speed asynchronous lines and high-speed tightly organized synchronous line. See Fig. 12.3.

It can have other devices also attached to the same line. The IBM 1006 has up to 30 low-speed lines entering it, each attached to one terminal. The high-speed line uses synchronous transmission, and the low-speed lines use start-stop transmission. The network therefore looks like that in Fig. 12.1.

This concentrator can handle several different line speeds, as shown in Fig. 12.3. Different terminal devices can be attached to it, and on airline systems it commonly handles a mixture of telegraph equipment and elaborate airline "agent sets." Where there are many terminals at one location, as at airports, or a sales office on Fifth Avenue or Piccadilly, the 1006 will probably be in a back room in the same building.

The machine has 4000 characters of core storage, each having seven bits of BCD code plus one parity bit. It addresses one bit or one character at a time with an 11.5 microsecond cycle time. (Remember the IBM 1401?)

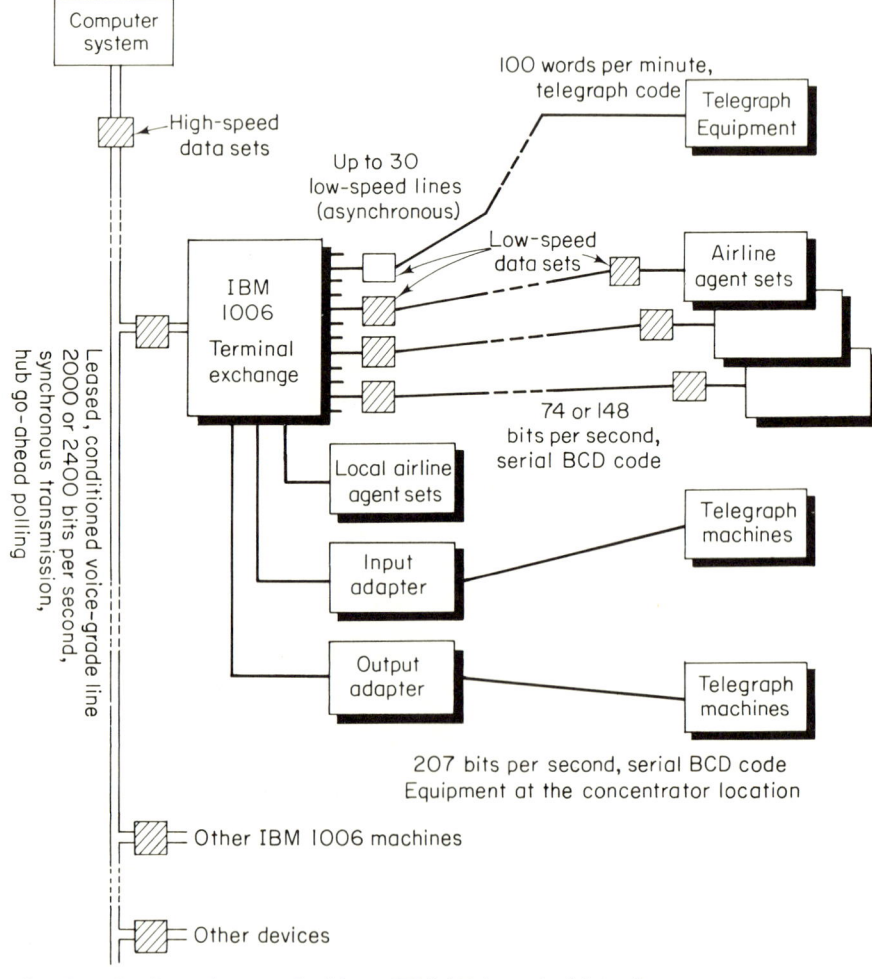

Fig. 12.3. Configurations used with an IBM 1006 terminal interchange in airline reservation systems.

The storage is divided into 40 blocks of 100 characters. Thirty-nine of these are used as dynamically allocatable buffers for data, and the other is for controls. When a terminal wants to transmit data, the concentrator selects one of the 39 blocks of core that is unused at that time and allocates it to the terminal. The data from the terminal are stored in it character by character. If it fills up, another block will be allocated and chained to the first. Each block must contain the address of the terminal that it refers to (one character). As all of the transmission is of variable-length blocks, it must also have space for an end-of-message character. Therefore, 98 of the 100 characters are available for storing data. (97 on long messages, when one character is used for chaining between blocks.)

When the computer polls the concentrator, the data for it in the blocks are transmitted synchronously at 2000 or 2400 bits per second. Six-bit BCD coding is used with no parity check but with an efficient cyclic checking character. Hub go-ahead polling is used as in the illustration on p. 167. The number of bits traveling on the high-speed line is thus cut down to a minimum.

Out-going data flow from the computer to the concentrator is six-bit BCD. It is stored in any of the 39 blocks that is free. From there the concentrator transmits to the terminal.

Five processes can take place simultaneously in the IBM 1006:

1. Scanning all of the low-speed lines for reception of incoming bits.
2. Transmission of bits on the low-speed lines.
3. Reception of bits from the high-speed line.
4. Transmission of bits on the high-speed line.
5. Processing of the characters a bit at a time, writing them in or reading them from the appropriate place in storage.

These operations are sufficiently fast, for all possible character activity can be handled during the real-time life of any character being transmitted or received on any line. The concentrator processes one bit of one message, then one bit of another, and so on.

There is a control word for each line that governs the storing of bits, the parity checking, the buffer assignation, bit timing (different lines have different bit speeds), special timing functions such as delays after type-head carrier returns on output, and code conversion, including the translation of the multiple-character "end of transmission" on telegraph messages.

SIMPLE LINE CONTROL

Hub go-ahead polling is used on the high-speed line like the example on p. 167. The line control is therefore fairly simple. A typical sequence is shown in Table 12.1.

Table 12.1

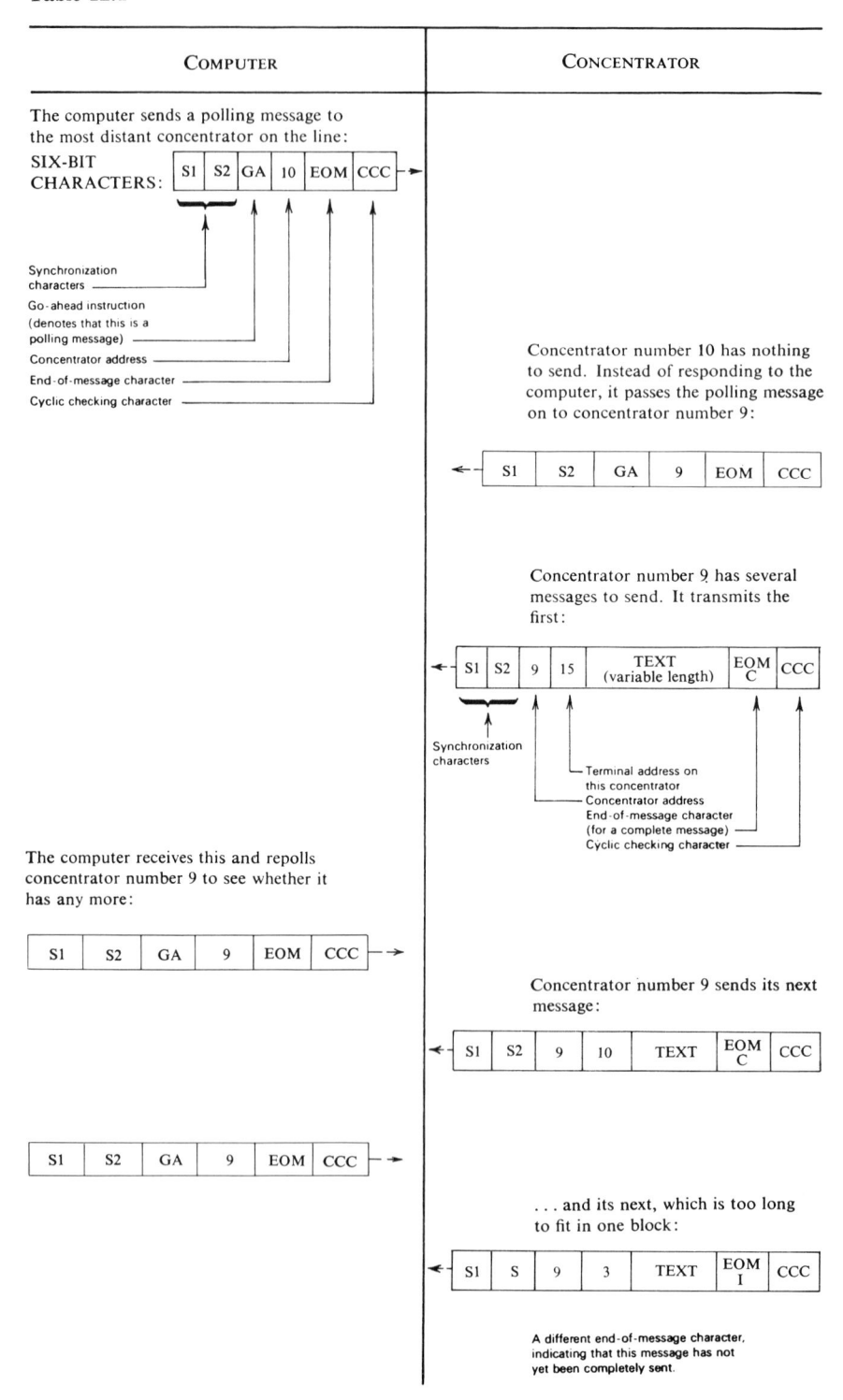

COMPUTER	CONCENTRATOR

The computer sends a polling message to the most distant concentrator on the line:

SIX-BIT CHARACTERS: | S1 | S2 | GA | 10 | EOM | CCC | →

- Synchronization characters ————— (S1, S2)
- Go-ahead instruction (denotes that this is a polling message) ————— (GA)
- Concentrator address ————— (10)
- End-of-message character ————— (EOM)
- Cyclic checking character ————— (CCC)

Concentrator number 10 has nothing to send. Instead of responding to the computer, it passes the polling message on to concentrator number 9:

← | S1 | S2 | GA | 9 | EOM | CCC |

Concentrator number 9 has several messages to send. It transmits the first:

← | S1 | S2 | 9 | 15 | TEXT (variable length) | EOM C | CCC |

- Synchronization characters
- Terminal address on this concentrator
- Concentrator address
- End-of-message character (for a complete message)
- Cyclic checking character

The computer receives this and repolls concentrator number 9 to see whether it has any more:

| S1 | S2 | GA | 9 | EOM | CCC | →

Concentrator number 9 sends its next message:

← | S1 | S2 | 9 | 10 | TEXT | EOM C | CCC |

| S1 | S2 | GA | 9 | EOM | CCC | →

... and its next, which is too long to fit in one block:

← | S1 | S | 9 | 3 | TEXT | EOM I | CCC |

A different end-of-message character, indicating that this message has not yet been completely sent.

HOLD-AND-FORWARD CONCENTRATORS

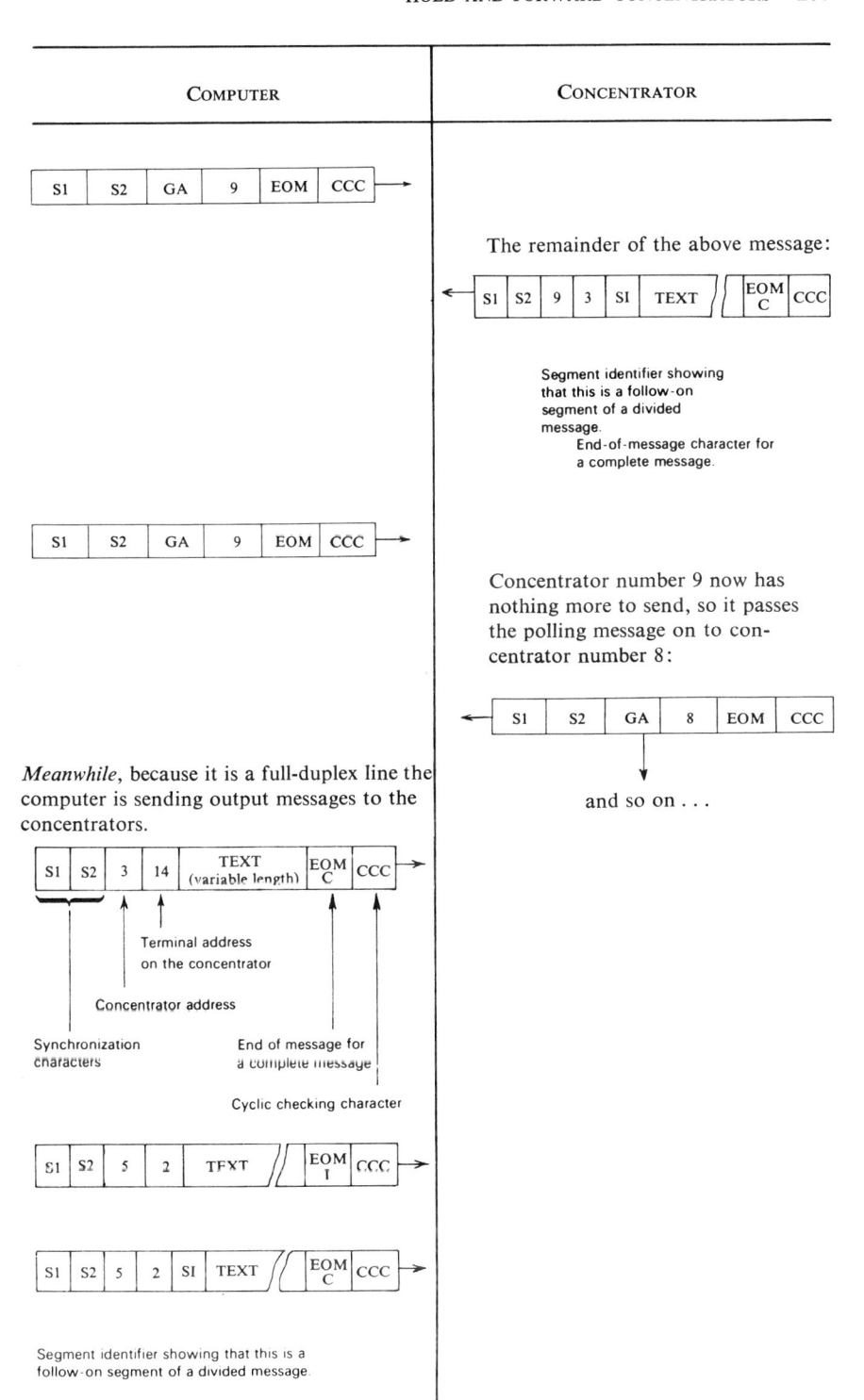

COMPUTER	CONCENTRATOR

| S1 | S2 | GA | 9 | EOM | CCC | →

The remainder of the above message:

← | S1 | S2 | 9 | 3 | SI | TEXT // EOMC | CCC |

        Segment identifier showing
        that this is a follow-on
        segment of a divided
        message.
            End-of-message character for
            a complete message.

| S1 | S2 | GA | 9 | EOM | CCC | →

Concentrator number 9 now has nothing more to send, so it passes the polling message on to concentrator number 8:

← | S1 | S2 | GA | 8 | EOM | CCC |

and so on . . .

Meanwhile, because it is a full-duplex line the computer is sending output messages to the concentrators.

| S1 | S2 | 3 | 14 | TEXT (variable length) | EOMC | CCC | →

- Synchronization characters
- Concentrator address
- Terminal address on the concentrator
- End of message for a complete message
- Cyclic checking character

| S1 | S2 | 5 | 2 | TEXT // EOMI | CCC | →

| S1 | S2 | 5 | 2 | SI | TEXT // EOMC | CCC | →

Segment identifier showing that this is a
follow-on segment of a divided message.

**RESPONSE
TIME REQUIREMENTS**

It would be possible for a concentrator to send *all* of the messages it contained at one instant instead of sending them one at a time as in the above illustration. Some machines do this. In the above illustration if a concentrator had ten messages to send to the computer, ten (six-character) polling messages and 20 line turnarounds with resynchronization would be needed. If the ten messages were sent in one block only one polling message and two line turnarounds would be needed.

The reason that this is not done is that it could degrade the terminal response time. On applications such as airline reservations it is considered very important to have a low response time. The contract of some such systems says that 90 percent of the transactions must have a response in less than three seconds—which requires a mean response time of less than two seconds. (Response time is here measured as being the time from when the terminal operator presses the last key of his input to when the first character of the reply from the computer is printed or displayed at the terminal.) If the concentrator had, in this case, transmitted ten full blocks together, this would have needed 1004 six-bit characters (assuming only one cyclic checking character). This would take more than three seconds on a 2000-bit-per-second line. None of the ten messages could have achieved a three-second response time. Transmitting them one at a time requires a total of 1100 characters traveling on the line, but most of them would have a good chance of meeting the three-second response time requirement.

Some concentrators compromise by sending several messages in a block but not all of them if the block would then exceed a given size. In some the computer can vary the amount the concentrator transmits at any time.

As in other aspects of network design, there is a tradeoff between overall throughput efficiency and response time.

**ERROR
CONTROL**

The line procedures in the above illustration were simplified somewhat by avoiding any positive/negative acknowledgments of correct transmission. The cyclic checking characters were sent and examined, giving a high probability of detecting any transmission errors. Checking is also carried out on the low-speed lines. But any messages found to be erroneous are *not* automatically retransmitted. Instead, a warning of the error is given to the terminal operator, and he resubmits the transaction to obtain a valid answer to his query or to ensure that the file he is concerned with has, in fact, been updated correctly. This does not happen very often. If the average length of a message and its response is 50 characters, and each of the four segments of transmission (low-speed line in, high-speed line in, high-speed line out, low-speed line out) has an error rate of 1 bit in

100,000, then less than one response in 100 will contain an error and need a new input message.

On airline reservation systems, and many other systems in which the operator is seeking information, this approach will be satisfactory. On other systems, however, especially those handling financial transactions or other numeric quantities that need careful control, tighter line error procedures are worthwhile. Automatic retransmission of data found to be in error can lessen the chance of falsely updating a file—failing to make the entry or inadvertently updating it twice.

MULTIDROP LOW-SPEED LINES Figure 12.4 shows concentrators with multidrop low-speed lines. With such an arrangement a large number of low-speed terminals scattered over a large area can be connected to a computer by using a simple voice line. Some commercial systems in the United States have more than 200 terminals permanently connected to a long-distance, leased, voice line in this way.

Suppose that the terminals in question operate at 10 characters per second and that the average number of characters in a message to the computer and its response is 50. Suppose that the operator sends one such message, on the average, every 100 seconds. The remainder of the time is "think" time or time when he is discussing the transaction with a client. In 100 seconds the high-speed line carries 50 characters of data per terminal. Suppose that there are eight bits per character; this becomes 400 bits per terminal. The total number of bits might be closer to 440 when we include those due to synchronization, addresses, end-of-message characters, and error checking. In one second, then, we average 4.4 bits per terminal. Suppose that the line speed is 2400 bits per second (on some in use today it is twice this). Because of queuing considerations we will not allow the line to be more than 60 percent utilized. We can therefore attach to one such voice line:

$$\frac{2400 \times 0.6}{4.4} = 327 \text{ terminals}$$

Needless to say, this reduces the line cost of a geographically dispersed system very greatly. Such a network design brings a scheme that would otherwise have been prohibitively expensive into the realms of practicality.

The character transmission time on the low-speed lines would be $\frac{50}{10} = 5$ seconds. However, part of this is input from a human operator, and as the terminal is not buffered, the line will be tied up for the length of time it takes him to type his transaction. Suppose that 15 of the 50 characters are input keyed in by the operator. (This average may be made

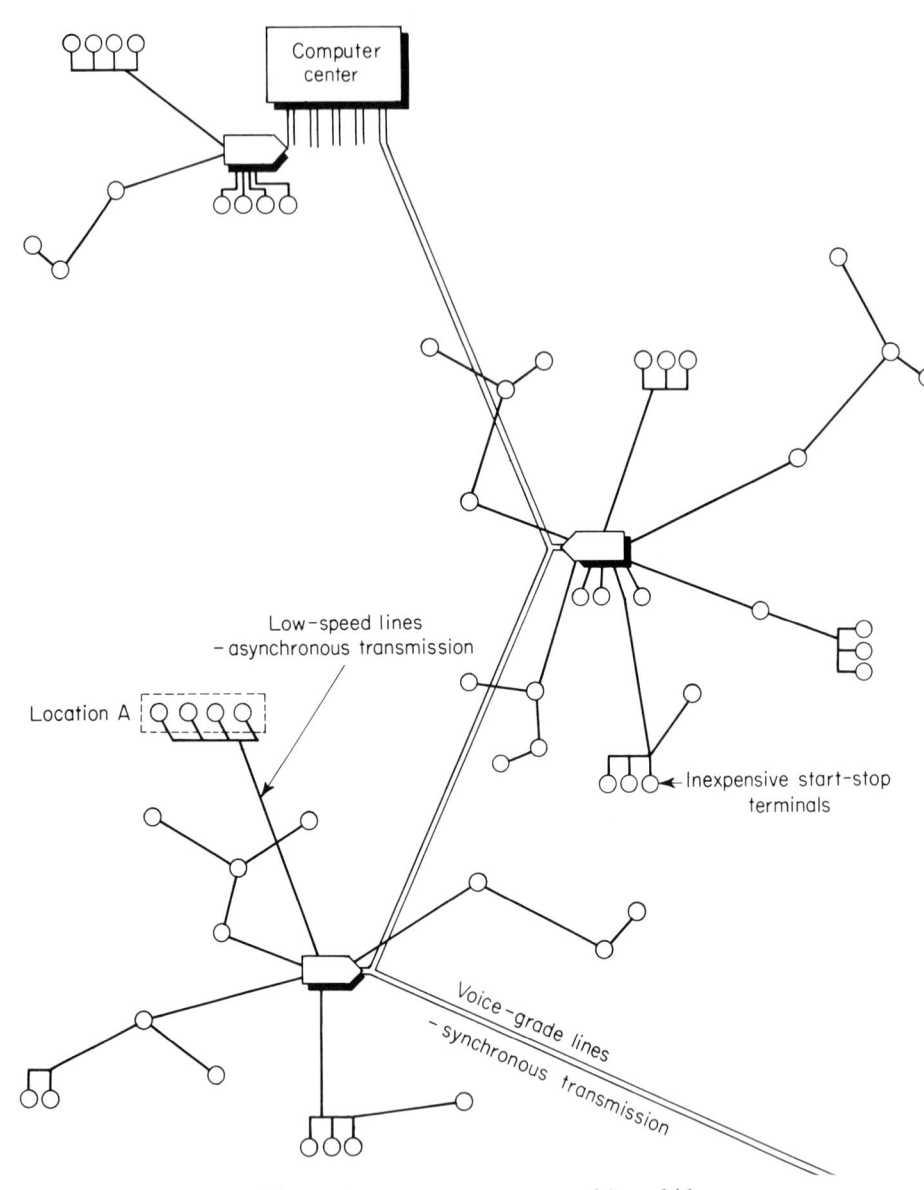

Fig. 12.4. Hold-and-forward remote concentrators with multidrop low-speed lines.

fairly low by the fact that a high proportion of the operator inputs are very brief.) If he types at the somewhat slow rate of two characters per second, then the total time this transaction occupies the low-speed line is $\frac{1.5}{2} + \frac{3.5}{10} =$ 11 seconds. We will again not allow the low-speed line to be more than 60 percent utilized. (100 × 0.6)/11 = 5.4: We can put five terminals on one low-speed line.

If the terminals handle larger transactions than this, the number that we can have on one line (without buffering) drops accordingly. On some systems it would be advisable to have only one or two on one line; otherwise a severe degradation of response time occurs.

Often, however, a reason for multidrop operation is that for a large proportion of the day the terminals are *unused*. Various persons have a terminal in their office and use it only when they need it. A number of such terminals may be connected to a private leased line. The cost is less than the cost of using public dial-up lines. This is only reasonable provided that no user transmits very long continuous messages. If he does, he will seriously degrade the response time of the other users. Such a user could be made to chop his message up, transmitting one line of print at a time. Alternatively, as we discussed in Chapter 5, terminals with buffers would solve this problem.

COMPLEX LINE CONTROL   As we commented above, with multiple terminals on the concentrator's low-speed line, the polling of these can be carried out either by the computer or by the concentrator itself. If the concentrator does it, then the high-speed line discipline need not be significantly more complex than that above. One extra addressing character is likely to be necessary to say which low-speed line the terminal is attached to. Also, there are extra messages for functions such as changing the concentrator's polling list. However, the moment-by-moment high-speed line control need not be basically different.

In some systems in actuality, however, the concentrator does not do the polling. The low-speed line control messages must be embedded in the messages traveling on the high-speed lines. This can become complicated especially if positive/negative response to error conditions is sent and automatic retransmission of messages in error takes place.

Let us illustrate such a complex control by describing that on the IBM 2905 concentrator. This machine is referred to in IBM terminology as a "remote multiplexor." A voice line may link many such machines, each with multidrop low-speed lines. The configuration could, then, be of the type represented by Fig. 12.4. Different-speed low-speed lines can be used, serving different types of terminals (one type per line) and transmitting different types of coding.

Let us suppose that we have IBM 1050 terminals on such a device, using a low-speed line polling discipline like that on p. 164. We will use the example on p. 164 again, showing what it would look like with this type of concentrator.

Suppose that the computer has some data that it wants to send to a terminal on one of the multidrop low-speed lines. The address of the terminal is F, and other details of the example are the same as those on p. 164, except for the intervention of the concentrator. The computer must now address the correct concentrator and indicate which low-speed line it is addressing. Instead of sending text to the concentrator, it will send a polling message for that low-speed line.

The sequence of operations is as shown in Table 12.2.

Table 12.2

COMPUTER	CONCENTRATOR 5	TERMINAL F ON LINE 14
The computer begins the sequence by sending an addressing message for the terminal in question: | S1 | S2 | 5 | 4 | © | F | 1 | EOM | CCC | Synchronization characters Concentrator address Line address Addressing message to terminal F End-of-message character Cyclic checking character	Concentrator 5 sends the terminal addressing message down line 14: | © | F | 1 | End-of-transmission character causes the terminals on line 14 to reset. The terminal address A code that selects the required component of the terminal, printer 1.	Terminal F receives the message, and is ready to receive the data. It therefore sends a positive response to the concentrator:

The computer eventually polls concentrator 5 again:

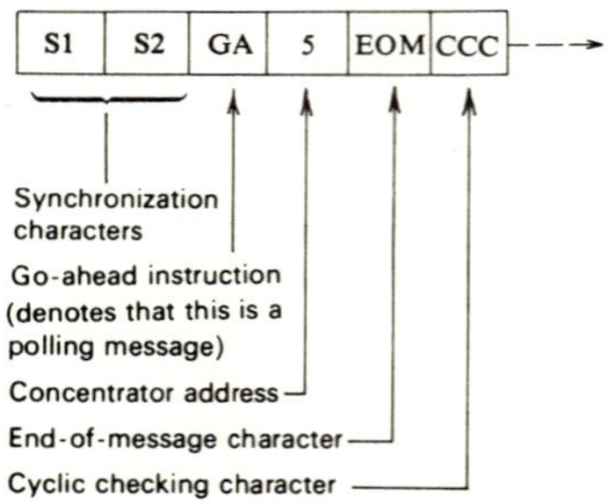

| S1 | S2 | GA | 5 | EOM | CCC |

- Synchronization characters
- Go-ahead instruction (denotes that this is a polling message)
- Concentrator address
- End-of-message character
- Cyclic checking character

The concentrator stores the positive response. It cannot send it to the computer until it is again polled by the computer.

When the concentrator receives this, it sends the terminal's positive response to the computer, along with an instruction to go ahead and transmit:

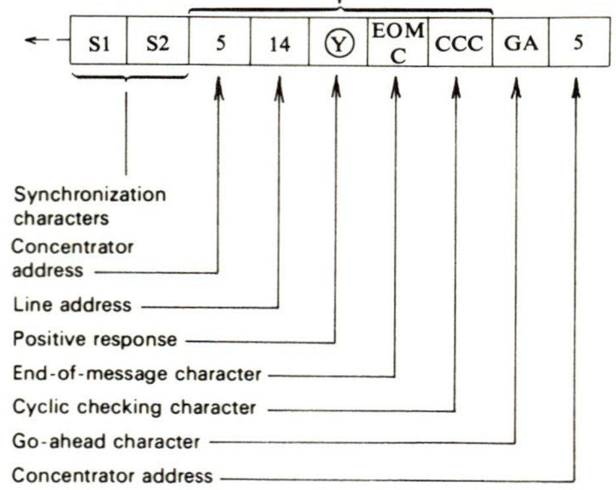

| S1 | S2 | 5 | 14 | Y | EOM | CCC | GA | 5 |

- Synchronization characters
- Concentrator address
- Line address
- Positive response
- End-of-message character
- Cyclic checking character
- Go-ahead character
- Concentrator address

(The last two characters are, in effect, an end-of-transmission signal telling the computer to go ahead and respond to concentrator 5.)

Note: This section could be repeated several times, giving responses from other low-speed lines and terminals.

The computer now receives the positive answer from the terminal and so transmits its data:

| S1 | S2 | 5 | 14 | Ⓓ | TEXT ⫽ | Ⓑ | EOM | CCC |

- Synchronization characters
- Line address
- Concentrator address
- Data message to send on to the terminal
- End-of-message character
- Cyclic checking character

The concentrator relays the message down line 14:

| Ⓓ | TEXT ⫽ | Ⓑ | LRC |

- End-of-block character
- End-of-address character means that text follows
- Longitudinal redundancy check

The concentrator again stores the positive response until polled by the computers:

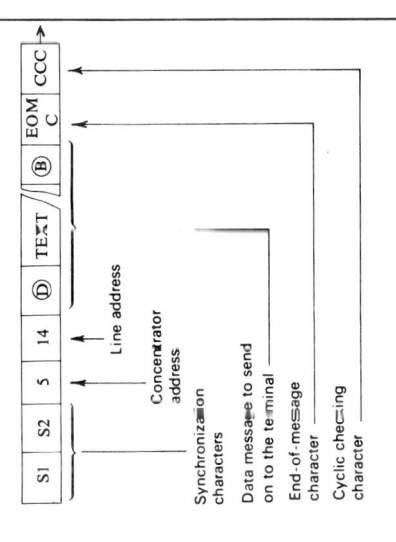

The terminal checks the longitudinal redundancy check character and finds it to be correct. It signals that it has received the data correctly:

Ⓨ

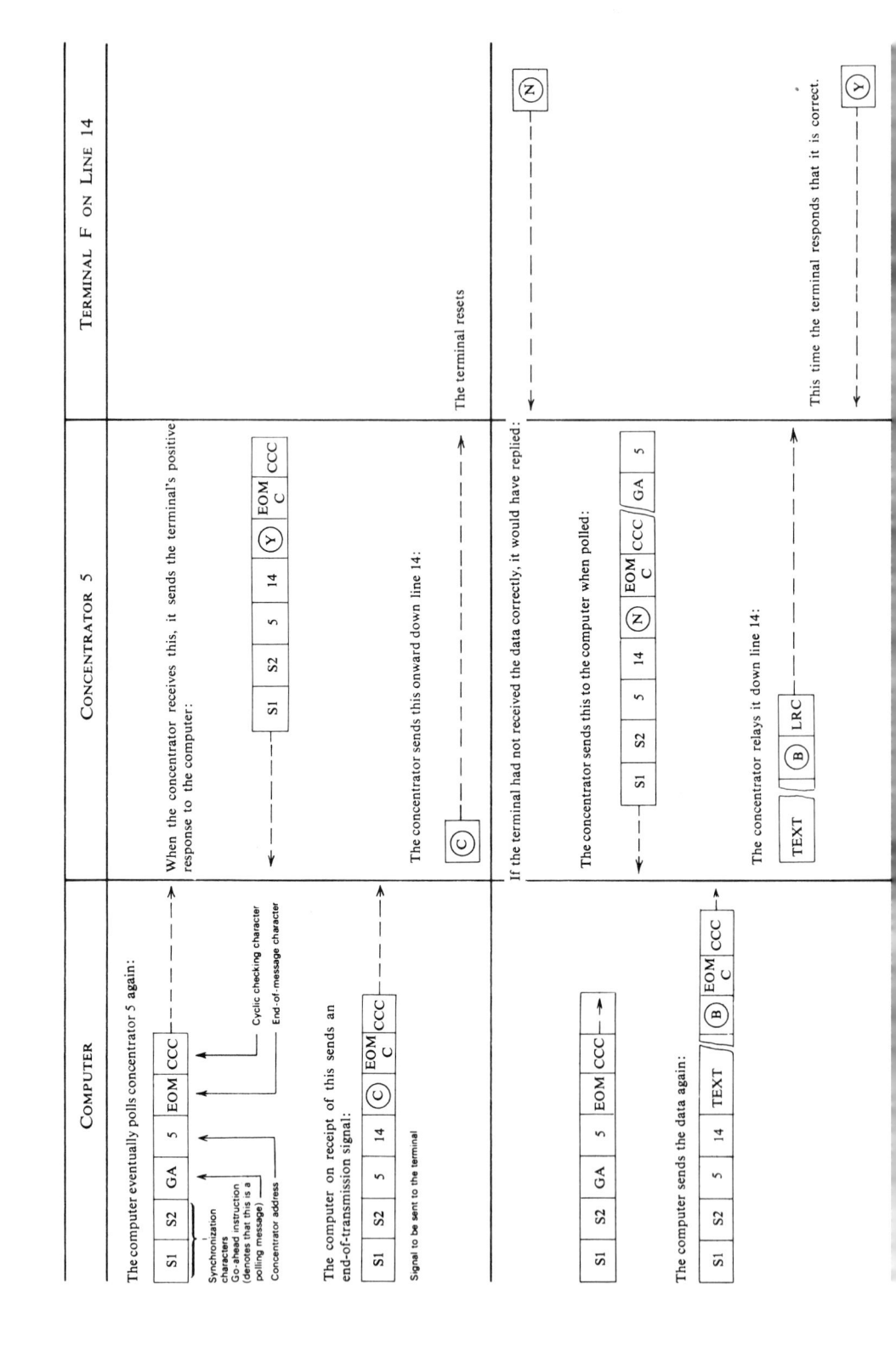

COMPUTER	CONCENTRATOR 5	TERMINAL F ON LINE 14
The computer eventually polls concentrator 5 **again**:		
| S1 | S2 | GA | 5 | EOM | CCC |	When the concentrator receives this, it sends the terminal's positive response to the computer:	
Synchronization characters — Go-ahead instruction (denotes that this is a polling message) — Concentrator address — Cyclic checking character — End-of-message character	| S1 | S2 | 5 | 14 | Y | EOM/C | CCC |	
	The concentrator sends this onward down line 14:	
The computer on receipt of this sends an end-of-transmission signal:		
| S1 | S2 | 5 | 14 | C | EOM/C | CCC |	C	
Signal to be sent to the terminal		The terminal resets
	If the terminal had not received the data correctly, it would have replied:	
		N
	The concentrator sends this to the computer when polled:	
| S1 | S2 | GA | 5 | EOM | CCC |	| S1 | S2 | 5 | 14 | N | EOM/C | CCC // GA | 5 |	
The computer sends the data again:	The concentrator relays it down line 14:	
| S1 | S2 | 5 | 14 | TEXT // B | EOM/C | CCC |	| TEXT // B | LRC |	This time the terminal responds that it is correct.
		Y

A somewhat similar sequence occurs when a terminal has something to send to the computer. Again, we follow the example on p. 164. A terminal operator wants to key in a long message at terminal F on line 14.

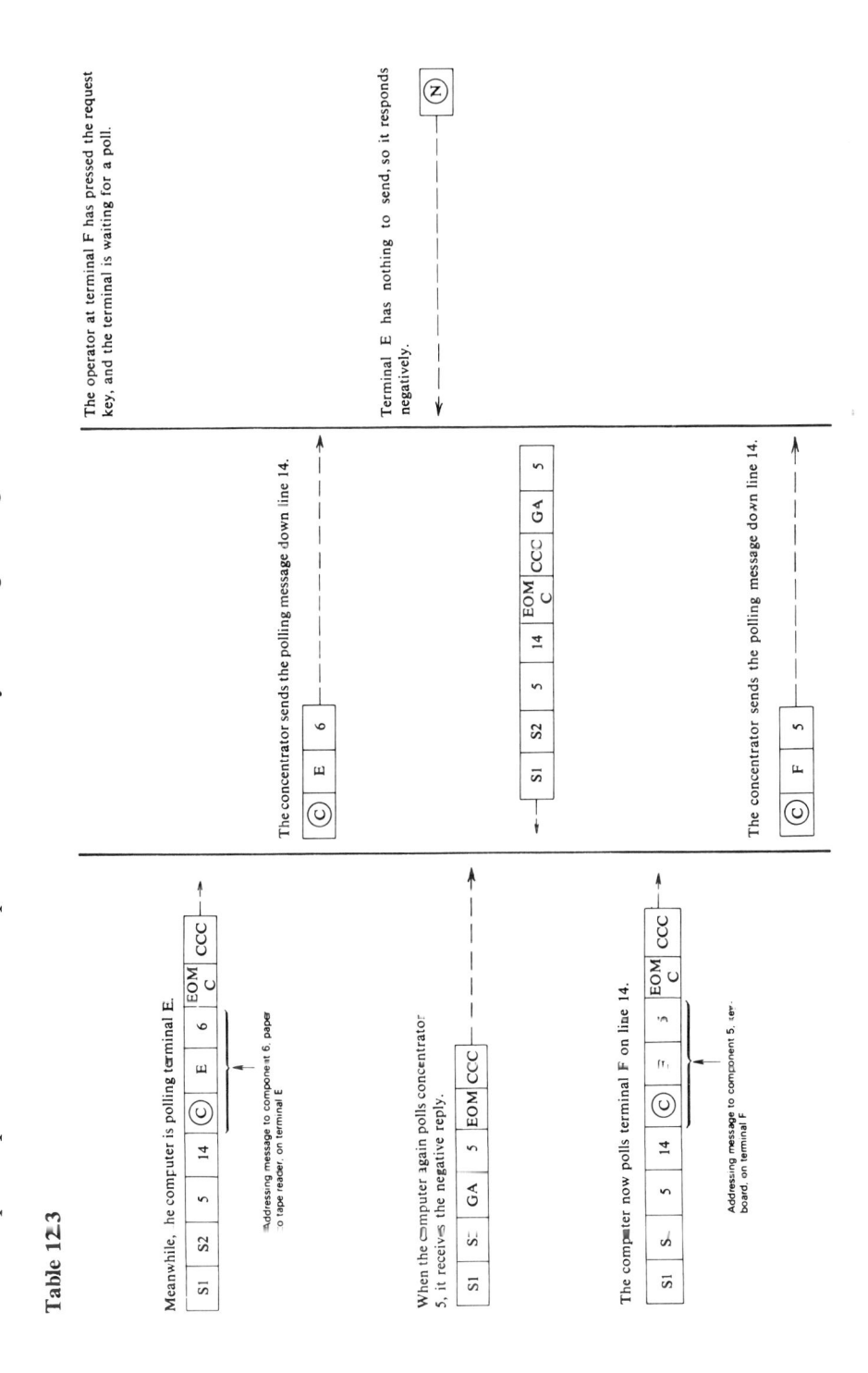

Table 12-3

Meanwhile, the computer is polling terminal E.

S1	S2	5	14	C	E	6	EOM	CCC

Addressing message to component 6, paper tape reader, on terminal E

The concentrator sends the polling message down line 14.

| C | E | 6 |

The operator at terminal F has pressed the request key, and the terminal is waiting for a poll.

When the computer again polls concentrator 5, it receives the negative reply.

| S1 | 5 | GA | 5 | EOM | CCC |

| S1 | S2 | 5 | 14 | EOM C | CCC | GA | 5 |

Terminal E has nothing to send, so it responds negatively.

| N |

The computer now polls terminal F on line 14.

| S1 | 5 | 14 | C | F | 5 | EOM C | CCC |

Addressing message to component 5, keyboard, on terminal F

The concentrator sends the polling message down line 14.

| C | F | 5 |

COMPUTER	CONCENTRATOR 5	TERMINAL F ON LINE 14
When the computer polls concentrator 5, it sends this data. | S1 | S2 | GA | 5 | EOM | CCC | →		

	If the data exceeds the concentrator block size, it is divided as follows:
	
	← | S1 | S2 | 5 | 14 | D | TEXT // EOM I | CCC |
	  | S1 | S2 | 5 | 14 | TEXT | EOM I | CCC |
	  | S1 | S2 | 5 | 14 | TEXT // B | EOM U | CCC | GA | 5 |
		The terminal responds positively by sending the data.
		
		← | D | TEXT // B | LRC |

The computer responds positively. | S1 | S2 | 5 | 14 | Y | EOM C | CCC | →		
	Ⓨ	
| S1 | S2 | 5 | 14 | EOM | CCC | →	← | S1 | S2 | 5 | 14 | Ⓒ | EOM C | CCC // GA | 5 |	The terminal acknowledges the positive response and so forces the computer to address the next terminal on this low-speed line. ← Ⓒ
The computer now goes on to poll the next terminal on its polling list for this line. | S1 | S2 | 5 | 14 | Ⓒ | A | 5 | EOM C | CCC | →		

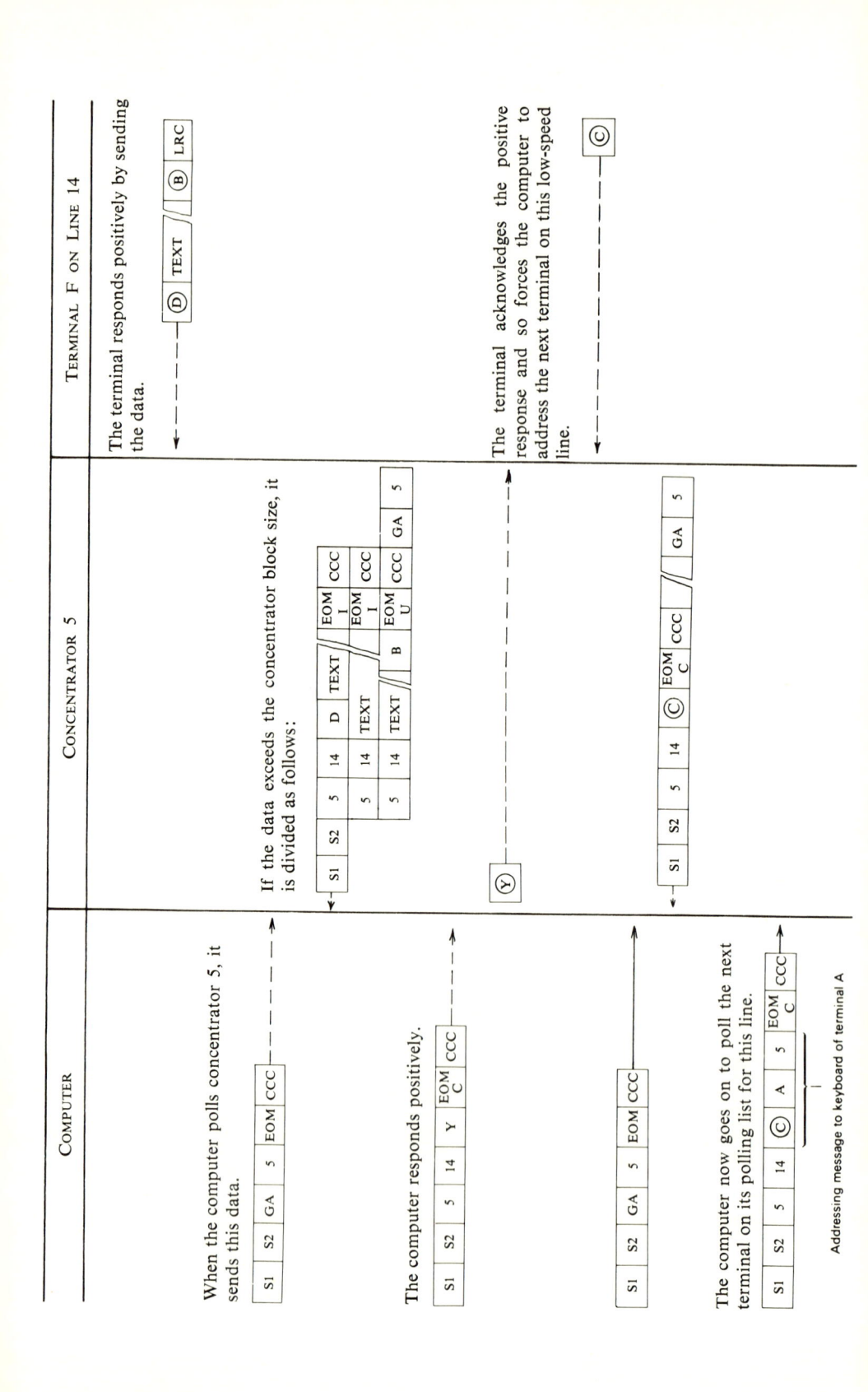

Addressing message to keyboard of terminal A

NUMBER OF CONTROL CHARACTERS

It will be seen that, in the above example, the line control procedure has become much more complex. This does not necessarily matter if it achieves our objective of lowering the network cost. It does, of course, increase the complexity of the communication line control program, but almost all of our efforts to organize computer systems more efficiently result in more complex software, and this is not much additional burden. It may, however, result in an excessive number of control characters flowing on the high-speed lines, and this would tend to defeat our objective.

On many systems the responses from the terminal operator to the computer are short. The computer sometimes gives an explanation of what it wants the operator to do, or how the operator might proceed with his inquiry, and the operator makes brief entries, one at a time. This will probably become increasingly so as terminals are used more and more for casual operations in which the operator does not spend long at the terminal but wants to cope quickly with a wide variety of applications. The computer leads him step by step through the necessary procedures. His responses are mostly very brief.

If this is so, then we could have far more control characters flowing on the high-speed line in the above example than data characters from the terminal operator. This situation can be resolved by further increasing the logic power in the concentrator. The machine from which the above illustration was taken, the IBM 2905, can itself have the ability to poll the low-speed lines (at extra cost). This function entails a considerable amount of logic most of which is in the form of a read-only storage microprogram. In the above illustration we used a terminal that gives a negative response to a poll when it has nothing to transmit (as is usually the case). If we had had 25 such terminals on the low-speed line from the concentrator, the computer would have had to send 25 polling messages of nine characters each. If they had had nothing to send, it would have received 25 negative responses, each needing five characters and residing in a block which needed four characters for synchronization and end of message, and which itself had to be read with a polling message. However, when the concentrator does the polling, this can be reduced to one polling message and response.

Taking the polling function away from the computer, wherever it is done, is referred to as "auto-polling." Auto-polling is further complicated if a mixture of terminals with differing polling procedures is attached to the device. Often, for example, telegraph equipment is intermixed with other more specialized terminals. These are polled differently but may be handled by the same auto-polling machine.

13 REMOTE LINE COMPUTERS

In our discussion of multiplexing and concentrating equipment we have been steadily increasing the logical complexity, until now we have a machine that is almost as complex as, and probably as expensive as, a small stored-program computer. Therefore, why not use a stored-program computer and benefit from its increased flexibility? A computer can do many operations that the special-purpose concentrator cannot, and we can use it for processing data when the network is not in use.

Figure 13.1 shows a network with a computer center on the eastern side of the United States and links to various locations across the country.

Fig. 13.1. Concentrator network.

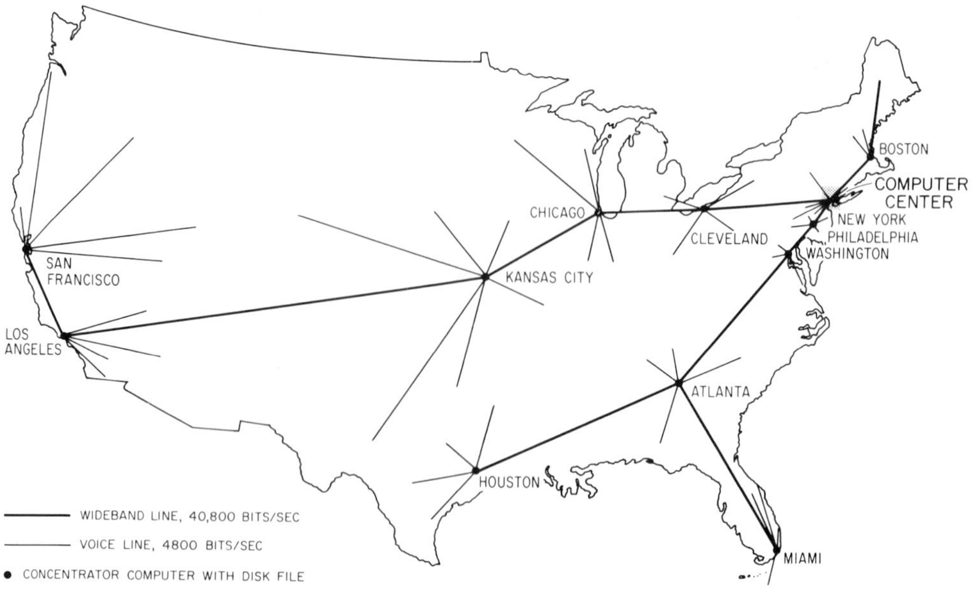

Broadband lines go to concentrators, and from these leased voice-grade lines go the branch locations. There are more branch locations than those shown in this simplified diagram, and at each such location there may be several visual display terminals and one or more printing terminals. The concentrators used here are small general-purpose computers. They are equipped with a disk file and high-speed printer and are located at regional centers of the location in question. When the real-time usage of the branch terminals ends in the evening, the central computer transmits listings and other information to the regional computers. The regional computers can then do processing of their own.

Figure 13.2 shows a line computer that is not permanently connected to the central computer. Here the object of the system is to collect data from a large number of peripheral locations. There is no need for real-time processing of this data. The regional computers here carry on the collection of data, checking accuracy and responding to the terminals. They store the data on their disk files, and later transmit them to the central computer. Leased subvoice-grade lines connect the terminals to the peripheral computers. The amount of transmission between the peripheral and central computers is not enough to warrant a leased line here. It is cheaper for the central computer to dial up each peripheral computer periodically and receive the data it has stored.

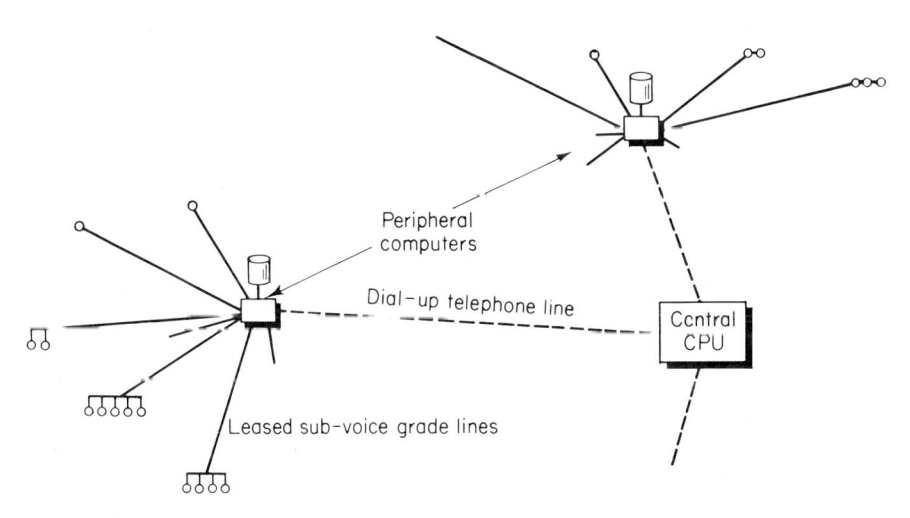

Figure 13.2

Figure 13.3 shows a converse configuration, with a leased link between the central computer and the peripheral machines, but a dial-up connection between the terminals and the peripheral computer. Here the terminals are used for time-sharing or as an enquiry device. Terminals may be in individuals' offices and much of the time are not in use. When they become operational their user dials his local computer. Some of the time the local computer will not be able to handle the processing requested, either because of load or because some facility is needed that is only on the central machine

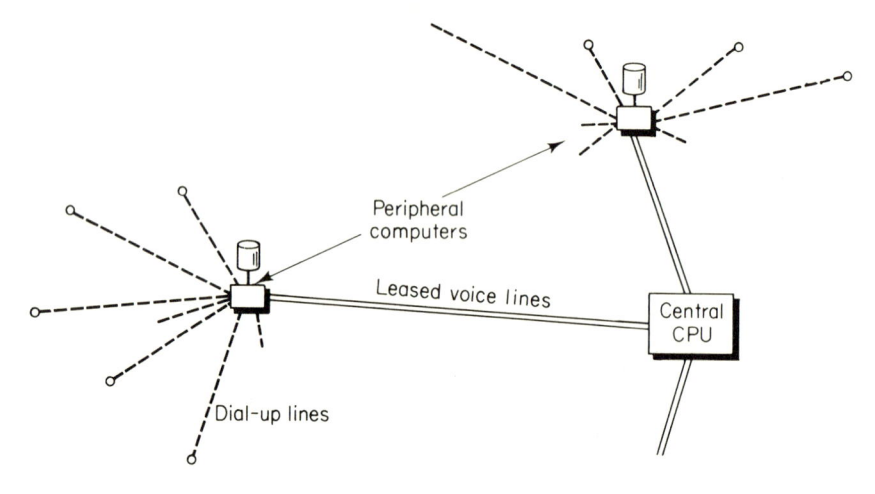

Figure 13.3

In the latter case the transaction is sent to the central computer on a leased line—leased because here the proportion of time it is in use makes this cheaper than a dial-up connection.

CANNED RESPONSES It is usually desirable that a terminal give responses to an operator in easily understood English. Also displays such as that in Fig. 15.2 require English words and headings. Such responses have been generated in most systems by the central computer. A considerable amount of verbiage then has to be sent down the communication lines. An alternative that involves less transmission time is to send only a coded reference to the responses and have the lengthy English phrases generated by the terminal control unit or concentrator.

The device may, for example, store a vocabulary of slightly less than 128 words, and when the central computer sends a stream of seven-bit characters in "coded text" mode, each of these (with the exception of a few, which are used for control purposes) would generate one of the words in the vocabulary. With eight-bit characters almost 256 words could be generated. With two seven-bit characters representing one word, a vocabulary of about 15,000 words becomes possible, and so on.

Alternatively, a character or group of characters may cause a stock sentence or phrase to be generated. For many applications 128 phrases would be plenty. One character could be given the meaning: "The next character is a phrase-generating character." The next character would then generate one from a list of 128 (or 256) phrases.

It is probably on commercial systems or systems with programs for carrying out highly specialized functions that one is most likely to find a need for canned responses. However, even on general-purpose systems where the operator writes programs at the keyboard, or where the terminal functions like an elaborate desk calculator, a set of fixed responses make the device easier to use. JOSS, the Rand Corporation time-sharing system that became famous as one of the earliest conversational computing facilities, had 40 canned responses that made it considerably easier to use. Figure 13.4 shows 13 of these being typed—for a somewhat browbeaten operator!

In Fig. 13.5 a much higher proportion of the transmission is in the form of canned printouts or phrases. To avoid long-distance transmission of the lengthy preamble and the repeated phases such as "MAX STRESS IN CONCRETE = ... KSI" would save most of the transmission time. Sometimes the printout is even more repetitive than that in Fig. 13.5. This, for example, is typed at the terminal in a similar program:

```
AT TIME  = 0.    THE OXYGEN DEFICIT IS 2.400 PPM
AT TIME  = 0.25  THE OXYGEN DEFICIT IS 3.291 PPM
AT TIME  = 0.50  THE OXYGEN DEFICIT IS 3.774 PPM
AT TIME  = 0.75  THE OXYGEN DEFICIT IS 3.987 PPM
AT TIME  = 1.00  THE OXYGEN DEFICIT IS 4.023 PPM
AT TIME  = 1.25  THE OXYGEN DEFICIT IS 3.946 PPM
AT TIME  = 1.50  THE OXYGEN DEFICIT IS 3.801 PPM
AT TIME  = 1.75  THE OXYGEN DEFICIT IS 3.614 PPM
AT TIME  = 2.00  THE OXYGEN DEFICIT IS 3.407 PPM
CRITICAL TIME = 0.94 CRITICAL DEFICIT = 4.027 PPM
```

However, whereas all of JOSS's canned responses could be stored in about 2000 characters, we are going to need a large storage if we are to store the verbiage from many programs like that in Fig. 13.5. The question arises: Would it be economically viable to have a disk file or other large storage at the concentrator, or terminal control computer? In some cases it would. It is interesting to note that for most such purposes the disk could be considerably smaller and slower than the slowest disks and drums on the market at the time of writing. For such purposes it might be worth making a very inexpensive disk unit, with a slow seek mechanism and disks the size of a 45 rpm phonograph record with no high precision packing.

1.1 Set x(1) = [− b + sqrt(b*2 − 4·a·c)]/[2·a].
1.2 Set x(2) = [− b − sqrt(b*2 − 4·a·c)]/[2·a].
1.3 Type x(1), x(2) in form 1.
Set b = − 1.
Set c = − 6.
Do part 1.1.
Eh?
Do part 1.1.
*Error above: Part number must be integer and 0 part 10*9.*
Do part 1.
Error at step 1.1: a is undefined.
Set a = 0.
Go.
Error at step 1.1: I have a zero divisor.
Set a = − 1.
Go.
Error at step 1.1: I have a negative argument for sqrt.
Set a = 1.
Go.
Error at step 1.3: I can't find the required form.
Form 1:
a =_____b =_____c =_____roots:_____._____.
Go.
Error at step 1.3: I have too few values for the form.
1.3 Type a, b, c, x(1), x(2) in form 1.
Go.
Error at step 1.3: I can't express value in your form.
Form 1:
a =_____b =_____c =_____roots:_____._____.
Go.
Error at step 1.3: Your fields run together in the form.
Form 1:
a =_____b =_____c =_____ roots:_____._____.
Go.
a = 1   b = − 1   c = − 6   roots: 3.0000   − 2.0000
Do part 1 for c = − 1(1) − 4.
a = 1   b = − 1   c = − 1   roots: 1.6180   − .6180
a = 1   b = − 1   c = − 2   roots: 2.0000   − 1.0000
a = 1   b = − 1   c = − 3   roots: 2.3028   − 1.3028
a = 1   b = − 1   c = − 4   roots: 2.5616   − 1.5616
2.1 Do step 2.1 for i = 1(1)100.
Do step 2.1.
I need more storage space.
Delete step 2.1.
Go.
It's a mess. Let's start over.
Go.
Error above: You haven't told me to do anything yet.
Do part 1 for c = − 5.
a = 1   b = − 1   c = − 5   roots: 2.7913   − 1.7913
Please wind up your work and turn off as soon as possible.
Delete all.

Fig. 13.4. Conversational computing with the RAND Corporation's JOSS system. Responses from the computer are in italics. Joss has 40 canned responses of which these are a sample. *Taken from RAND Corporation report P-3149.*

REMOTE LINE COMPUTERS   223

```
101. -READY     LOAD(BEAM)
191. +READY     START(0)
        • REINFORCED CONCRETE BEAM DESIGN PROGRAM
        • UNLESS OTHERWISE SPECIFIED, DATA IS TO BE ENTERED AS A SERIES OF NUM-
        • BERS SEPARATED BY SLASHES.
        • THE PROGRAM WILL FIRST DESCRIBE ALL INPUT VARIABLES. SUBSEQUENTLY,
        • WHEN A VALUE IS REQUIRED FOR A GIVEN VARIABLE, A REQUEST
        • FOR INPUT WILL BE GIVEN BY VARIABLE NAME AS IN THE EXAMPLE:
        •     ENTER: NAME1/NAME2./NAME3./NAME4
        •
        • A PERIOD ASSOCIATED WITH A VARIABLE NAME MEANS THAT A DECIMAL POINT
        • MUST BE INCLUDED IN THE VALUE GIVEN FOR THAT VARIABLE.
        •
        • THE DATA ENTERED IN RESPONSE TO THE ABOVE REQUEST MIGHT BE:
        •
        •     3/452./13.4/57
        •
        • ON SUBSEQUENT EXECUTIONS OF THIS PROGRAM, THE USER MAY
        •
        •   START(0)   TO BEGIN WITH THIS EXPLANATION
        •   START(2)   TO BEGIN WITH A LISTING OF THE INPUT VARIABLES
        •   START(1)   TO BEGIN WITH THE FIRST REQUEST FOR INPUT
        •
        • AN "XEQER" MESSAGE DURING EXECUTION PROBABLY SIGNIFIES INCONSISTENT
        • DATA INPUT.  CHECK YOUR INPUT, AND BE SURE YOU HAVE INCLUDED THE
        • DECIMAL POINT WHERE (AND ONLY WHERE) IT IS REQUIRED.
        • THEN BEGIN AGAIN (I.E., TAB AND TYPE "START(1)" )
        •
        • M  = BENDING MOMENT (KIP-FEET)
        • B  = BEAM WIDTH FOR RECTANGULAR BEAM (IN)
        •    = FLANGE WIDTH FOR T BEAM
        • B1 = B - BEAM WIDTH FOR RECTANGULAR BEAM (IN)
        •    = STEM WIDTH FOR T BEAM
        • T  = FLANGE THICKNESS FOR T BEAM (IN)
        • D  = DEPTH TO TENSION STEEL (IN)
        • AS = CROSS SECTIONAL AREA OF TENSION STEEL (IN2)
        • N  = MODULAR RATIO
        • D1 = DEPTH TO COMPRESSION STEEL (IN)
        • AS1 = CROSS SECTIONAL AREA OF COMPRESSION STEEL (IN2)
        • R  = MULTIPLIER FOR COMPRESSION STEEL MODULAR RATIO
        •
        • ENTER: M.
125. -1 00    76.8
        • RECTANGULAR, OR T-BEAM? (ENTER 0 FOR RECTANGLE; OR 1 FOR T-BEAM)
134. -1 00    0
        • RECTANGULAR BEAM.  ENTER: B.
142. -1 00    12.
        • IF TENSION STEEL ONLY, ENTER 0; IF COMPRESSION STEEL, ENTER 1
147. -1 00    0
        • ENTER: D./AS./N.
155. -1 00    8.0/3.0/10.
169. -0 12        MAX STRESS IN CONCRETE   =     5.439 KSI
171. -0 13        STRESS IN TENSION STEEL  =    46.783 KSI
173. -0 14        COMPRESSION STEEL STRESS =     0.    KSI
175. -0 15        BEAM WEIGHT PER FOOT     =   125.    LBS
        • WHEN THE TERMINAL TYPES "+READY" YOU MAY CHANGE THE VALUE OF ANY
        • VARIABLES YOU DESIRE BY HITTING "TAB", TYPING IN "NAME=XXX.XX", AND
        • HITTING EOB FOR EACH CHANGE. THEN TAB AND TYPE "START".  E.G.:
        •
        •     T=4.
        •     K1=1.7
        •     START
        • NOTE THAT ALL VARIABLES WILL RETAIN THEIR VALUES UNLESS YOU
        • SPECIFICALLY REASSIGN THEM.  IF YOU DESIRE TO WORK WITH A
        • DIFFERENT TYPE BEAM (E.G. T-BEAM INSTEAD OF RECT.), TAB
        • AND TYPE IN "START(1)".
        •
188. =PAUSE
200. +READY   D=10.5
201. +READY   AS=4.5
202. +READY   START
169. -0 12        MAX STRESS IN CONCRETE   =     3.057 KSI
171. -0 13        STRESS IN TENSION STEEL  =    23.985 KSI
173. -0 14        COMPRESSION STEEL STRESS =     0.    KSI
175. -0 15        BEAM WEIGHT PER FOOT     =   156.    LBS
188. =PAUSE
202. +READY   AS=5.0
203. +READY   START
169. -0 12        MAX STRESS IN CONCRETE   =     2.984 KSI
171. -0 13        STRESS IN TENSION STEEL  =    21.747 KSI
173. -0 14        COMPRESSION STEEL STRESS =     0.    KSI
175. -0 15        BEAM WEIGHT PER FOOT     =   156.    LBS
```

```
188.  =PAUSE
203.  +READY      B=11.5
204.  +READY      D=11.0
205.  +READY      START
169.  =O 12              MAX STRESS IN CONCRETE =      2.748 KSI
171.  =O 13              STRESS IN TENSION STEEL =    20.691 KSI
173.  =O 14              COMPRESSION STEEL STRESS =       0.  KSI
175.  =O 15              BEAM WEIGHT PER FOOT =         163. LBS
188.  =PAUSE
205.  +READY      AS=5.25
206.  +READY      START
169.  =O 12              MAX STRESS IN CONCRETE =      2.717 KSI
171.  =O 13              STRESS IN TENSION STEEL =    19.774 KSI
173.  =O 14              COMPRESSION STEEL STRESS =       0.  KSI
175.  =O 15              BEAM WEIGHT PER FOOT =         163. LBS
188.  =PAUSE
206.  +READY      START(1)
                  • ENTER: M.
125.  =I 00       90.0 FOOT KIPS
                  • RECTANGULAR, OR T-BEAM? (ENTER 0 FOR RECTANGLE; OR 1 FOR T-BEAM)
134.  =I 00       1 I WANT A T BEAM
                  • T-BEAM. ENTER: B./B1./T.
138.  =I 00       8.0 INCHES IS FLANGE WIDTH/4.0 IN. STEM WIDTH/4.0 IN. FLANGE THICKNESS
                  • IF TENSION STEEL ONLY, ENTER 0; IF COMPRESSION STEEL, ENTER 1
147.  =I 00       1 WNT CMPR STL
                  • ENTER: D./AS./N./D1./AS1./R.
151.  =I 00       12./4./10./3./2./11.
169.  =O 12              MAX STRESS IN CONCRETE =      1.516 KSI
171.  =O 13              STRESS IN TENSION STEEL =    28.871 KSI
173.  =O 14              COMPRESSION STEEL STRESS =   45.641 KSI
175.  =O 15              BEAM WEIGHT PER FOOT =          75. LBS
```

Figure 13.5

Looking at the example in Fig. 13.5, we could now take this argument a step further and say: If the concentrator or terminal control unit is a stored-program computer capable of storing a variety of responses for different applications, why not have a machine capable of executing the program needed for Fig. 13.5? Perhaps the peripheral machine will not be able to store all of the large numbers of programs that a user would need. Probably also it would not be able to execute all of them—not those needing the speed or memory size of a powerful computer. A network might, therefore, consist of local, relatively inexpensive computers that can do some of the computing needed locally, but which will be able to pass on to a powerful distant machine work that they cannot handle. The small computers may not have the program in Fig. 13.5 when a user requests it. If, as in this case, it is a small program that *can* be executed in the peripheral machine, the local computer will request that it be transmitted and temporarily stored on the file of the local machine. The local machine will then execute it and carry on the conversation with the terminal user. Transmitting a program such as that in Fig. 13.5 may take about half a minute on a conditioned leased, voice line—two or three times as long as on a dial-up line.

It will be seen that as soon as we use remote stored-program computers as part of our line organization, we open up a range of possibilities that extends from simple concentration functions to the complete execution of application programs. In many operating examples at the time of writing the remote line computer is being used in a relatively simple fashion for concentration, editing, formatting, the printing and reading of remote jobs or generation of graphics. However, as the software improves, and when we allow the application programer to utilize the remote machine, there will be more scope for argument as to what program function is done in what machine.

COMMERCIAL SYSTEMS

The case for elaborately programed line control computers may be stronger in geographically dispersed *commercial* systems than in those for scientific or general-purpose computing. A fundamental issue in the design of such systems is deciding whether the concentrators contain *application* programs as opposed to control programs which merely manipulate or store the characters flowing on the network. Once application programs are introduced at the concentrator level, new problems arise in their debugging, cutover, and maintenance. Often such programs remain relatively simple and so these problems are not severe. Otherwise, their effect on network cost and system feasibility can be very great.

An interesting example of this is demonstrated by the banking systems installed in Britain. London has several large banks with more than a thousand branches throughout the British Isles handling checking accounts. In these branches, it is desirable to have terminals connected to a computer in London that handles all customer records. The terminals are used to update the records, answer inquiries, and produce up-to-date statements on request giving all transaction details. They print exception reports and management information from the computer center. In addition, they must have the facilities of a multiregister accounting machine to "prove" batches of work and to enable the elimination of clerical errors.

It is clearly desirable to have a network of concentrators, especially in view of the very large number of branch locations. Because of the expense and difficulty of obtaining wideband lines in England, it was desirable to use voice lines to connect the concentrators to the computer center. Leased voice lines operating at 1200 bits per second (no more) were available along with dial-up lines operating at 600 bits per second. The network used by IBM for this situation was of the type shown in Fig. 12.1 (except that asynchronous transmission was used on the voice-grade lines). In addition to the leased connections, some concentrators could dial up to the computer center.

To reduce the network cost, it was desirable for a variety of computing functions to be carried out at the concentrator, and for a minimal number of characters to be transmitted over the lines to the computers. To achieve

226 REMOTE LINE COMPUTERS

this, a concentrator was designed based on the IBM 1130 with 8,192 16-bit words of core and a program loaded from paper tape. In addition to providing the normal concentrator functions discussed in Chapter 12, this machine was programed for arithmetic and editing that were unique to this application.

Let us illustrate this with an example of how a bank clerk uses his terminal. Figure 13.6 shows the keyboard of the terminal, which is located under the printing mechanism like a typewriter keyboard. On the extreme

Fig. 13.6. IBM 3982 terminal keyboard.

right are a set of status keys that can place the terminal in different statuses. For example "ENQ" places it in "Enquiry" status for sending enquiries to the main computer. "LCL TYPE" converts it into a local typewriter. "PRF" converts it, in effect, to an accounting machine with 16 sterling currency or decimal registers and a fully printed audit. When it is in this status it is used for "proving" batches of work as an accounting machine would be. All calculations are made in the concentrator and no transactions are transmitted to the main computer. "SEL" places the terminal in "Selective Transmit" status, and it then has the same functions as in "PRF" status, but in addition the terminal sends certain transactions on to the computer, their format depending upon the transaction type.

The operation in "SEL" status might look as follows:

REMOTE LINE COMPUTERS 227

(Two or three characters in a circle below refer to the pressing of one key, e.g. (TOT) for the "Total" key.)

KEYED DATA	PRINTED OUTPUT
The operator presses the "SEL" key, placing the terminal in "Selective Transmit" status.	SELCT
He then checks that the "accounting machine registers" he is going to use are clear:	
(TOT)(EOE) ↑ ↑ Total Key End-of-Entry Key	0.00.0 TOTAL DEBITS 0.00.0 TOTAL CREDITS 0.00.0 TOTAL HEAD OFFICE DEBITS 0.00.0 TOTAL HEAD OFFICE CREDITS 0.00.0 TOTAL LOCAL OFFICE REMS Zero Sterling amounts Description of registers
He then presses the proof key to check that the proof register is zero:	
(PRF)(EOE) ↑ ↑ Proof Key End-of-Entry Key	PRF 0.00.0 Zero Sterling amount printed on the right-hand side for ease of checking
Next he enters the items:	
(CHQ) 618030 6 (00)(000)(EOE) ↑ ↑ ↑ ↑ Check (cheque) Account number £600.0s.0d. End-of-Entry Key	618030 600.00.0 CHQ DR
(CSH) 40 (000)(DR)(EOE) ↑ ↑ ↑ Cash £400.0s.0d. Debit	40.00.0 CSH DR
(HOC) 600983 560 (000) EOE ↑ ↑ Head office credit Branch bank code	60–0983 640.00.0 HC
(PRF)(EOE)	PRF 0.00.0 Zero Sterling amount printed showing a correct zero-proof operation.

228 REMOTE LINE COMPUTERS

KEYED DATA	PRINTED OUTPUT

Transfer of money
(TRF) **618050** **46146** (CR)(EOE)
         ↑
     £46.14s.6d.
 	
The operator has made an error in keying in the account number.
 	
The operator corrects the error:
	618050 ERROR IN ACCOUNT NUMBER
	
	The error is found because self-checking account numbers are used. The concentrator checks the modulus of the account number and reports the error.

(TRF) **618030** **46146** (CR)(EOE)	618030 46.14.6 TRF CR
(TRF) **950861** **46146** (DR)(EOE)	950861 46.14.6 TRF DR
(PRF)(EOE)	PRF 0.00.0
(SDS) **397605** **128019** (CR)(EOE) ↑ Sundry deposits	397605 128.01.9 SDS CR
(CHQ) **218507** 1(**00**)(**000**)(EOE)	218507 100.00.0 CHQ DR

(HOD) **600983** **10** (**000**)(EOE)
         ↑
    Branch bank code
↑
Head office debit
	60–0983 10.00.0 HD

(LOR) **200788** **18109** (EOE) ↑ Local office remnants	20–0788 18.10.9 LR
(PRF)(EOE)	PRF 0.09.0 ERROR
(REV) **200788** **18109** (EOE) ↑ Reverse the above erroneous action.	20–0788 18.10.9 LR REV
(LOR) **200788** **18019** (EOE)	20–0788 18.01.9 LR
(PRF)(EOE)	PRF 0.00.0
(TOT)(EOE) ↑ This causes the totaling routine to be executed and the amounts are cross-footed into the appropriate registers to prove the accuracy of the work.	786.14.6 TOTAL DEBITS 174.16.3 TOTAL CREDITS 10.00.0 TOTAL HEAD OFFICE DEBITS 640.00.0 TOTAL HEAD OFFICE CREDITS 18.01.9 TOTAL LOCAL OFFICE REMS.
(PRF)(EOE)	PRF 0.00.0

NOTE THAT DURING THE ABOVE OPERATIONS ONLY THE DATA IN BOLD TYPE ARE TRANSMITTED FROM THE CONCENTRATOR TO THE MAIN COMPUTER.

In many other types of system the user carries on a conversation with a terminal either to obtain information or to give information to the system, which it stores in a central data base and uses for some form of control or decision-making. The system is usually dedicated to a relatively narrow, precisely specified set of functions. Often, however, the information is much more complex than on a banking system (e.g., with systems for airline bookings, sales order handling, providing management information, etc.).

SYSTEMS WITH COM- During the input of information the computer
PLEX CONVERSATIONS must ensure that all of the facts are entered, and, as far as is possible, must check them for validity. Several different methods are used for doing this. In many cases it is considered desirable for the computer to tell the operator what to enter next. Man-machine conversations of differing degrees of complexity have been used. We are interested in the question: To what extent need this conversation tie up the long communication links? On the majority of systems at the time of writing, the entire conversation has to be transmitted to and from the main computer. This has, in some cases, led to a very expensive network, which could have been much lower in cost if peripheral computers had handled those parts of the man-machine conversation that did not need transmission to the central computer.

Figure 13.7 shows a hypothetical example of one type of data entry. The terminal screen presents the operator a "form," which he has to "fill in." He uses the typewriter keyboard to move the cursor to appropriate places on the screen and type the requisite data. If our prime concern is in minimizing the data transmitted over the long lines, we would like not only to store the "form" in the peripheral machines, but also to edit the data entered before sending it to the main computer. There is no worth in transmitting "blank" characters. The editing of input data generally needs to be programed uniquely for each application.

A technique more commonly seen is to ask questions of the terminal user one by one rather than presenting him with a "form" such as that in Fig. 13.7. The following is a typical sequence in a system in which it is necessary to make terminal operation easy for the user:

```
SCREEN:  ENTER CUSTOMER NUMBER IF KNOWN. IF NOT TYPE "N"
USER:    N
SCREEN:  TYPE FIRST TWO LETTERS OF CUSTOMER'S NAME
USER:    GO
SCREEN:  WHICH CUSTOMER? ENTER LINE NUMBER
         1.  GOLDBERG SIDNEY & CO.
         2.  GOLDEN KAUFMAN SALES CO.
         3.  GOLDSMITH WALTER & CO.
         4.  GOODMAN ABE
         5   NONE OF THE ABOVE
```

230   REMOTE LINE COMPUTERS

USER: 3
SCREEN: CUSTOMER NUMBER: 17-8341
          GOLDSMITH WALTER & CO.
          745, OLD STONE HIGHWAY, SPRINGS,
          EAST HAMPTON, N. Y., 11937
          REPRESENTATIVE E. S. KOPLEY #543271

          PRESS ENTER IF CORRECT
USER: Presses "Enter" key.
SCREEN: WHICH INFORMATION IS TO BE CHANGED?
          ENTER LINE NUMBER
          1. NEW ORDER   9. INVOICE TO ADDRESS
          2. CANCEL ORDER   10. SHIP TO ADDRESS
          3. ORDER QUANTITY   11. CUSTOMER ADDRESS
          4. DELIVERY DATE   12. CUSTOMER NAME
          5. PRIORITY CODE   13. MISC. INSTRUCTIONS
          6. ORDER DETAILS   14. REPRESENTATIVE DETAILS
          7. DISCOUNT   15. BRANCH OFFICE
          8. TAX DETAILS   16. OTHER
USER: 4
SCREEN: WHICH ORDER IS TO BE CHANGED?
          ENTER LINE NUMBER

ORDER#		ITEMS			ORDER DATE	DEL. DATE
1	573824	87289	00241	00029	4/1/69	4/1/70
		00310	77139	04126		
		00900				
2	575931	00391	00700	00421	10/1/69	4/1/71
3	575828	00750	00900		12/1/69	12/1/70
4	579003	02107			6/1/69	8/1/70

USER: 2
SCREEN: ORDER NUMBER 575931

ITEM #	DESCRIPTION	QTY.
00391	MAGNETIC D EXTRACTOR 7 BR	5
00700	LINK ADAPTOR ADDL	5
00421	CTC DIRECT CONTROL	5

          ORDER DATE 10/1/69   DELIV DATE 4/1/71

          ENTER NEW REQUIRED DELIV DATE (XX/XX/XX)
USER: 10/01/70
SCREEN: EARLIEST DELIV DATE POSSIBLE AT PRIORITY 3 IS
          2/1/70
          IF THIS IS REQUESTED ENTER "Y"
USER: Y

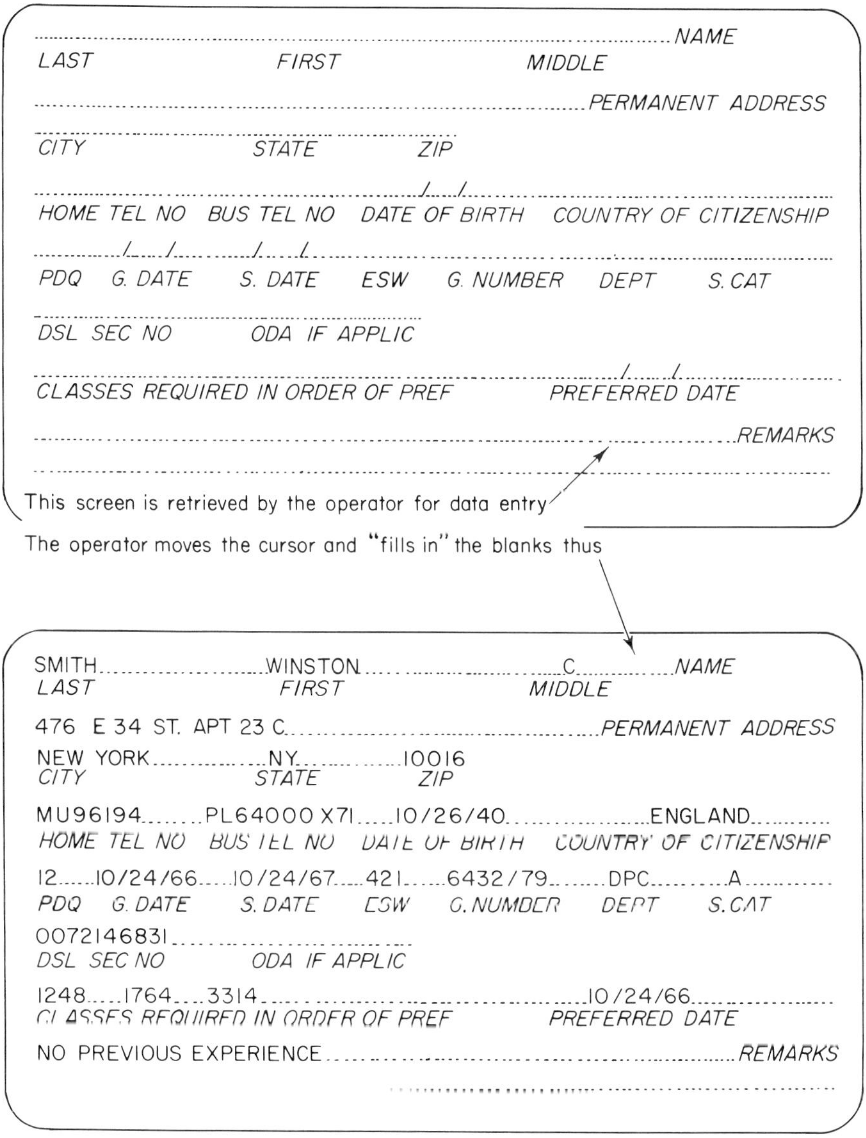

```
..................................................................................NAME
LAST              FIRST            MIDDLE
..........................................................................PERMANENT ADDRESS
CITY            STATE        ZIP
................................................./...../.............................................
HOME TEL NO  BUS TEL NO  DATE OF BIRTH  COUNTRY OF CITIZENSHIP
.............../...../............/......./...................................................................
PDQ    G. DATE    S. DATE    ESW    G. NUMBER    DEPT    S. CAT
----------------------------------------
DSL SEC NO       ODA IF APPLIC
............................................................................../...../........
CLASSES REQUIRED IN ORDER OF PREF       PREFERRED DATE
..................................................................................REMARKS
```

This screen is retrieved by the operator for data entry

The operator moves the cursor and "fills in" the blanks thus

```
SMITH.................WINSTON...............C........NAME
LAST              FIRST            MIDDLE
476 E 34 ST. APT 23 C.........................................PERMANENT ADDRESS
NEW YORK..........NY..........10016
CITY            STATE        ZIP
MU96194......PL64000 X71.....10/26/40...............ENGLAND..........
HOME TEL NO  BUS TEL NO  DATE OF BIRTH  COUNTRY OF CITIZENSHIP
12......10/24/66.....10/24/67.....421......6432/79........DPC..........A............
PDQ    G. DATE    S. DATE    ESW    G. NUMBER    DEPT    S. CAT
0072146831.................................
DSL SEC NO       ODA IF APPLIC
1248.....1764....3314...............................10/24/66................
CLASSES REQUIRED IN ORDER OF PREF       PREFERRED DATE
NO PREVIOUS EXPERIENCE.........................................REMARKS
```

Fig. 13.7. If the operator uses a technique such as this for filling in information, how much editing of data will be carried out before they are transmitted to a distant computer?

In terminal use such as this we generally have a large number of "canned" responses. There is no requirement for the central computer to prepare every response. Some of them need information from its data base, but most need neither its data nor its power. Much of the conversation could be carried out in the peripheral computer without any long-distance transmission of data. The central machine would be contacted only when necessary.

Consider the conversation that takes place when an airline journey is booked on a far distant computer, for example. It is first necessary to determine which flights between the cities requested have seats available. Thus information from the central seat inventory files is needed. Then, when the flights to be booked have been determined, details about the passenger are collected: first his name, then his address, his home telephone number, possibly also his work telephone number, the date by which he will pay for his ticket (some make bookings and do not fly), the fact that his wife is traveling with him, and his small baby (who needs a carry-cot), and his English sheepdog (no more than two dogs are allowed on one plane), that he wants a car booking at San Francisco and a hotel in Los Angeles, and general information about him. All of this data could be collected by the peripheral computer in a manner designed to give rise to the least number of operator errors, and then transmitted in a block on the long-distance lines.

In fact, in most airline systems in operation at the time of writing a peripheral computer is not used in this manner. The above entries about the passenger are transmitted, and responded to, one at a time. The format of the conversation is consequently designed to minimize the number of characters used. Lengthy English statements telling the operator what to do, as in the above example, are avoided. The operator needs an extensive period of training in how to use the terminal language. Many of the responses are highly coded.[1] This is satisfactory for an airline reservation system, because the operator spends most of his working life using the terminal for that one application. In an administrative or management information system, however, such coding would rarely be usable, because the operators use the terminal only periodically, and for several different applications. They cannot be given such intensive training and practice, so terminal operations must be more self-explanatory and verbose, as in the case above.

---

[1] James Martin, *Design of Real-Time Computer Systems*. Prentice-Hall, Inc., Englewood Cliffs, N.J., 1967 (p. 96).

14 MESSAGE SWITCHING

For many years in conventional telegraphy it has been considered vital to saturate the capacity of the costly long-distance lines as fully as possible. This was achieved by using message-switching centers to relay messages, and using *multidrop* telegraph lines from these centers.

As we discussed in Chapter 7, message switching is an alternative to line switching as a means for interconnecting many terminal locations. Messages to be transmitted from one terminal to another are sent to the switching center, where they are temporarily stored and then relayed to the terminal for which they were intended. There is no physical circuit interconnection as in line switching.

With line switching, an interconnection exists between the communicating machines for the duration of the call, so the operators can respond to each other in real time as over the telephone. This is not possible with message switching. The reason for using message switching is that better line utilization can be obtained, thus reducing the overall line cost. The relay center transmits to the terminals at the maximum speed—better than the speed obtained by an operator. Several terminals can be used on one line, and the messages can be packed together to make good use of the line.

If an operator were trying to contact another operator on a system with *line* switching there could sometimes be a considerable delay. In the line configuration in Fig. 14.1, if terminal A tries to contact terminal G, it might have to wait some time because the line to G is occupied by traffic to F, G, H, I, or J, or by traffic *from* these terminals, or because its own line is busy with traffic to or from B. With a *message*-switching system, A stands a much better chance of being able to transmit to the switching center without delay, and the switching center will store the transaction. If A cannot obtain its own line immediately, the transaction may wait, punched in paper tape (or cards), at the terminal until the switching center polls that terminal. Today when *data* are transmitted a terminal may send large quantities of data at a time, and when this happens congestion problems may be made worse than with conventional telegraph traffic.

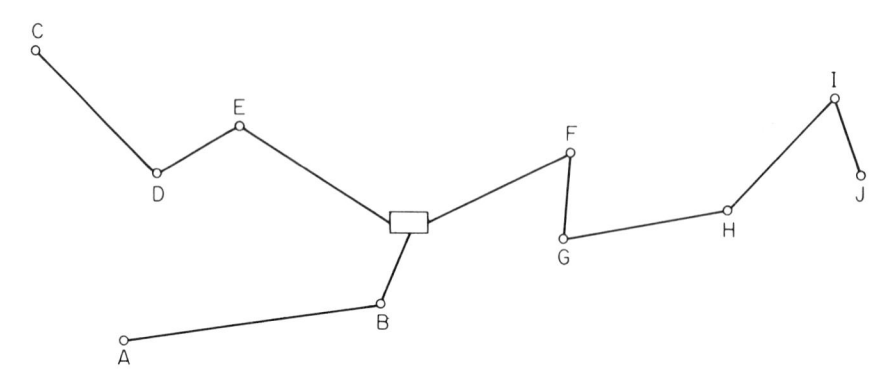

Figure 14.1

In the United States, message-switching systems have had a greater advantage over line-switching exchanges than in Europe, because of the greater distances between cities and the lesser population density. The trunk circuits were much longer and so more expensive. Consequently, Telex-type line-switching systems, with relatively inefficient use of trunk circuits, were more practical in Europe, and message-switching systems were more economical in the United States. However, the cost per mile of telegraph circuits has dropped drastically with the advent of carrier techniques and transmission media with bandwidths orders of magnitude higher.

Many private message-switching centers are in operation today in firms, and certain related *groups* of firms such as airlines, to lower their costs for message transmission.

TORN-TAPE SWITCHING CENTERS

The earliest of such systems were manually operated. Incoming lines were connected to paper-tape punches and paper tape was the storage medium used. Many such systems are still in operation today. They are referred to as *torn-paper-tape switching centers* because the operators tear off transactions punched in tape to retransmit them. Figure 14.2 shows a torn-tape center and its lines. The messages punched into tape are preceded by the address or addresses to which they are to be sent. The pieces of paper tape are stored on a rack awaiting retransmission. Girls operating this type of switching center tear off the messages received, read the address to which they are sent, and place them on the appropriate rack or transmitting paper-tape reader. The operator transmitting the message, having selected the requisite output line, then selects the appropriate station of the multidrop line by pressing a button on the transmitter console.

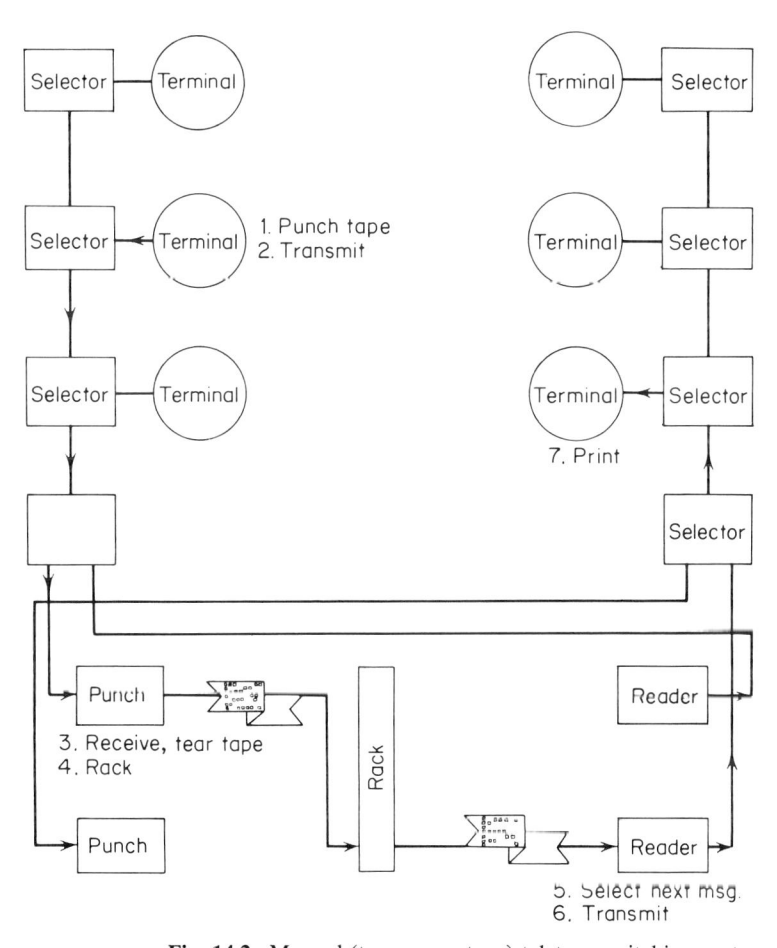

Fig. 14.2. Manual (torn-paper-tape) teletype switching center.

Queues of transactions develop in such a system. If a large number of messages are received in a short period on one line, these sit in a wire basket waiting for the girls to inspect them. Similarly, a temporary high load on an outgoing line results in a number of pieces of torn paper tape

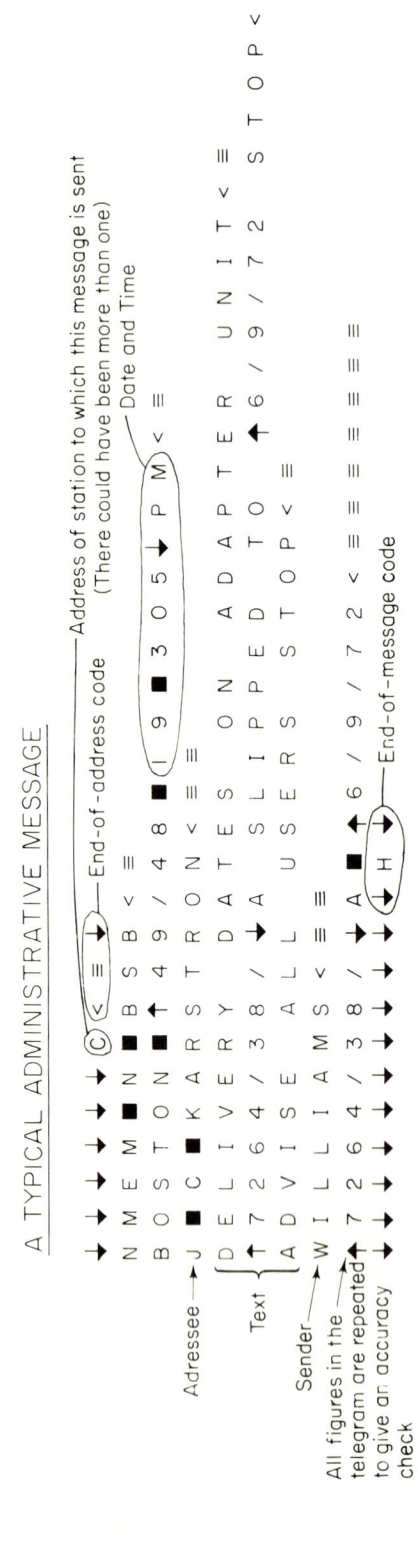

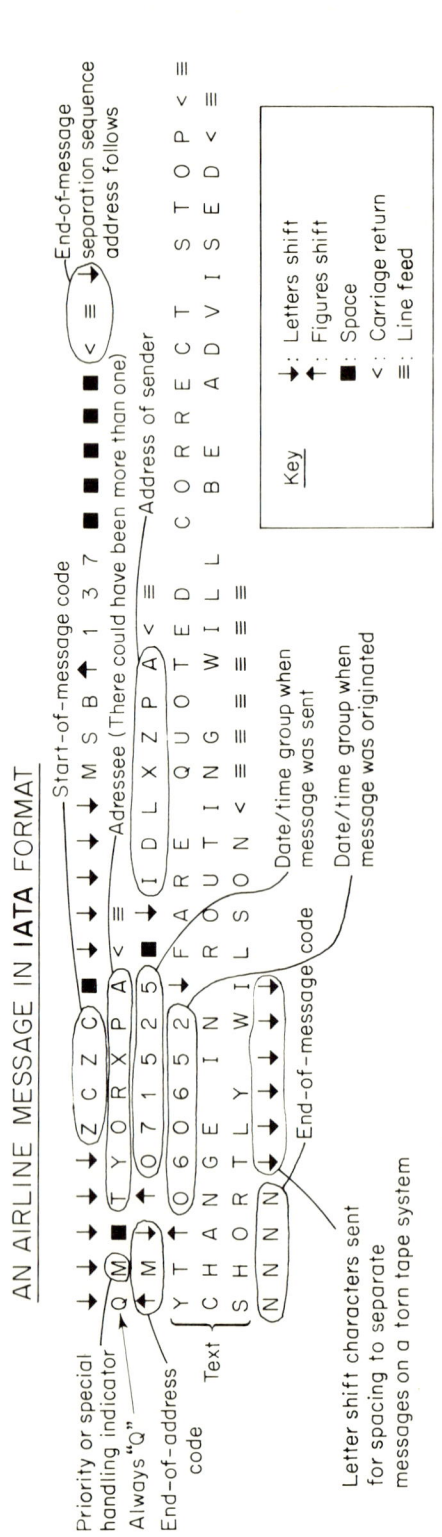

Fig. 14.3. Format of typical teletype messages sent to message-switching centers.

waiting to be transmitted on that line. A large switching center of this type may employ as many as sixty girls to keep it in operation.

Figure 14.3 shows the format of typical teletype messages transmitted to message-switching centers.

SEMIAUTOMATIC SWITCHING CENTERS

A somewhat more mechanized system than the torn-tape center is shown in Fig. 14.4. Here the messages are still stored in punched paper tape, but now it is not torn by the operators, and the operators do not physically carry it to the outgoing readers. Instead, an unbroken paper tape loop is developed. The message resides

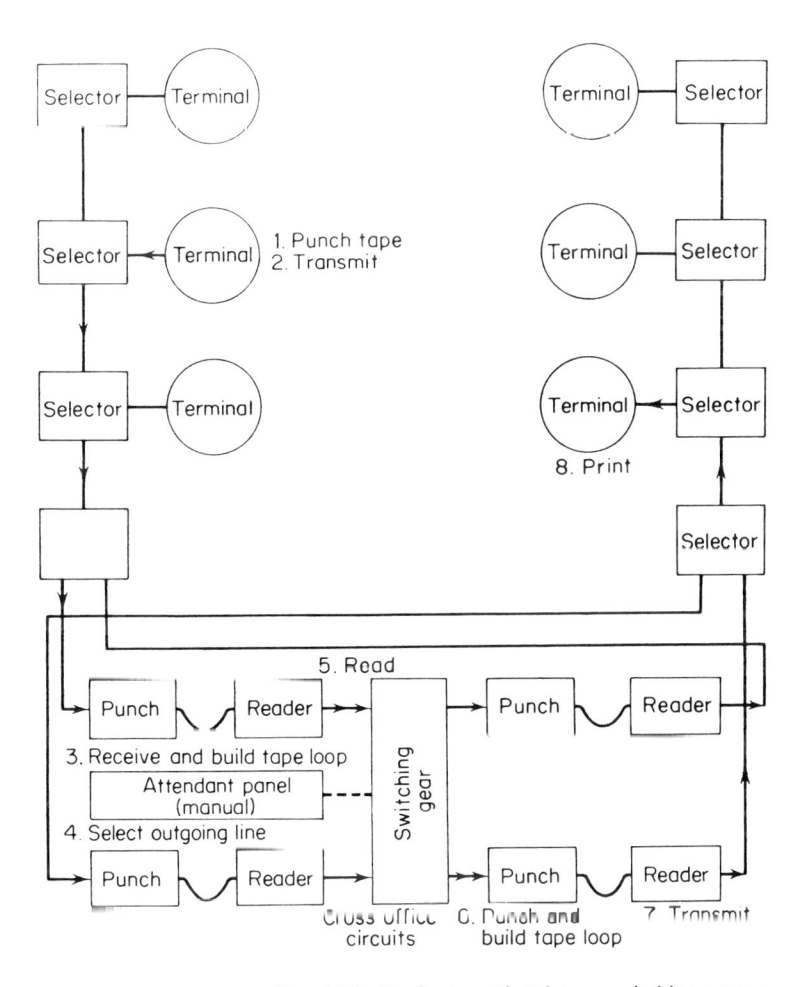

Fig. 14.4. Semiautomatic teletype switching center.

in this loop until the operator has time to service it. By reading the address in the message and pushing the correct buttons on the console, the operator can set up a connection to another reader, this time one associated with an outgoing line. The message is then read and transmitted from a second paper tape loop as shown.

In some systems the job of the console operator in Fig. 14.4 is mechanized, thus giving a fully automatic switching center, still using paper tape storage.

COMPUTERS FOR MESSAGE SWITCHING

Today computer systems are in common use for message switching, and the messages are stored on disks or whatever the storage medium of the computer is. In general, any switching center that handles 6000 messages per day would probably be best operated with a computer. Some networks with a much *lower* traffic volume than this are switched by computer.

One of the largest commercial message-switching centers is the Collins Radio System at Cedar Rapids, Iowa. This uses 10 computers, 24 hours of every day to switch airline messages. Its volume totals more than 200,000 messages per day.

A computerized switching center can be considered to be an automation of the manual and semiautomatic systems in Figs. 14.2 and 14.4. Its advantages over these are

1. Messages are delivered more rapidly (much faster than the torn-tape system).
2. Operating errors are eliminated.
3. Priority indicators in the messages can be easily recognized and acted upon.
4. Transmission errors can be handled with automatic retransmission.
5. Messages can be analyzed and various automatic actions taken if required.
6. Routing a message to several different destinations can be done easily. The computer can book address lists or send broadcast messages.
7. Messages can be stored for later retrieval if desired.
8. Processing of certain message types can be carried out.
9. For message volumes above a certain limit it becomes the cheapest system.

The *single* computer may have one disadvantage over the above methods. If a computer failure occurs, the entire exchange is out of action until it is repaired. Most of the likely failures in the other methods put out only one line or, at worst, a few lines. The mean time to repair the waiting to be transmitted on that line. A large switching center of this type may employ as many as sixty girls to keep it in operation.

Figure 14.3 shows the format of typical teletype messages transmitted to message-switching centers.

SEMIAUTOMATIC SWITCHING CENTERS

A somewhat more mechanized system than the torn-tape center is shown in Fig. 14.4. Here the messages are still stored in punched paper tape, but now it is not torn by the operators, and the operators do not physically carry it to the outgoing readers. Instead, an unbroken paper tape loop is developed. The message resides

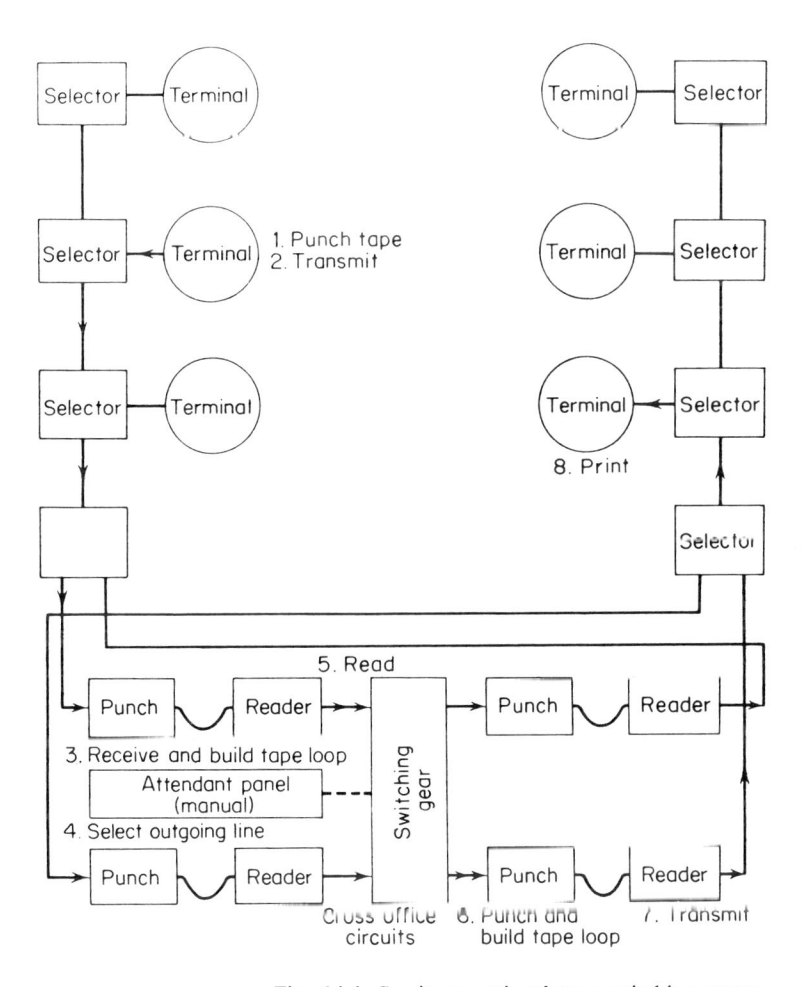

Fig. 14.4. Semiautomatic teletype switching center.

in this loop until the operator has time to service it. By reading the address in the message and pushing the correct buttons on the console, the operator can set up a connection to another reader, this time one associated with an outgoing line. The message is then read and transmitted from a second paper tape loop as shown.

In some systems the job of the console operator in Fig. 14.4 is mechanized, thus giving a fully automatic switching center, still using paper tape storage.

COMPUTERS FOR MESSAGE SWITCHING

Today computer systems are in common use for message switching, and the messages are stored on disks or whatever the storage medium of the computer is. In general, any switching center that handles 6000 messages per day would probably be best operated with a computer. Some networks with a much *lower* traffic volume than this are switched by computer.

One of the largest commercial message-switching centers is the Collins Radio System at Cedar Rapids, Iowa. This uses 10 computers, 24 hours of every day to switch airline messages. Its volume totals more than 200,000 messages per day.

A computerized switching center can be considered to be an automation of the manual and semiautomatic systems in Figs. 14.2 and 14.4. Its advantages over these are

1. Messages are delivered more rapidly (much faster than the torn-tape system).
2. Operating errors are eliminated.
3. Priority indicators in the messages can be easily recognized and acted upon.
4. Transmission errors can be handled with automatic retransmission.
5. Messages can be analyzed and various automatic actions taken if required.
6. Routing a message to several different destinations can be done easily. The computer can book address lists or send broadcast messages.
7. Messages can be stored for later retrieval if desired.
8. Processing of certain message types can be carried out.
9. For message volumes above a certain limit it becomes the cheapest system.

The *single* computer may have one disadvantage over the above methods. If a computer failure occurs, the entire exchange is out of action until it is repaired. Most of the likely failures in the other methods put out only one line or, at worst, a few lines. The mean time to repair the computer when it fails may be of the order of two hours. On some networks a break of two hours is tolerable. On others it is not. If it is not, two computers must be used, giving a *duplexed* system, so that if one fails the other takes over. On some systems the duplexing may not be as wasteful as it sounds, because the standby computer may run data-processing or other programs. Some organizations have deliberately placed the message exchange next to a data-processing center for this reason. The data-processing computer may interrupt its work only for the occasional period when the message switching has failed. Certain computer operating systems permit message switching to go on at the same time as other data processing on the same machine.

Some message-switching systems have special no-break power supplies to maintain operation when mains power fails. A direct-current motor-driven alternator may be used, driven off a direct-current battery that is kept fully charged by being floated across the mains. Such a system can provide at least two hours of electricity supply when the mains fail. This gives ample time to start up a secondary diesel electric set to keep the battery charged.

The facilities desirable on a message-switching computer are a small random-access file for the temporary storing of messages being routed, and a serial-access file, often magnetic tape, for logging messages if they are to be kept for a long period of time. It must, of course, have the facilities for handling all the various types of communication lines used. The core size usually varies from about 40,000 to 200,000 characters, or the equivalent in binary words. Figure 14.5 sketches typical configurations.

In a data-processing system, message switching may be thought of along with the other work and regarded as one element of the complete system. In this case it may be done by a separate line control computer that feeds messages also to the main computers, or one computer may handle message routing and data processing at the same time, in parallel. When message switching was first performed by computer, it was usually done by stand-alone machines dedicated to this function. However as multiprogramming and supervisory program techniques are improving, more and more systems combine message switching with some data processing. If a system handles, say, 2000 messages per day, even a small computer is left with much idle time that can be absorbed in other work.

FUNCTIONS OF THE SYSTEM To explain in detail what can be involved in message switching, the functions that a message-switching computer performs are listed below:

1. The system accepts messages from distant terminals. The terminals are often teleprinters and paper-tape readers, but other devices

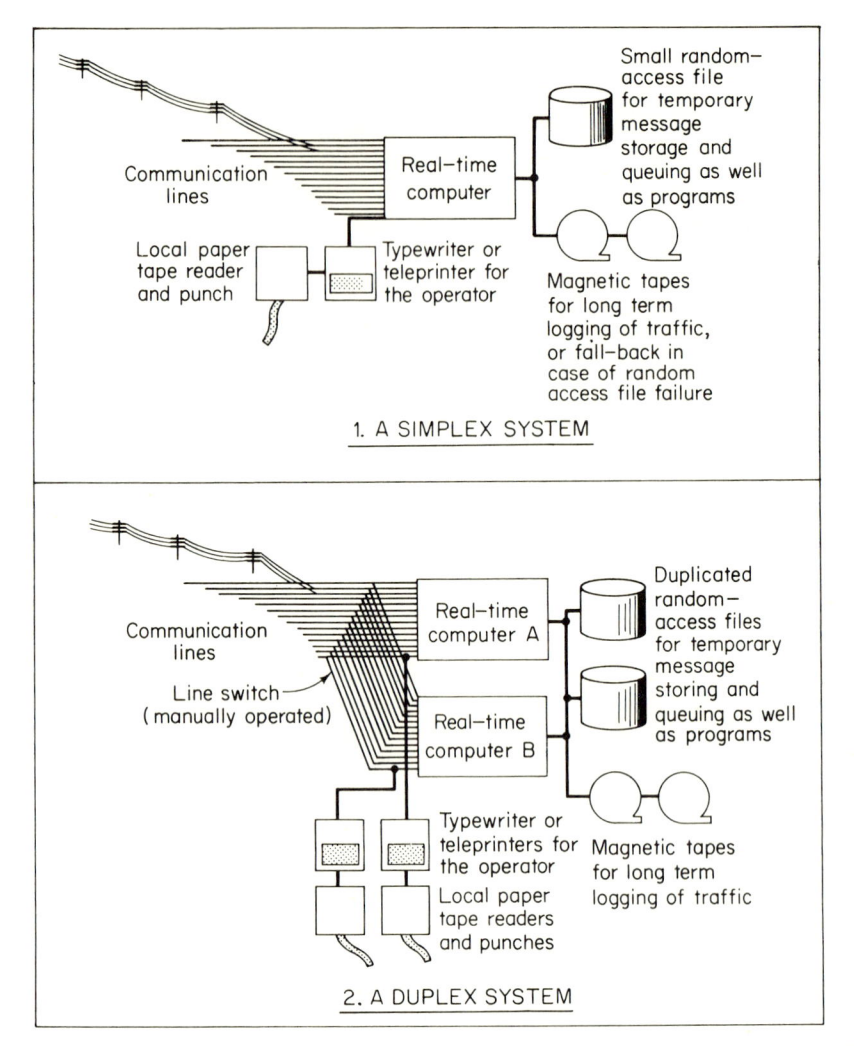

Fig. 14.5. Configurations for computer message switching.

may be used, such as card readers and special input keyboards. The system may also accept messages from other computers.

2. On receipt of a message it analyzes the message's header to determine the destination or destinations to which the message must be sent.

3. The system may analyze the header for a priority indication. This will tell the program that certain messages are urgent. They must

jump any queues of messages and be sent to their destination immediately.
4. It may analyze the header for an indication that some processing of the message is necessary; for example, statistical information from the message may be gathered by the system.
5. The system detects any errors in transmission of the incoming message and requests a retransmission of faulty messages. This retransmission may be automatic.
6. It detects format errors in incoming messages as far as possible. Types of format errors that may be picked up include the following:
    (a) Address invalid. The address to which the message is to be sent is not included in the computer's directory.
    (b) Excessive addresses. There are more than the given maximum number of addresses allowed.
    (c) Incorrect format. An invalid character, for example, the control character, appears in the message in an incorrect location.
    (d) A priority indicator is invalid.
    (e) Originator code error. The address of the originator is not included in the computer's list.
    (f) Incorrect character counts.
7. The system stores all the messages arriving and protects them from possible subsequent damage.
8. It takes messages from one store and transmits them to the desired addresses. One message may be sent to many different addresses. In doing this, it does not destroy the message held in the store. The store is thus a queuing area for messages received and messages waiting to be sent, as well as a file in which messages are retained.
9. The system redirects messages from the store and sends them to the terminals requesting them. It may, for example, be asked to resend all messages from a given serial number or to resend a message with a specified serial number.
10. Systems in use store messages in this manner for several hours or, on some systems, several days. Any message in the store is immediately accessible for this period of time.
11. The system may also maintain a permanent log of messages received. This will probably be done on a relatively inexpensive medium, such as magnetic tape, and not on a random-access file.
12. If messages are sent to a destination at which the terminal is temporarily inoperative, the system intercepts these messages. It may automatically reroute them to alternative terminals that are operative. On the other hand, it may store them until the inoperative terminal is working again.

13. It may intercept messages for other reasons. For example, the system may be programed to send a message to the location of an important person, although he may be moving from one place to another. The person in question leaves his current location with the computer, and the computer diverts messages for him to that location. The system may handle messages on a priority basis. There may be one urgent priority level so that these messages are sent before any others. Some systems have more than one level of priority, priority level 1 being transmitted before priority level 2, priority level 2 being transmitted before priority level 3, and so on. The system may notify the operator in the event that any priority queue becomes too great.

    A simple system may have no priority scheme, messages being handled on a first-in, first-out basis.
14. The system maintains an awareness of the status of lines and terminals. It is programed to detect faulty operation on terminals where possible, to make a log of excessive noise on lines, and to notify its operator when a line goes out. The system maintains records of any faults it detects.
15. On a well-planned system the messages should be given serial numbers by the operator sending them. The computer checks the serial numbers and places new serial numbers on the outgoing messages. When serial numbers are used, the system can be designed to avoid the loss of any message. This is especially important in the event of a computer failure or of a switchover in a duplex system.
16. At given intervals, perhaps once an hour, the system may send a message to each terminal, quoting the serial number of the last message it received from that terminal. The terminal's operator then knows that the switching system is still on the air.
17. The system may conduct a statistical analysis of the traffic that it is handling.
18. It may be programed to bill the users for the messages sent. It may, for example, make a small charge per character sent from each terminal and bill the terminal location appropriately.
19. It produces periodic reports of its operation for its operator. These may include reports on the status of all facilities, error statistics, reports giving the number of messages in each queue, message counts, and so on.

The computer is programed in such a way that the operator may make a modification in its action, for example, change the routing of messages to certain destinations in the event of line outages.

The program to carry out these functions resides mainly in core, although exception routines may be on a drum or disk to be called in when wanted, for example, when emergencies or line outages are being handled. Also in core is a set of tables giving the addresses of the terminals and the lines they are on. This enables the computer to find the correct terminal for a message with a certain destination code. Other tables indicate the status of each line and whether there are messages waiting to be transmitted down it.

The organization of the files of messages is a major design question in this type of system. The queues of messages that are kept on the files vary from one period to another. But messages must be written on the files and retrieved from them in the minimum time. On a disk file in which the read/write heads are physically moved to seek a record, the data must be placed so that the seek times—relatively long in terms of computer speeds—are not too long.

A common method of organizing such files is to allocate areas for each output terminal. As messages are received, they are written sequentially in these areas. These are rather like the pigeonholes for letters in a club or college common room. When the attendant receives letters, he places them in the pigeonholes of the persons who are to receive them. If a pigeonhole becomes full because many letters are received for one person, an overflow area is available. The computer sends messages from these areas to where they are required. If the retrieval of any message is demanded, the machine can search the appropriate "pigeonhole" for the message.

The active area of the file at any one instant is relatively small. As the day proceeds and messages are sent, the active area may move across the file in such a way that the seek times within that area are always small.

VERY LARGE NETWORKS

Where the originating points and destinations of messages are widely separated geographically, the cost of lines for the system will be high. Any means of cutting down the line cost should be considered. For networks above a certain size it becomes economical to have more than one computer for message routing. A large organization is likely to have computers at several distant locations. Commonly these are at centers of the organization's operation and so could conveniently handle message routing. Some firms have used such points as relay stations for message routing. In Fig. 14.6 computer A is the main message-switching machine with the capacity to store a day's traffic on a random-access file and to perform all the functions outlined above. If computer A had communication lines to all the terminals shown in the diagram with, say, not more than three locations on one line, to avoid overloading the lines, the lines would cost much more than those shown. Computers B, C, D, and E are needed in places shown for other work. They are, therefore, made to perform a dual function, and part of the day they relay message traffic as well as doing their other work.

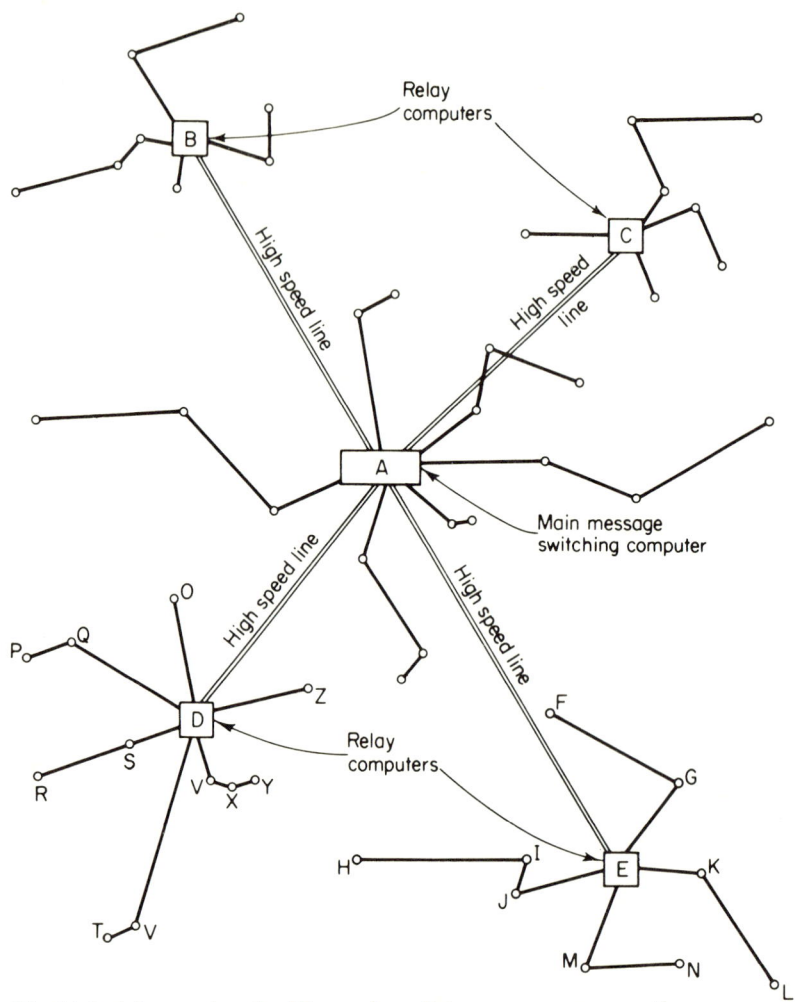

Fig. 14.6. A large network with very long distances may use more than one computer for message routing.

Computer A is a system dedicated to message switching. Computers B, C, D, and E are not dedicated and do not carry out all the functions that A performs. They merely pass on communication-line traffic as a secondary job, just as a computer may do a tape-to-printer or card-to-tape job secondary to its main processing. Any retrieval of past messages or interception of messages is done by A, not by B, C, D, and E. With this system, if location K wishes to send a message to location Z, the message is passed by computer E down the high-speed line to computer A. A determines its routing from that point on and carries out any processing that may be needed on it. A sends it down the high-speed line to computer D, which routes it to terminal Z. If location K wishes to send a message to location G on this system, the message may also have to go via A. This gives the high-speed lines a higher load than if E itself switched the message to G. For this reason, it may be economical to use larger programs in E that carry out all the functions described above if the distances involved are very large, as, for example, on a worldwide network. This will probably increase the cost of E, as it will need more core storage and a random-access file capable of storing the traffic handled by E for a period of some hours. On the other hand, the most economical solution may be to have a simple routing program in E and send the message to A and back.

Communication links of this type between the data-processing centers in an organization can have many uses. When one computer becomes overloaded, data or programs can be transmitted to an alternative machine. One machine may keep a central file of data that can be used by other machines. Persons at any of the terminals can have direct access to distant central files or distant computers. A location with no computer can transmit programs to be executed to a distant machine.

If the message-switching machine is duplexed, the off-line computer of the pair may be used to give a service to the organization and may be used as a standby in case of failure of other distant computers. If a computer fails and cannot perform some vital work, the job may be transmitted to the switching center.

Data-processing systems and message-switching facilities can thus become very much interlinked. Message-switching facilities can be an integral part of larger real-time systems or can operate independently of other data processing. It is apparent that many large organizations need an integrated network of computers and data-transmission facilities.

A PUBLIC NETWORK

An ingenious form of public transmission facility has been designed in England using message switching computers. This system would enable entirely incompatible machines to communicate. Any machine on the network could send data to any other. The data would be formatted into packages which are passed through the network in milliseconds, from one mode to another like a hot potato. This concept could also be used in corporate transmission networks. It is described in *Future Developments in Telecommunications* by James Martin, Prentice-Hall Inc., 1971.

15 DISTRIBUTED INTELLIGENCE

In a telecommunication-based system, as we have seen, many operations must be performed on data besides the execution of the application programs that process it. A variety of housekeeping and system-control operations have been listed. There are various places in the system network where these operations could be performed. A commonly used approach has been to let the central computer control the entire network and perform all, or almost all, of the housekeeping operations. This is certainly not the only approach, and it may not be the best approach, particularly when the networks in question become larger or more complex.

It may not even be desirable, as we have seen, to carry out all of the *application* programs on the central computer. On a large network it may be more economical to have small computers at peripheral or junctional points, and use these to reduce drastically the amount of data that need be sent to and from the distant center. There are several ways in which the "intelligence" of the system can be distributed throughout the network, rather than all residing in the central computer. This chapter discusses these, and their pros and cons.

LOCATIONS FOR SUBSYSTEMS Subsystems in which logic operations are performed could be in a variety of different places in the network. Figure 15.1 shows six places on a typical large network where "intelligence" can reside:

1. *In the Central Computer, A*

On many systems today the central computer handles everything. The network may then contain no other expensive components. Relatively simple and inexpensive terminals may be used. The lines will be constantly

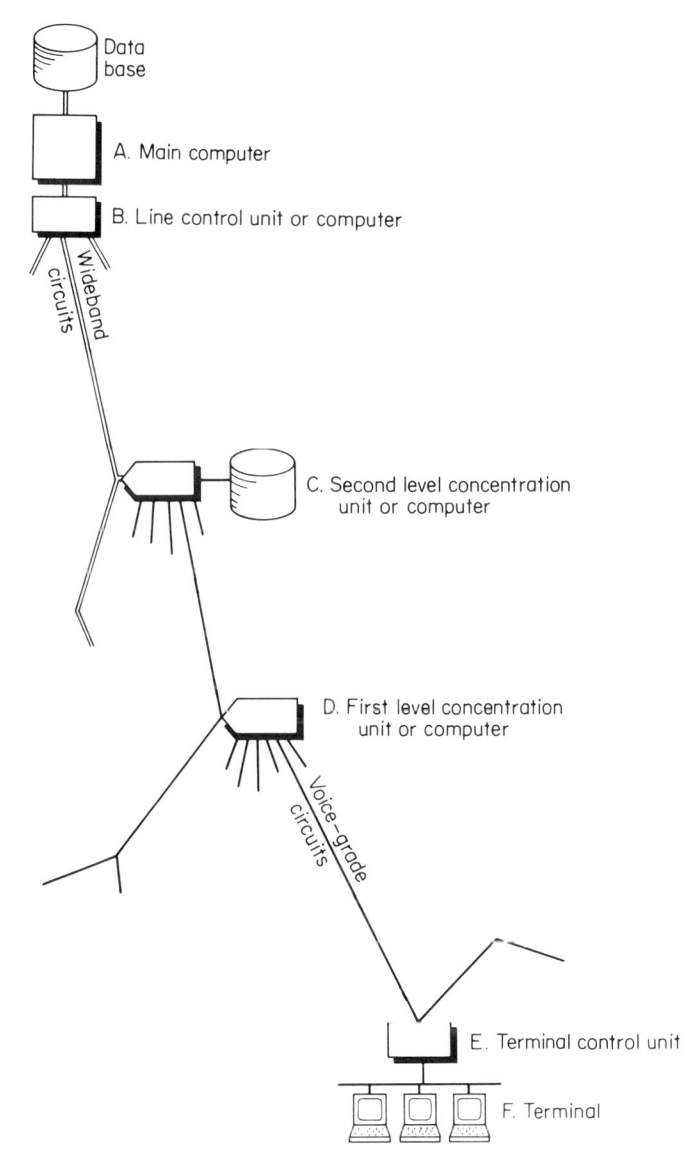

Fig. 15.1. Six places where "intelligence" can reside in a large network.

in use scanning the terminals. There may be only one terminal per line. Where the distances are very short, this may be the most economical design.

2. *In a Line Control Unit or Line Control Computer, B*

Many functions are necessary to control a terminal network. If the host computer performs all of these itself, it will be constantly interrupting its main processing, and many machine cycles will be needed for line control. Some of the line control functions may be performed by a separate line control unit. In some systems all of them are performed by a separate and specialized computer. The balance of which functions are performed by a line control unit, which by the host computer hardware, and which by its software, vary widely from system to system. Some *application* programs could be performed by the subsystem computer, for example, accuracy checking and message logging.

3. *In the Network Concentrators, C and D*

The concentrators may take a variety of different forms, as discussed in Chapter 12. They may be relatively simple machines with unchangeable logic. They may have wired-in logic, part or all of which can be changed by an engineer. They may be microprogramed and hence flexible in possible means of use. Or they may be stored-program computers, sometimes designed solely for concentration but sometimes also capable of other operations and equipped with files, high-speed printers, and other input-output equipment.

Sometimes a computer positioned in a network similarly to the concentrators in Fig. 15.1 may be a real-time system itself, or at least an on-line data-processing system. One may, for example, have a terminal network in which this computer handles most of the responses to terminals, and occasionally passes transactions onward to the main computer. It may pass large jobs on to the main computer, or it may pass transactions on when they need to obtain information from, or update, the files kept centrally at the main computer.

4. *In the Terminal Control Unit, E*

Again, terminal control units differ greatly in their logical complexity. They range from a simple "stunt box" to a stored-program computer. They may control one terminal or many. They may contain elaborate logic capable of editing, formatting, error correction, generating curves on graphic terminals, and so on, or they may do no more than convert the characters sent into appropriate mechanical actions, including carriage returns and page skips. They may contain a large storage for buffering and manipulating data, or they may have no storage at all.

As with the concentrator, the terminal control unit may be constructed to carry out many logical functions that would otherwise be done at, or closer to, the center of the network. If it is microprogramed or a stored-program machine, it may carry out functions unique to that application. It may generate messages to be displayed at the terminal. It may operate conversational routines to collect information from the terminal operator and transmit it only when a complete transaction has been assembled. Like the concentrator, it may have a small file unit for assisting in this.

5. *In the Terminal, F*

The terminal itself may carry out certain logic operations ranging from simple functions, such as accumulating totals in a terminal handling financial transactions, to operations involving intricate logic. In the years to come we will hear much discussion of "intelligent terminals." Where several terminals share a control unit, it will normally be better to put all common functions in the control unit.

DUPLICATION OF HARDWARE Distributing the logic functions throughout the network increases the amount of hardware required. The lower in Fig. 15.1 a function is performed, from B to F, the more duplication of hardware is necessary. The least expensive approach, if line costs were ignored, would be to have very simple terminals linked directly to the computer center, where the main computer or line control computer handles all of the necessary logic. Many systems have one teletypewriter per line linked to the computer center in this way.

As we have seen, distributing the logic throughout the network can lower the communication line cost. If there is only one teleprinter per line, the total network cost will be higher than with several per line, but in the latter case the added line control functions will increase the hardware cost. Some large commercial systems with transactions utilizing a control data base have very costly line requirements, needing a wideband link from the concentrators to computer center as in Fig. 15.1. When one examines their transactions and their responses, it is often the case that only a small fraction of them need either the power of the central machine or its data base. Every element in their man-machine conversation involves transmission to the computer center and a response from there. If this relatively simple conversation handling could be programed at the concentrator level, the expensive wideband link might not be needed.

The length of the lines determines the best way to organize them. In a geographically dispersed system the lines are expensive, so there is a strong argument for distributed intelligence to lower the line cost. If, on the other hand, the lines are short, the most economical design may be that with simple terminals and the logic of line control and housekeeping being done at the computer center. A system with terminals in the Wall Street area only may have one terminal per line with no elaborate network logic. The same system with terminals throughout the greater New York area may have multidrop lines and terminal control units that buffer, edit, and error-check the transactions. The same system with terminals across several states may use concentrators and possibly stored-program logic at either the concentrator or the terminal control unit to lessen the data flowing on the lines, and hence to lower the line requirements. It is interesting to note that to meet all of these possibilities a manufacturer would need to have different sets of data transmission hardware and software in his product line.

In the few years ahead the cost of both fixed and stored-program logic can be expected to drop rapidly, especially with the increasing use of *large-scale integration* circuitry. The cost per bit of data transmission will drop at a much slower rate. The economic balance is, therefore, changing in favor of logic distributed throughout the network. (Data transmission costs may drop drastically with new tariffs based on pulse code modulation carriers, but even considering this, the logic cost should be still much lower.) Large-scale integration becomes particularly worthwhile when a very large number of one kind of machine can be sold. This is likely to be true in the terminal and control unit field and so adds to the argument for distributed logic.

FUNCTIONS TO BE PERFORMED   Let us review the functions that will be performed on a data transmission network and consider which of them might be distributed:

1. *Generation of Bits to be Sent and Interpretation of Bits Received*

The terminal generates the bits to be sent, including start and stop bits if asynchronous transmission is being used. It translates the bits received into appropriate mechanical action. In the simplest form of network this would be the only function (other than data set functions) carried on away from the computer center.

2. *Functions Associated with Multidrop Operation*

When there are several terminals on one line, the functions discussed in Chapter 9 become necessary. In a polled system the terminal or control unit must recognize its address on messages sent to it, and ignore others. It must respond logically to polling, go-ahead, and end-of-transmission signals. These functions can be carried out by a unit that services one terminal or several.

3. *Functions Associated with Error Control and Retransmission*

For the system to recognize transmission errors, the transmitting unit must compose redundant bit patterns or characters that can be checked by the receiving unit. The receiving unit responds, saying whether or not it detected an error. The transmitting unit must recognize this response and have a means of retransmitting the item if there was an error. In order to retransmit the item it is usually desirable to store it until it is known to have been received correctly. Some terminals simply notify their operator when an error is detected and the operator re-enters the item. However, many retransmit automatically. If it was transmitted from paper tape or card, this may involve merely rereading the item. Otherwise, some means of storing the item is needed.

4. *Functions Associated with Buffering, Message Assembly, and Synchronous Transmission*

Storage for the transaction may be needed for error retransmission. It is also used for buffering transactions so that they may be transmitted at a high line speed. Suppose that the line speed is 4800 bits per second, and that we want to put as many terminals as possible on one line. Neither the keyboard operator nor the various reading devices function at this speed; therefore, a buffer is used in which the transaction is assembled and a header and end-of-message indicator added. The transaction is sent from the buffer when the line control machine gives the instruction. Usually synchronous transmission is used, in which case the sending unit must also transmit the required synchronization pattern. The converse functions are needed when the unit receives data. It must recognize the synchronization pattern and so store the item correctly, then cause the information part of the transaction to be printed or displayed.

Again, the buffering unit may handle several terminals in one location. If this is so, then storage allocation mechanisms are needed. It becomes economical to allocate storage dynamically, as was described in Chapter 7. The complexity is building up.

5. *Functions Associated with Concentration*

The 4800-bit-per-second line above may not extend to the terminal location itself, but only to a concentrator. If a wideband line is used on a system without high-speed terminals (such as magnetic tape or disk transmission units), then it usually extends only as far as the concentration device. The functions above in 4, and possibly those in 2, may be performed only in the concentrator. Often, to keep the terminal cost down, the terminals are unbuffered and asynchronous transmission is used to the concentrator. The concentrator then assembles the asynchronous characters into messages that can be transmitted onward synchronously. The concentrator is now attending to two different types of lines, both of which need control for incoming and outgoing transactions.

Its logic is further complicated if the lines are multidrop. If the line from the computer to the concentrator is multidrop, the concentrator must be enabled to handle the functions in 2, above. If the lines on the terminal side of the concentrator are multidrop, this complicates its operation considerably because it will normally have to handle polling on these lines. It must have a list of the sequence in which terminals are to be polled, and must receive and send all of the messages needed for polled line control. It must be possible to change the polling list when terminals are added or go out of operation.

6. *Functions Associated with Security*

The two main security functions are first, to lessen the likelihood of effective wire-tapping, and second, to prevent unauthorized access to the system or files. The former may require the scrambling and unscrambling of transmitted data at the terminal control unit. The safer the security must be, the greater the logic needed for this. Prevention of unauthorized access may need a physical lock on the terminal or a programed lock in the system. The operator may have to type in a password, insert an identity card, or, in years to come, use a voice-print. The logic operations associated with these functions may be at the terminal control unit or at the computer center. For a complex operation such as voice-print recognition, a specialized unit separate from the main computer, though probably at the computer center, may be the best solution.

7. *Functions Associated with Message Switching*

In some networks not all of the messages traveling on the lines need to go to the central computer. A message-switching machine may separate them, store them, and route them to the requisite locations, as described in Chapter 14. Sometimes this is done at the computer center. Sometimes, however, it is economical to locate the message-switching function in other parts of the network. Airlines, for example, have message-switching systems that in some cases are located close to the computer handling reservations, and in other cases are in different locations, filtering out the reservation messages and transmitting them to the reservations computer.

8. *Functions Associated with Editing and Formatting*

A terminal operator may wish to edit his input at the terminal. He may, for example, make mistakes and want to backspace and correct them. He may want to insert words or figures into a format laid out on the terminal screen by the computer. On many systems the code for doing this, for example, two backspace characters followed by two replacement characters, travels all the way to the central computer and this computer does the editing. Some terminal control units, however, have editing logic in them. The edited message may, for example, appear on a terminal screen and be checked by the operator before it is transmitted. Editing functions could also take place in a concentrator as part of the work of assembling a message for synchronous transmission.

Editing a message before transmission reduces slightly the number of characters to be transmitted. The reduction is rarely enough to pay for the added control unit logic that is required. On the other hand, logic for formatting the messages *from* the computer may bring a very substantial saving. Suppose, for the sake of example, that it is desired to display on the terminal screen tables or statements such as that in Fig. 15.2. If the terminal or control unit has no formatting logic, then many *blank* characters

	A R JENKINS, 397 E 34 ST, APT 19B, NEW YORK 10017 073-2-037948			
DATE	PARTICULARS	PAYMENT	RECEIPTS	BALANCES
9.24	OPENING BALANCE			2338.66
9.26			956.60	3295.26
9.30	INTEREST CHARGES	63.85		3231.68
10.5	INVESTMENTS BOUGHT	1218.00		2013.68
10.5	ALREADY ADVISED		993.87	3007.55
10.15		265.00		2742.55
10.16		44.00		2698.55
10.16		2600.00		98.55
10.22		100.00		1.45
10.23	CHARGES	2.10		3.55

Figure 15.2

must be transmitted in order to format the display in a readable manner. Figure 15.2 would need more blanks than other characters. If its data could be sent in a compressed form, with formatting characters but no blanks, then the number of characters transmitted would be almost halved. At the terminal or control unit (or possibly concentrator) the formatting characters would be used to expand the data into a neat display format. Here again, we have a tradeoff between distributed logic hardware and communication line costs.

Formatting logic saves buffer storage as well as transmission time. The Sanders 620 Data Display System terminal, for example, has 2000 character positions on its screen, but only 1000 characters of buffer storage. Formatting characters among the 1000 stored cause the display to be tabulated neatly on the 2000 available screen positions.

9. *The Generation of Responses*

As we discussed in Chapter 13, some applications have many standard responses that could be generated peripherally rather than being transmitted over long lines.

Today such a scheme could be programed into a stored-program terminal control or concentrator. With this and good formatting logic a relatively small number of characters transmitted can produce an elaborate display. Perhaps in the future it may be built into the hardware. It may make sense to use a small, cheap, slow disk storage at the control unit or concentrator location. Another approach that has been used effectively is to have a means of displaying frames of colored film at the terminal. A code from the computer selects the frame that is to be displayed.

10. *Functions Associated with Man-Machine Conversation*

For the use of real-time terminals to become widely accepted we must design conversations as easy to use and unconfusing as possible; however this is generally in conflict with the desire to minimize communication costs. In some commercial systems with terminals throughout the United States it has been found that the design of a good effective conversational language has led to exceedingly high telecommunication costs.

A solution to this dilemma on large networks is to use small remote computers at the concentrator or control unit locations to handle the bulk of the man-machine conversation.

The conversation may, for example, involve the updating or modifying of a record in a centrally kept data base: perhaps the modifying of a complex machine order, or the changing of a passenger's airline itinerary. The peripheral computer in this case needs to communicate with the main, distant computer twice: first to obtain the record in question from the data base, and second to send back the modified record. It communicates with the terminal user many times, however, asking for information one simple step at a time and making accuracy checks as it does it. Once we take this approach, we can afford to design a conversational language that conforms to whatever our behavioral psychologists or simulator experiments indicate.

11. *Specialized Subsidiary Processors*

In the rapidly evolving technology of time-sharing systems there are schools of thought in favor of general-purpose systems and in favor of highly specialized systems and software. Some of the smaller and highly specialized systems have proved to be very efficient and to give an impressively fast response time. It seems likely that we will see more of them and that their cost per user will remain competitive with the larger, more general systems.

A corporation may have a number of such computers in its various locations, capable of meeting a local need. Small specialized systems are relatively easy to implement. These systems may be designed to be capable of passing transactions that they cannot handle onwards to a larger data-processing system. Typical of such systems are ones providing on-line computation with languages such as APL and BASIC, ones providing secretaries with text editing and letter-typing facilities, and small commercial enquiry and data entry systems. Figure 15.3 shows a possible arrangement with a line from the main data-processing center wandering around several small real-time systems that are self-contained for their own functions. General-purpose terminals are used. The computer at the center polls the peripheral computers periodically to see whether they have any work for the main computer. The terminals will often be on dial-up lines to their local computer, though a permanent connection is used if the terminal has sufficiently heavy use. Similarly, the connection from peripheral to main computer will be dial-up if the usage is low.

The network, in fact, may not have a distinguishable main computer. An organization may operate several computer centers primarily dedicated to different types of work, but capable of interchanging some jobs or transactions. The centers may have different files of information available to them and different programs. Transactions are routed to the appropriate location in this complex.

AVAILABILITY Cost and ease of implementation are major considerations in the decision as to what extent logic functions should be dispersed throughout the network. A further consideration, which on some systems is vital, is that of availability.

A much higher availability may be built into the central computer system, by duplexing, than into the concentrators. It is often considered too expensive to duplicate the concentrators, or, if this were done a large part of their economic justification would be lost. Terminals and their control units, on the other hand, are often duplicated because the location needs more than one for its normal load anyway. In this case if a terminal fails

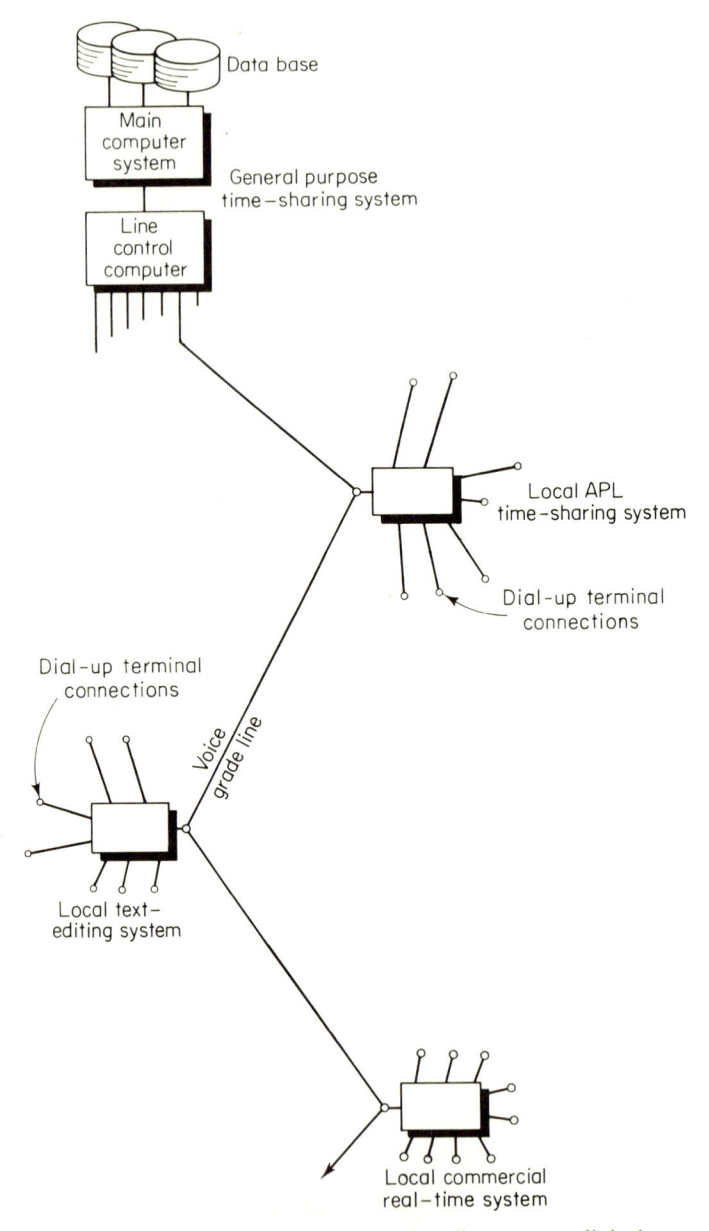

Fig. 15.3. Small, local, special-purpose, on-line systems linked to a large central system with wider capabilities and a large data base. The nodes of the network have become self-contained systems. If the path to the main computer is used only a small part of the day, the peripheral computers may dial up the main center rather than use a leased line.

the location is not cut off because the other terminals are used. The concentrator, then, might be the weak link in the chain. For this reason some means of bypassing the concentrator might be devised. This may be a reason for some systems not putting too many functions into the concentrator. It may be desirable to make it a simple, and hence reliable, machine.

There is a strong argument for designing concentrators with far higher reliability than today's typical data-processing hardware. Highly reliable machines can be built when the cycle speed used is not the fastest, as has been proved by such machines as the Bell System ESS 1.

The concentrator, then, may not have a high level of "intelligence" built into it in a system in which reliability is vital. The power to handle the man-machine conversation may be provided at the terminal location, where it is easier to bypass it in an emergency. The same functions may be performed at the central computer with a temporary penalty in line loading.

GLOSSARY

There is little point in redefining the wheel, and where useful the definitions in this glossary have been taken from other recognized glossaries.

A suffix "2" after a definition below indicates that it is the CCITT definition, published in *List of Definitions of Essential Telecommunication Terms*, International Telecommunication Union, Geneva.

A suffix "1" after a definition below indicates that the definition is taken from the *Data Communications Glossary*, International Business Machines Corporation, Poughkeepsie, 1967 (Manual number C20-1666).

Address. A coded representation of the destination of data, or of their originating terminal. Multiple terminals on one communication line, for example, must have unique addresses. Telegraph messages reaching a switching center carry an address before their text to indicate the destination of the message.

Alphabet (telegraph or data). A table of correspondence between an agreed set of characters and the signals which represent them. (2).

Alternate routing. An alternative communications path used if the normal one is not available. There may be one or more possible alternative paths.

Amplitude modulation. One of three ways of modifying a sine wave signal in order to make it "carry" information. The sine wave, or "carrier," has its amplitude modified in accordance with the information to be transmitted.

Analog data. Data in the form of *continuously variable* physical quantities. (Compare with **Digital data**.) (1).

Analog transmission. Transmission of a continuously variable signal as opposed to a discretely variable signal. Physical quantities such as temperature are continuously variable and so are described as "analog." Data characters, on the other hand, are coded in discrete separate pulses or signal levels, and are referred to as "digital." The normal way of transmitting a telephone, or voice, signal has been analog; but now digital encoding (using PCM) is coming into use over trunks.

Application program. The working programs in a system may be classed as *application programs* and *supervisory programs*. The application programs are the main data-processing programs. They contain no input-output coding except in the form of macroinstructions that transfer control to the supervisory programs. They are usually unique to one type of application, whereas the supervisory programs could be used for a variety of different application types. A number of different terms are used for these two classes of program.

ARQ (Automatic Request for Repetition). A system employing an error-detecting code and so conceived that any false signal initiates a repetition of the transmission of the character incorrectly received. (2).

ASCII (American Standard Code for Information Interchange). Usually pronounced "ask'-ee." An eight-level code for data transfer adopted by the American Standards Association to achieve compatibility between data devices. (1).

Asynchronous transmission. Transmission in which each information character, or sometimes each word or small block, is individually synchronized, usually by the use of start and stop elements. The gap between each character (or word) is not of a necessarily fixed length. (Compare with **Synchronous transmission.**) Asynchronous transmission is also called *start-stop transmission*.

Attended operation. In data set applications, individuals are required at both stations to establish the connection and transfer the data sets from talk (voice) mode to data mode. (Compare **Unattended operation.**) (1).

Attenuation. Decrease in magnitude of current, voltage, or power of a signal in transmission between points. May be expressed in decibels. (1).

Atenuation equalizer. (*See* **Equalizer.**)

Audio frequencies. Frequencies that can be heard by the human ear (usually 30 to 20,000 cycles per second). (1).

Automatic calling unit (ACU). A dialing device supplied by the communications common carrier, which permits a business machine to automatically dial calls over the communication networks. (1).

Automatic dialing unit (ADU). A device capable of automatically generating dialing digits. (Compare with **Automatic calling unit.**) (1).

Bandwidth. The range of frequencies available for signaling. The difference expressed in cycles per second (hertz) between the highest and lowest frequencies of a band.

Baseband signaling. Transmission of a signal at its original frequencies, i.e., a signal not changed by modulation.

Baud. Unit of signaling speed. The speed in bauds is the number of discrete conditions or signal events per second. (This is applied only to the actual signals on a communication line.) If each signal event represents only one bit condition, baud is the same as bits per second. When each signal event represents other than one bit (e.g., see **Dibit**), baud does not equal bits per second. (1).

Baudot code. A code for the transmission of data in which five equal-length bits represent one character. This code is used in most DC teletypewriter machines where 1 start element and 1.42 stop elements are added. (See page 109.) (1).

Bel. Ten decibels, q.v.

BEX. Broadband exchange, q.v.

Bias distortion. In teletypewriter applications, the uniform shifting of the beginning of all marking pulses from their proper positions in relation to the beginning of the start pulse. (1).

Bias distortion, asymmetrical distortion. Distortion affecting a two-condition (or binary) modulation (or restitution) in which all the significant conditions have longer or shorter durations than the corresponding theoretical durations. (2).

Bit. Contraction of "binary digit," the smallest unit of information in a binary system. A bit represents the choice between a mark or space (one or zero) condition.

Bit rate. The speed at which bits are transmitted, usually expressed in bits per second. (Compare with **Baud**.)

Broadband. Communication channel having a bandwidth greater than a voice-grade channel, and therefore capable of higher-speed data transmission. (1).

Broadband exchange (BEX). Public switched communication system of Western Union, featuring various bandwidth FDX connections. (1).

Buffer. A storage device used to compensate for a difference in rate of data flow, or time of occurrence of events, when transmitting data from one device to another. (1).

Cable. Assembly of one or more conductors within an enveloping protective sheath, so constructed as to permit the use of conductors separately or in groups. (1).

Carrier. A continuous frequency capable of being modulated, or impressed with a second (information carrying) signal. (1).

Carrier, communications common. A company which furnishes communications services to the general public, and which is regulated by appropriate local, state, or federal agencies. The term strictly includes truckers and movers, bus lines, and airlines, but is usually used to refer to telecommunication companies.

Carrier system. A means of obtaining a number of channels over a single path by modulating each channel on a different carrier frequency and demodulating at the receiving point to restore the signals to their original form.

Carrier telegraphy, carrier current telegraphy. A method of transmission in which the signals from a telegraph transmitter modulate an alternating current. (2).

Central office. The place where communications common carriers terminate customer lines and locate the switching equipment which interconnects those lines. (Also referred to as an *exchange, end office,* and *local central office.*)

Chad. The material removed when forming a hole or notch in a storage medium such as punched tape or punched cards.

Chadless tape. Perforated tape with the chad partially attached, to facilitate interpretive printing on the tape.

Channel. 1. (CCITT and ASA standard) A means of one-way transmission. (Compare with **Circuit.**)
2. (Tariff and common usage) As used in the tariffs, a path for electrical transmission between two or more points without common-carrier-provided terminal equipment. Also called *circuit, line, link, path,* or *facility.* (1).

Channel, analog. A channel on which the information transmitted can take any value between the limits defined by the channel. Most voice channels are analog channels.

Channel, voice-grade. A channel suitable for transmission of speech, digital or analog data, or facsimile, generally with a frequency range of about 300 to 3400 cycles per second.

12-channel group (of carrier current system). The assembly of 12 telephone channels, in a carrier system, occupying adjacent bands in the spectrum, for the purpose of simultaneous modulation or demodulation. (2).

Character. Letter, figure, number, punctuation or other sign contained in a message. Besides such characters, there may be characters for special symbols and some control functions. (1).

Characteristic distortion. Distortion caused by transients which, as a result of the modulation, are present in the transmission channel and depend on its transmission qualities.

Circuit. A means of both-way communication between two points, comprising associated "go" and "return" channels. (1).

Circuit, four-wire. A communication path in which four wires (two for each direction of transmission) are presented to the station equipment. (1).

Circuit, two-wire. A metallic circuit formed by two conductors insulated from each other. It is possible to use the two conductors as either a one-way transmission path, a half-duplex path, or a duplex path. (1).

Common carrier. (*See* **Carrier, communications common.**)

Compandor. A compandor is a combination of a compressor at one point in a communication path for reducing the volume *range* of signals, followed by an expandor at another point for restoring the original volume range. Usually its purpose is to improve the ratio of the signal to the interference entering in the path between the compressor and expandor. (2).

Compressor. Electronic device which compresses the volume range of a signal, used in a compandor (q.v.). An "expandor" restores the original volume range after transmission.

Conditioning. The addition of equipment to a leased voice-grade channel to provide minimum values of line characteristics required for data transmission. (1).

Contention. This is a method of line control in which the terminals request to transmit. If the channel in question is free, transmission goes ahead; if it is not free, the terminal will have to wait until it becomes free. The queue of contention requests may be built up by the computer, and this can either be in a prearranged sequence or in the sequence in which the requests are made.

Control character. A character whose occurrence in a particular context initiates, modifies, or stops a control operation—e.g., a character to control carriage return. (1).

Control mode. The state that all terminals on a line must be in to allow line control actions, or terminal selection to occur. When all terminals on a line are in the control mode, characters on the line are viewed as control characters performing line discipline, that is, polling or addressing. (1).

Cross-bar switch. A switch having a plurality of vertical paths, a plurality of horizontal paths, and electromagnetically operated mechanical means for interconnecting any one of the vertical paths with any of the horizontal paths.

Cross-bar system. A type of line-switching system which uses cross-bar switches.

Cross talk. The unwanted transfer of energy from one circuit, called the *disturbing* circuit, to another circuit, called the *disturbed* circuit. (2).

Cross talk, far-end. Cross talk which travels along the disturbed circuit in the same direction as the signals in that circuit. To determine the far-end cross talk between two pairs, 1 and 2, signals are transmitted on pair 1 at station A, and the level of cross talk is measured on pair 2 at station B. (1).

Cross talk, near-end. Cross talk which is propagated in a disturbed channel in the direction opposite to the direction of propagation of the current in the distrubing channel. Ordinarily, the terminal of the disturbed channel at which the near-end cross talk is present is near or coincides with the energized terminal of the disturbing channel. (1).

Dataphone. Both a service mark and a trademark of AT & T and the Bell System. As a service mark it indicates the transmission of data over the telephone network. As a trademark it identifies the communications equipment furnished by the Bell System for data communications services. (1).

Data set. A device which performs the modulation/demodulation and control functions necessary to provide compatibility between business machines and communications facilities. (*See also* **Line adapter, Modem,** *and* **Subset**.) (1).

Data-signaling rate. It is given by $\sum_{i=1}^{m} \frac{1}{T_i} \log_2 n_i$, where m is the number of parallel channels, T is the minimum interval for the ith channel, expressed in seconds, n is the number of significant conditions of the modulation in the ith channel. Data-signaling rate is expressed in bits per second. (2).

Dataspeed. An AT&T marketing term for a family of medium-speed paper tape transmitting and receiving units. Similar equipment is also marketed by Western Union. (1).

DDD. (*See* **Direct distance dialing,** q.v.)

Decibel (db). A tenth of a bel. A unit for measuring relative strength of a signal parameter such as power, voltage, etc. The number of decibels is ten times the logarithm (base 10) of the ratio of the measured quantity to the reference level. The reference level must always be indicated, such as 1 milliwatt for power ratio. (1). See Fig. 9.5.

Delay distortion. Distortion occurring when the envelope delay of a circuit or system is not constant over the frequency range required for transmission.

Delay equalizer. A corrective network which is designed to make the phase delay or envelope delay of a circuit or system substantially constant over a desired frequency range. (*See* **Equalizer**.) (1).

Demodulation. The process of retrieving intelligence (data) from a modulated carrier wave; the reverse of modulation. (1).

Diagnostic programs. These are used to check equipment malfunctions and to pinpoint faulty components. They may be used by the computer engineer or may be called in by the supervisory programs automatically.

Diagnostics, system. Rather than checking one individual component, system diagnostics utilize the whole system in a manner similar to its operational running. Programs resembling the operational programs will be used rather than systematic programs that run logical patterns. These will normally detect overall system malfunctions but will not isolate faulty components.

Diagnostics, unit. These are used on a conventional computer to detect faults in the various units. Separate unit diagnostics will check such items as arithmetic circuitry, transfer instructions, each input-output unit, and so on.

Dial pulse. A current interruption in the DC loop of a calling telephone. It is produced by the breaking and making of the dial pulse contacts of a calling telephone when a digit is dialed. The loop current is interrupted once for each unit of value of the digit. (1).

Dial-up. The use of a dial or pushbutton telephone to initiate a station-to-station telephone call.

Dibit. A group of two bits. In four-phase modulation, each possible dibit is encoded as one of four unique carrier phase shifts. The four possible states for a dibit are 00, 01, 10, 11.

Differential modulations. A type of modulation in which the choice of the significant condition for any signal element is dependent on the choice for the previous signal element. (2).

Digital data. Information represented by a code consisting of a sequence of discrete elements. (Compare with **Analog data**.) (1).

Digital signal. A discrete or discontinuous signal; one whose various states are discrete intervals apart. (Compare with **Analog transmission**.) (1).

Direct distance dialing (DDD). A telephone exchange service which enables the telephone user to call other subscribers outside his local area without operator assistance. In the United Kingdom and some other countries, this is called *Subscriber Trunk Dialing* (STD).

Disconnect signal. A signal transmitted from one end of a subscriber line or trunk to indicate at the other end that the established connection should be disconnected. (1).

Distortion. The unwanted change in waveform that occurs between two points in a transmission system. (1).

Distributing frame. A structure for terminating permanent wires of a telephone central office, private branch exchange, or private exchange, and for permitting the easy change of connections between them by means of cross-connecting wires. (1).

Double-current transmission, polar direct-current system. A form of binary telegraph transmission in which positive and negative direct currents denote the significant conditions. (2).

Drop, subscriber's. The line from a telephone cable to a subscriber's building. (1).

Duplex transmission. Simultaneous two-way independent transmission in both directions. (Compare with **Half-duplex transmission**. Also called *full-duplex transmission*.) (1).

Duplexing. The use of duplicate computers, files or circuitry, so that in the event of one component failing an alternative one can enable the system to carry on its work.

Echo. An echo is a wave which has been reflected or otherwise returned with sufficient magnitude and delay for it to be perceptible in some manner as a wave distinct from that directly transmitted.

Echo check. A method of checking data transmission accuracy whereby the received data are returned to the sending end for comparison with the original data.

Echo suppressor. A line device used to prevent energy from being reflected back (echoed) to the transmitter. It attenuates the transmission path in one direction while signals are being passed in the other direction. (1).

End distortion. End distortion of start-stop teletypewriter signals is the shifting of the end of all marking pulses from their proper positions in relation to the beginning of the start pulse.

End office. (*See* **Central office**.)

Equalization. Compensation for the attenuation (signal loss) increase with frequency. Its purpose is to produce a flat frequency response while the temperature remains constant. (1).

Equalizer. Any combination (usually adjustable) of coils, capacitors, and/or resistors inserted in transmission line or amplifier circuit to improve its frequency response. (1).

Equivalent four-wire system. A transmission system using frequency division to obtain full-duplex operation over only one pair of wires. (1).

Error-correcting telegraph code. An error-detecting code incorporating sufficient additional signaling elements to enable the nature of some or all of the errors to be indicated and corrected entirely at the receiving end.

Error-detecting and feedback system, decision feedback system, request repeat system, ARQ system. A system employing an error-detecting code and so arranged that a signal detected as being in error automatically initiates a request for retransmission of the signal detected as being in error. (2).

Error-detecting telegraph code. A telegraph code in which each telegraph signal conforms to specific rules of construction, so that departures from this construction in the received signals can be automatically detected. Such codes necessarily require more signaling elements than are required to convey the basic information.

ESS. (Electronic Switching System). Bell System term for computerized telephone exchange. ESS 1 is a central office. ESS 101 gives private branch exchange (PBX) switching controlled from the local central office.

Even parity check (odd parity check). This is a check which tests whether the number of digits in a group of binary digits is even (even parity check) or odd (odd parity check). (2).

Exchange. A unit established by a communications common carrier for the administration of communication service in a specified area which usually embraces a city, town, or village and its environs. It consists of one or more central offices together with the associated equipment used in furnishing communication service. (This term is often used as a synonym for "central office," q.v.)

Exchange, classes of. Class 1 (*see* **Regional center**); class 2 (*see* **Sectional center**); class 3 (*see* **Primary center**); class 4 (*see* **Toll center**); class 5 (*see* **End office**).

Exchange, private automatic (PAX). A dial telephone exchange that provides private telephone service to an organization and that does *not* allow calls to be transmitted to or from the public telephone network.

Exchange, private automatic branch (PABX). A private automatic telephone exchange that provides for the transmission of calls to and from the public telephone network.

Exchange, private branch (PBX). A manual exchange connected to the public telephone network on the user's premises and operated by an attendant supplied by the user. PBX is today commonly used to refer also to an automatic exchange.

Exchange, trunk. An exchange devoted primarily to interconnecting trunks.

Exchange service. A service permitting interconnection of any two customers' stations through the use of the exchange system.

Expandor. A transducer which for a given amplitude range or input voltages produces a larger range of output voltages. One important type of expandor employs the information from the envelope of speech signals to expand their volume range. (Compare **Compandor**.) (1).

Facsimile (FAX). A system for the transmission of images. The image is scanned at the transmitter, reconstructed at the receiving station, and duplicated on some form of paper. (1).

GLOSSARY

Fail softly. When a piece of equipment fails, the programs let the system fall back to a degraded mode of operation rather than let it fail catastrophically and give no response to its users.

Fall-back, double. Fall-back in which two separate equipment failures have to be contended with.

Fall-back procedures. When the equipment develops a fault the programs operate in such a way as to circumvent this fault. This may or may not give a degraded service. Procedures necessary for fall-back may include those to switch over to an alternative computer or file, to change file addresses, to send output to a typewriter instead of a printer, to use different communication lines or bypass a faulty terminal, etc.

FCC. Federal Communications Commission, q.v.

FD or FDX. Full duplex. (*See* **Duplex.**)

FDM. Frequency-division multiplex, q.v.

Federal Communications Commission (FCC). A board of seven commissioners appointed by the President under the Communication Act of 1934, having the power to regulate all interstate and foreign electrical communication systems originating in the United States. (1).

Figures shift. A physical shift in a teletypewriter which enables the printing of numbers, symbols, upper-case characters, etc. (Compare with **Letters shift.**) (1).

Filter. A network designed to transmit currents of frequencies within one or more frequency bands and to attenuate currents of other frequencies. (2).

Foreign exchange service. A service which connects a customer's telephone to a telephone company central office normally not serving the customer's location. (Also applies to TWX service.) (1).

Fortuitous distortion. Distortion resulting from causes generally subject to random laws (accidental irregularities in the operation of the apparatus and of the moving parts, disturbances affecting the transmission channel, etc.). (2).

Four-wire circuit. A circuit using two pairs of conductors, one pair for the "go" channel and the other pair for the "return" channel. (2).

Four-wire equivalent circuit. A circuit using the same pair of conductors to give "go" and "return" channels by means of different carrier frequencies for the two channels. (2).

Four-wire terminating set. Hybrid arrangement by which four-wire circuits are terminated on a two-wire basis for interconnection with two-wire circuits.

Frequency-derived channel. Any of the channels obtained from multiplexing a channel by frequency division. (2).

Frequency-division multiplex. A multiplex system in which the available transmission frequency range is divided into narrower bands, each used for a separate channel. (2).

Frequency modulation. One of three ways of modifying a sine wave signal to make it "carry" information. The sine wave or "carrier" has its frequency modified in accordance with the information to be transmitted. The frequency function of the modulated wave may be continuous or discontinuous. In the latter case, two or more particular frequencies may correspond each to one significant condition.

Frequency-shift signaling, frequency-shift keying (FSK). Frequency modulation method in which the frequency is made to vary at the significant instants. 1. By smooth transitions: the modulated wave and the change in frequency are continuous at the significant instants. 2. By abrupt transitions: the modulated wave is continuous but the frequency is discontinuous at the significant instants. (2).

FSK. Frequency-shift keying, q.v.

FTS. Federal Telecommunications System.

Full-duplex (FD or FDX) transmission. (*See* **Duplex transmission.**)

Half-duplex (HD or HDX) circuit.
  1. CCITT definition: A circuit designed for duplex operation, but which, on account of the nature of the terminal equipments, can be operated alternately only.
  2. Definition in common usage (the normal meaning in computer literature): A circuit designed for transmission in either direction but not both directions simultaneously.

Handshaking. Exchange of predetermined signals for purposes of control when a connection is established between two data sets.

Harmonic distortion. The resultant presence of harmonic frequencies (due to non-linear characteristics of a transmission line) in the response when a sinusoidal stimulus is applied. (1)

HD or HDX. Half duplex. (*See* **Half-duplex circuit.**)

Hertz (Hz). A measure of frequency or bandwidth. The same as cycles per second.

Home loop. An operation involving only those input and output units associated with the local terminal. (1).

In-house. *See* **In-plant system.**

In-plant system. A system whose parts, including remote terminals, are all situated in one building or localized area. The term is also used for communication systems spanning several buildings and sometimes covering a large distance, but in which no common carrier facilities are used.

International Telecommunication Union (ITU). The telecommunications agency of the United Nations, established to provide standardized communications procedures and practices including frequency allocation and radio regulations on a world-wide basis.

Interoffice trunk. A direct trunk between local central offices.

Intertoll trunk. A trunk between toll offices in different telephone exchanges. (1).

ITU. International Telecommunication Union, q.v.

Keyboard perforator. A perforator provided with a bank of keys, the manual depression of any one of which will cause the code of the corresponding character or function to be punched in a tape. (2).

Keyboard send/receive. A combination teletypewriter transmitter and receiver with transmission capability from keyboard only.

KSR. Keyboard send/receive, q.v.

Leased facility. A facility reserved for sole use of a single leasing customer. (*See also* **private line.**) (1).

Letters shift. A physical shift in a teletypewriter which enables the printing of alphabetic characters. Also, the name of the character which causes this shift. (*Compare* with **Figures shift.**) (1).

Line switching. Switching in which a circuit path is set up between the incoming and outgoing lines. Contrast with message switching (q.v.) in which no such physical path is established.

Link communication. The physical means of connecting one location to another for the purpose of transmitting and receiving information. (1).

Loading. Adding inductance (load coils) to a transmission line to minimize amplitude distortion. (1).

Local exchange, local central office. An exchange in which subscribers' lines terminate. (Also referred to as *end office*.)

Local line, local loop. A channel connecting the subscriber's equipment to the line terminating equipment in the central office exchange. Usually metallic circuit (either two-wire or four-wire). (1).

Longitudinal redundancy check (LRC). A system of error control based on the formation of a block check following preset rules. The check formation rule is applied in the same manner to each character. In a simple case, the LRC is created by forming a parity check on each bit position of all the characters in the block (e.g., the first bit of the LRC character creates odd parity among the one-bit positions of the characters in the block).

Loop checking, message feedback, information feedback. A method of checking the accuracy of transmission of data in which the received data are returned to the sending end for comparison with the original data, which are stored there for this purpose. (2).

LRC. Longitudinal redundancy check.

LTRS. Letters shift, q.v. (*See* **Letters shift.**)

Mark. Presence of signal In telegraph communications a mark represents the closed condition or current flowing. A mark impulse is equivalent to a binary 1. (*See*

Mark-hold. The normal no-traffic line condition whereby a steady mark is transmitted. (Compare with **Space-hold.**) (1).

Mark-to-space transition. The transition, or switching from a marking impulse to a spacing impulse.

Master station. A unit having control of all other terminals on a multipoint circuit for purposes of polling and/or selection. (1).

Mean time to failure. The average length of time for which the system, or a component of the system, works without fault.

Mean time to repair. When the system, or a component of the system, develops a fault, this is the average time taken to correct the fault.

Message reference block. When more than one message in the system is being processed in parallel, an area of storage is allocated to each message and remains uniquely associated with that message for the duration of its stay in the computer. This is called the *message reference block* in this book. It will normally contain the message and data associated with it that are required for its processing. In most systems, it contains an area of working storage uniquely reserved for that message.

Message switching. The technique of receiving a message, storing it until the proper outgoing line is available, and then retransmitting. No direct connection between the incoming and outgoing lines is set up as in line switching (q.v.).

Microwave. Any electromagnetic wave in the radio-frequency spectrum above 890 megacycles per second. (1).

Modem. A contraction of "modulator-demodulator." The term may be used when the modulator and the demodulator are associated in the same signal-conversion equipment. (*See* **Modulation** *and* **Data set**.) (1).

Modulation. The process by which some characteristic of one wave is varied in accordance with another wave or signal. This technique is used in data sets and modems to make business machine signals compatible with communications facilities. (1).

Modulation with a fixed reference. A type of modulation in which the choice of the significant condition for any signal element is based on a fixed reference. (2).

Multidrop line. Line or circuit interconnecting several stations. (Also called *multipoint line*.) (1).

Multiplex, multichannel. Use of a common channel in order to make two or more channels, either by splitting of the frequency band transmitted by the common channel into narrower bands, each of which is used to constitute a distinct channel (frequency-division multiplex), or by allotting this common channel in turn, to constitute different intermittent channels (time-division multiplex). (2).

Multiplexing. The division of a transmission facility into two or more channels either by splitting the frequency band transmitted by the channel into narrower bands, each of which is used to constitute a distinct channel (frequency-division multiplex), or by allotting this common channel to several different information channels, one at a time (time-division multiplexing). (2).

Multiplexor. A device which uses several communication channels at the same time, and transmits and receives messages and controls the communication lines. This device itself may or may not be a stored-program computer.

Multipoint line. (*See* **Multidrop line.**)

Neutral transmission. Method of transmitting teletypewriter signals, whereby a mark is represented by current on the line and a space is represented by the absence of current. By extension to tone signaling, neutral transmission is a method of signaling employing two signaling states, one of the states representing both a space condition and also the absence of any signaling. (Also called *unipolar*. Compare with **Polar transmission**.) (1).

Noise. Random electrical signals, introduced by circuit components or natural disturbances, which tend to degrade the performance of a communications channel. (1).

Off hook. Activated (in regard to a telephone set). By extension, a data set automatically answering on a public switched system is said to go "off hook." (Compare with **On hook**.) (1).

Off line. Not in the line loop. In telegraph usage, paper tapes frequently are punched "off line" and then transmitted using a paper tape transmitter.

On hook. Deactivated (in regard to a telephone set). A telephone not in use is "on hook." (1).

On line. Directly in the line loop. In telegraph usage, transmitting directly onto the line rather than, for example, perforating a tape for later transmission. (*See also* **On-line computer system.**)

On-line computer system. An on-line system may be defined as one in which the input data enter the computer directly from their point of origin and/or output data are transmitted directly to where they are used. The intermediate stages such as punching data into cards or paper tape, writing magnetic tape, or off-line printing, are largely avoided.

Open wire. A conductor separately supported above the surface of the ground—i.e., supported on insulators.

Open-wire line. A pole line whose conductors are principally in the form of open wire.

PABX. Private automatic branch exchange. (*See* **Exchange, private automatic branch.**)

Parallel transmission. Simultaneous transmission of the bits making up a character or byte, either over separate channels or on different carrier frequencies on the channel. (1). The simultaneous transmission of a certain number of signal elements constituting the same telegraph or data signal. For example, use of a code according to which each signal is characterized by a combination of 3 out of 12 frequencies simultaneously transmitted over the channel. (2).

Parity check. Addition of noninformation bits to data, making the number of ones in a grouping of bits either always even or always odd. This permits detection of bit groupings that contain single errors. It may be applied to characters, blocks, or any convenient bit grouping. (1).

Parity check, horizontal. A parity check applied to the group of certain bits from every character in a block. (*See also* **Longitudinal redundancy check.**)

Parity check, vertical. A parity check applied to the group which is all bits in one character. (Also called *vertical redundancy check.*) (1).

PAX. Private automatic exchange. (*See* **Exchange, private automatic.**)

PBX. Private branch exchange. (*See* **Exchange, private branch.**)

PCM. (*See* **Pulse-code modulation.**)

PDM. (*See* **Pulse-duration modulation.**)

Perforator. An instrument for the manual preparation of a perforated tape, in which telegraph signals are represented by holes punched in accordance with a predetermined code. Paper tape is prepared off line with this. (Compare with **Reperforator.**) (2).

Phantom telegraph circuit. Telegraph circuit superimposed on two physical circuits reserved for telephony. (2).

Phase distortion. (*See* **Distortion, delay.**)

Phase equalizer, delay equalizer. A delay equalizer is a corrective network which is designed to make the phase delay or envelope delay of a circuit or system substantially constant over a desired frequency range. (2).

Phase-inversion modulation. A method of phase modulation in which the two significant conditions differ in phase by π radians. (2).

Phase modulation. One of three ways of modifying a sine wave signal to make it "carry" information. The sine wave or "carrier," has its phase changed in accordance with the information to be transmitted.

Pilot model. This is a model of the system used for program testing purposes which is less complex than the complete model, e.g., the files used on a pilot model may contain a much smaller number of records than the operational files; there may be few lines and fewer terminals per line.

Polar transmission. A method for transmitting teletypewriter signals, whereby the marking signal is represented by direct current flowing in one direction and the spacing signal is represented by an equal current flowing in the opposite direction. By extension to tone signaling, polar transmission is a method of transmission employing three distinct states, two to represent a mark and a space and one to represent the absence of a signal. (Also called *bipolar*. Compare with **Neutral transmission.**)

Polling. This is a means of controlling communication lines. The communication control device will send signals to a terminal saying, "Terminal A. Have you anything to send?" if not, "Terminal B. Have you anything to send?" and so on. Polling is an alternative to contention. It makes sure that no terminal is kept waiting for a long time.

Polling list. The polling signal will usually be sent under program control. The program will have in core a list for each channel which tells the sequence in which the terminals are to be polled.

PPM. (*See* **Pulse position modulation.**)

Primary center. A control center connecting toll centers; a class 3 office. It can also serve as a toll center for its local end offices.

Private automatic branch exchange. (*See* **Exchange, private automatic branch.**)

Private automatic exchange. (*See* **Exchange, private automatic.**)

Private branch exchange (PBX). A telephone exchange serving an individual organization and having connections to a public telephone exchange. (2).

Private line. Denotes the channel and channel equipment furnished to a customer as a unit for his exclusive use, without interexchange switching arrangements. (1).

Processing, batch. A method of computer operation in which a number of similar input items are accumulated and grouped for processing.

Processing, in line. The processing of transactions as they occur, with no preliminary editing or sorting of them before they enter the system. (1).

Propagation delay. The time necessary for a signal to travel from one point on a circuit to another.

Public. Provided by a common carrier for use by many customers.

Public switched network. Any switching system that provides circuit switching to many customers. In the U.S.A. there are four such networks: Telex, TWX, telephone, and Broadband Exchange. (1).

Pulse-code modulation (PCM). Modulation of a pulse train in accordance with a code. (2).

Pulse-duration modulation (PDM) (**pulse-width modulation**) (**pulse-length modulation**). A form of pulse modulation in which the durations of pulses are varied. (2).

Pulse modulation. Transmission of information by modulation of a pulsed, or intermittent, carrier. Pulse width, count, position, phase, and/or amplitude may be the varied characteristic.

Pulse-position modulation (PPM). A form of pulse modulation in which the positions in time of pulses are varied, without modifying their duration. (2).

Pushbutton dialing. The use of keys or pushbuttons instead of a rotary dial to generate a sequence of digits to establish a circuit connection. The signal form is usually multiple tones. (Also called *tone dialing, Touch-call, Touch-Tone.*) (1).

Real time. A real-time computer system may be defined as one that controls an environment by receiving data, processing them, and returning the results sufficiently quickly to affect the functioning of the environment at that time.

Reasonableness checks. Tests made on information reaching a real-time system or being transmitted from it to ensure that the data in question lie within a given range. It is one of the means of protecting a system from data transmission errors.

Recovery from fall-back. When the system has switched to a fall-back mode of operation and the cause of the fall-back has been removed, the system must be restored to its former condition. This is referred to as *recovery from fall-back*. The recovery process may involve updating information in the files to produce two duplicate copies of the file.

Redundancy check. An automatic or programmed check based on the systematic insertion of components or characters used especially for checking purposes. (1).

Redundant code. A code using more signal elements than necessary to represent the intrinsic information. For example, five-unit code using all the characters of International Telegraph Alphabet No. 2 is not redundant; five-unit code using only the figures in International Telegraph Alphabet No. 2 is redundant; seven-unit code using only signals made of four "space" and three "mark" elements is redundant. (2).

Reference pilot. A reference pilot is a different wave from those which transmit the telecommunication signals (telegraphy, telephony). It is used in carrier systems to facilitate the maintenance and adjustment of the carrier transmission system. (For example, automatic level reguSetAction, synchronization of oscillators, etc.) (2).

Regenerative repeater. (*See* **Repeater, regenerative.**)

Regional center. A control center (class 1 office) connecting sectional centers of the telephone system together. Every pair of regional centers in the United States has a direct circuit group running from one center to the other. (1).

Repeater.
1. A device whereby currents received over one circuit are automatically repeated in another circuit or circuits, generally in an amplified and/or reshaped form.
2. A device used to restore signals, which have been distorted because of attenuation, to their original shape and transmission level.

Repeater, regenerative. Normally, a repeater utilized in telegraph applications. Its function is to retime and retransmit the received signal impulses restored to their original strength. These repeaters are speed- and code-sensitive and are intended for use with standard telegraph speeds and codes. (Also called *regen.*) (1).

Repeater, telegraph. A device which receives telegraph signals and automatically retransmits corresponding signals. (2).

Reperforator (receiving perforator). A telegraph instrument in which the received signals cause the code of the corresponding characters or functions to be punched in a tape. (1).

Reperforator/transmitter (RT). A teletypewriter unit consisting of a reperforator and a tape transmitter, each independent of the other. It is used as a relaying device and is especially suitable for transforming the incoming speed to a different outgoing speed, and for temporary queuing.

Residual error rate, undetected error rate. The ratio of the number of bits, unit elements, characters or blocks incorrectly received but undetected or uncorrected by the error-control equipment, to the total number of bits, unit elements, characters or blocks sent. (2).

Response time. This is the time the system takes to react to a given input. If a message is keyed into a terminal by an operator and the reply from the computer, when it comes, is typed at the same terminal, response time may be defined as the time interval between the operator pressing the last key and the terminal typing the first letter of the reply. For different types of terminal, response time may be defined similarly. It is the interval between an event and the system's response to the event.

Ringdown. A method of signaling subscribers and operators using either a 20-cycle AC signal, a 135-cycle AC signal, or a 1000-cycle signal interrupted 20 times per second. (1).

Routing. The assignment of the communications path by which a message or telephone call will reach its destination. (1).

Routing, alternate. Assignment of a secondary communications path to a destination when the primary path is unavailable. (1).

Routing indicator. An address, or group of characters, in the heading of a message defining the final circuit or terminal to which the message has to be delivered. (1).

RT. Reperforator/transmitter, q.v.

Saturation testing. Program testing with a large bulk of messages intended to bring to light those errors which will only occur very infrequently and which may be triggered by rare coincidences such as two different messages arriving at the same time.

Sectional center. A control center connecting primary centers; a class 2 office. (1).

Seek. A mechanical movement involved in locating a record in a random-access file. This may, for example, be the movement of an arm and head mechanism that is necessary before a read instruction can be given to read data in a certain location on the file.

Selection. Addressing a terminal and/or a component on a selective calling circuit. (1).

Selective calling. The ability of the transmitting station to specify which of several stations on the same line is to receive a message. (1).

Self-checking numbers. Numbers which contain redundant information so that an error in them, caused, for example, by noise on a transmission line, may be detected.

Serial transmission. Used to identify a system wherein the bits of a character occur serially in time. Implies only a single transmission channel. (Also called *serial-by-bit*.) (1). Transmission at successive intervals of signal elements constituting the same telegraph or data signal. For example, transmission of signal elements by a standard teleprinter, in accordance with International Telegraph Alphabet No. 2; telegraph transmission by a time-divided channel. (2).

Sideband. The frequency band on either the upper or lower side of the carrier frequency within which fall the frequencies produced by the process of modulation. (2).

Signal-to-noise ratio (S/N). Relative power of the signal to the noise in a channel. (1).

Simplex circuit.
1. CCITT definition: A circuit permitting the transmission of signals in either direction, but not in both simultaneously.
2. Definition in common usage (the normal meaning in computer literature): A circuit permitting transmission in one specific direction only.

Simplex mode. Operation of a communication channel in one direction only, with no capability for reversing. (1).

Simulation. This is a word which is sometimes confusing as it has three entirely different meanings, namely:

Simulation for design and monitoring. This is a technique whereby a model of the working system can be built in the form of a computer program. Special computer languages are available for producing this model. A complete system may be described by a succession of different models. These models can then be adjusted easily and endlessly, and the system that is being designed or monitored can be experimented with to test the effect of any proposed changes. The simulation model is a program that is run on a computer separate from the system that is being designed.

Simulation of input devices. This is a program testing aid. For various reasons it is undesirable to use actual lines and terminals for some of the program testing. Therefore, magnetic tape or other media may be used and read in by a special program which makes the data appear as if they came from actual lines and terminals. Simulation in this sense is the replacement of one set of equipment by another set of equipment and programs, so that the behavior is similar.

Simulation of superivsory programs. This is used for program testing purposes when the actual supervisory programs are not yet available. A comparatively simple program to bridge the gap is used instead. This type of simulation is the replacement of one set of programs by another set which imitates it.

Single-current transmission, (inverse) **neutral direct-current system.** A form of telegraph transmission effected by means of unidirectional currents. (2).

Space. 1. An impulse which, in a neutral circuit, causes the loop to open or causes absence of signal, while in a polar circuit it causes the loop current to flow in a direction opposite to that for a mark impulse. A space impulse is equivalent to a binary 0. 2. In some codes, a character which causes a printer to leave a character width with no printed symbol. (1).

Space-hold. The normal no-traffic line condition whereby a steady space is transmitted. (Compare with **Mark-hold.**) (1).

Space-to-mark transition. The transition, or switching, from a spacing impulse to a marking impulse. (1).

Spacing bias. *See* **Distortion, bias.**

Spectrum. 1. A continuous range of frequencies, usually wide in extent, within which waves have some specific common characteristic. 2. A graphical representation of the distribution of the amplitude (and sometimes phase) of the components of a wave as a function of frequency. A spectrum may be continuous or, on the contrary, contain only points corresponding to certain discrete values. (2).

Start element. The first element of a character in certain serial transmissions, used to permit synchronization. In Baudot teletypewriter operation, it is one space bit. (1).

Start-stop system. A system in which each group of code elements corresponding to an alphabetical signal is preceded by a start signal which serves to prepare the receiving mechanism for the reception and registration of a character, and is followed by a stop signal which serves to bring the receiving mechanism to rest in preparation for the reception of the next character. (Contrast with **Synchronous system.**) (Start-stop transmission is also referred to as *asynchronous transmission*, q.v.)

Station. One of the input or output points of a communications system—e.g., the telephone set in the telephone system or the point where the business machine interfaces the channel on a leased private line. (1).

Status maps. Tables which give the status of various programs, devices, input-output operations, or the status of the communication lines.

Step-by-step switch. A switch that moves in synchronism with a pulse device such as a rotary telephone dial. Each digit dialed causes the movement of successive selector switches to carry the connection forward until the desired line is reached. (Also called *stepper switch*. Compare with **Line switching** and **Cross-bar switch.**) (1).

Step-by-step system. A type of line-switching system which uses step-by-step switches. (1).

Stop bit. (*See* **Stop element.**)

Stop element. The last element of a character in asynchronous serial transmissions, used to ensure recognition of the next start element. In Baudot teletypewriter operation it is 1.42 mark bits. (*See also* **Start-stop system.**) (1).

Store and forward. The interruption of data flow from the originating terminal to the designated receiver by storing the information enroute and forwarding it at a later time. (*See* **Message switching.**)

Stunt box. 1. A device to control the nonprinting functions of a teletypewriter terminal, such as carriage return and line feed; and 2. a device to recognize line control characters (e.g., DCC, TSC, etc.). (1).

Subscriber trunk dialing. (*See* **direct distance dialing.**)

Subscriber's line. The telephone line connecting the exchange to the subscriber's station. (2).

Subscriber's loop. (*See* **Local loop.**)

Subset. A subscriber set of equipment, such as a telephone. A modulation and demodulation device. (Also called *data set*, which is a more precise term.) (1).

Subvoice-grade channel. A channel of bandwidth narrower than that of voice-grade channels. Such channels are usually subchannels of a voice-grade line. (1).

Supergroup. The assembly of five 12-channel groups, occupying adjacent bands in the spectrum, for the purpose of simultaneous modulation or demodulation. (2).

Supervisory programs. Those computer programs designed to coordinate service and augment the machine components of the system, and coordinate and service application programs. They handle work scheduling, input-output operations, error actions, and other functions.

Supervisory signals. Signals used to indicate the various operating states of circuit combinations. (1).

Supervisory system. The complete set of supervisory programs used on a given system.

Support programs. The ultimate operational system consists of supervisory programs and application programs. However, a third set of programs are needed to install the system, including diagnostics, testing aids, data generator programs, terminal simulators, etc. These are referred to as *support programs*.

Suppressed carrier transmission. That method of communication in which the carrier frequency is suppressed either partially or to the maximum degree possible. One or both of the sidebands may be transmitted. (1).

Switch hook. A switch on a telephone set, associated with the structure supporting the receiver or handset. It is operated by the removal or replacement of the receiver or handset on the support. (*See also* **Off hook** *and* **On hook.**) (1).

Switching center. A location which terminates multiple circuits and is capable of interconnecting circuits or transferring traffic between circuits; may be automatic, semiautomatic, or torn-tape. (The latter is a location where operators tear off the incoming printed and punched paper tape and transfer it manually to the proper outgoing circuit.) (1).

Switching message. (*See* **Message switching.**)

Switchover. When a failure occurs in the equipment a switch may occur to an alternative component. This may be, for example, an alternative file unit, an alternative communication line or an alternative computer. The switchover process may be automatic under program control or it may be manual.

Synchronous. Having a constant time interval between successive bits, characters, or vents. The term implies that all equipment in the system is in step.

Synchronous system. A system in which the sending and receiving instruments are operating continuously at substantially the same frequency and are maintained, by means of correction, if necessary, in a desired phase relationship. (Contrast with **Start-stop system.**) (2).

Synchronous transmission. A transmission process such that between any two significant instants there is always an integral number of unit intervals. (Contrast with **Asynchronous.**) (1).

Tandem office. An office that is used to interconnect the local end offices over tandem trunks in a densely settled exchange area where it is uneconomical for a telephone company to provide direct interconnection between all end offices. The tandem office completes all calls between the end offices but is not directly connected to subscribers. (1).

Tandem office, tandem central office. A central office used primarily as a switching point for traffic between other central offices. (2).

Tariff. The published rate for a specific unit of equipment, facility, or type of service provided by a communications common carrier. Also the vehicle by which the regulating agencies approve or disapprove such facilities or services. Thus the tariff becomes a contract between customer and common carrier.

TD. Transmitter-distributor, q.v.

Teleprocessing. A form of information handling in which a data-processing system utilizes communication facilities. (Originally, but no longer, an IBM trademark.) (1).

Teletype. Trademark of Teletype Corporation, usually referring to a series of different types of teleprinter equipment such as tape punches, reperforators, page printers, etc., utilized for communications systems.

Teletypewriter exchange service (TWX). An AT&T public switched teletypewriter service in which suitably arranged teletypewriter stations are provided with lines to a central office for access to other such stations throughout the U.S.A. and Canada. Both Baudot- and ASCII-coded machines are used. Business machines may also be used, with certain restrictions. (1).

Telex service. A dial-up telegraph service enabling its subscribers to communicate directly and temporarily among themselves by means of start-stop apparatus and of circuits of the public telegraph network. The service operates world wide. Baudot equipment is used. Computers can be connected to the Telex network.

Terminal. Any device capable of sending and/or receiving information over a communication channel. The means by which data are entered into a computer system and by which the decisions of the system are communicated to the environment it affects. A wide variety of terminal devices have been built, including teleprinters, special keyboards, light displays, cathode tubes, thermocouples, pressure gauges and other instrumentation, radar units, telephones, etc.

TEX. (*See* **Telex service.**)

Tie line. A private-line communications channel of the type provided by communications common carriers for linking two or more points together.

Time-derived channel. Any of the channels obtained from multiplexing a channel by time division.

Time-division multiplex. A system in which a channel is established in connecting intermittently, generally at regular intervals and by means of an automatic distribution, its terminal equipment to a common channel. At times when these connections are not established, the section of the common channel between the distributors can be utilized in order to establish other similar channels, in turn.

Toll center. Basic toll switching entity; a central office where channels and toll message circuits terminate. While this is usually one particular central office in a city, larger cities may have several central offices where toll message circuits terminate. A class 4 office. (Also called "toll office" and "toll point.") (1).

Toll circuit (American). *See* **Trunk circuit** (British).

Toll switching trunk (American). *See* **Trunk junction** (British).

Tone dialing. (*See* **Pushbutton dialing.**)

Touch-call. Proprietary term of GT&E. (*See* **Pushbutton dialing.**)

Touch-tone. AT&T term for pushbutton dialing, q.v.

Transceiver. A terminal that can transmit and receive traffic.

Translator. A device that converts information from one system of representation into equivalent information in another system of representation. In telephone equipment, it is the device that converts dialed digits into call-routing information. (1).

Transmitter-distributor (TD). The device in a teletypewriter terminal which makes and breaks the line in timed sequence. Modern usage of the term refers to a paper tape transmitter.

Transreceiver. A terminal that can transmit and receive traffic. (1).

Trunk circuit (British), **toll circuit** (American). A circuit connecting two exchanges in different localities. *Note*: In Great Britain, a trunk circuit is approximately 15 miles long or more. A circuit connecting two exchanges less than 15 miles apart is called a *junction circuit*.

Trunk exchange (British), **toll office** (American). An exchange with the function of controlling the switching of trunk (British) [toll (American)] traffic.

Trunk group. Those trunks between two points both of which are switching centers and/or individual message distribution points, and which employ the same multiplex terminal equipment.

Trunk junction (British), **toll switching trunk** (American). A line connecting a trunk exchange to a local exchange and permitting a trunk operator to call a subscriber to establish a trunk call.

Unattended operations. The automatic features of a station's operation permit the transmission and reception of messages on an unattended basis. (1).

Vertical parity (redundancy) check. (*See* **Parity check, vertical.**)

VOGAD (Voice-Operated Gain-Adjusting Device). A device somewhat similar to a compandor and used on some radio systems; a voice-operated device which removes fluctuation from input speech and sends it out at a constant level. No restoring device is needed at the receiving end. (1).

Voice-frequency, telephone-frequency. Any frequency within that part of the audio-frequency range essential for the transmission of speech of commerical quality, i.e., 300–3400 c/s. (2).

Voice-frequency carrier telegraphy. That form of carrier telegraphy in which the carrier currents have frequencies such that the modulated currents may be transmitted over a voice-frequency telephone channel. (1).

Voice-frequency multichannel telegraphy. Telegraphy using two or more carrier currents the frequencies of which are within the voice-frequency range. Voice-frequency telegraph systems permit the transmission of up to 24 channels over a single circuit by use of frequency-division multiplexing.

Voice-grade channel. (*See* **Channel, voice-grade.**)

Voice-operated device. A device used on a telephone circuit to permit the presence of telephone currents to effect a desired control. Such a device is used in most echo suppressors. (1).

VRC. Vertical redundancey check. (*See also* **Parity check.**)

Watchdog timer. This is a timer which is set by the program. It interrupts the program after a given period of time, e.g., one second. This will prevent the system from going into an endless loop due to a program error, or becoming idle because of an equipment fault. The Watchdog timer may sound a horn or cause a computer interrupt if such a fault is detected.

WATS (Wide Area Telephone Service). A service provided by telephone companies in the United States which permits a customer by use of an access line to make calls to telephones in a specific zone in a dial basis for a flat monthly charge. Monthly charges are based on the size of the area in which the calls are placed, not on the number or length of calls. Under the WATS arrangement, the U.S. is divided into six zones to be called on a full time or measured-time basis. (1).

Word. 1. In telegraphy, six operations or characters (five characters plus one space). ("Group" is also used in place of "word.") 2. In computing, a sequence of bits or characters treated as a unit and capable of being stored in one computer location. (1).

WPM (Words per minute). A common measure of speed in telegraph systems.

INDEX

A

ASCII code 29, 30, 31, 34, 35, 36, 159, 176, 199
Acoustical coupling 22, 23, 38ff, 127
Application program 136, 225, 246, 248
Asynchronous transmission 43-50, 58, 60, 61, 199, 201, 251
  and multidrop lines 155, 156, 167
Automatic error correction 66-75

B

BCD 31, 32
Bandwidth 4-6, 9-11, 184, 186
Batch totals and errors 70, 71, 72
Batching:
  and multidrop systems 154, 157
  and multiplexing 194
Baudot 5-bit telegraph code 26, 27
Binary coded decimal (BCD) 31, 32
Binary-synchronous transmission 94
Bit stream:
  high-speed 186, 190, 192, 193
  low-speed 190-192
Block check 54-57, 70-71
Block retransmission, design factors 69-72
Block transmission 50, 51, 54, 69-72, 73-75

Block transmission (*cont.*)
  design factors 69-70
  error detection in 69-75
  odd-even check 73, 74
  parity check 104
  serial number check 74
Bose-Chaudhuri codes 83
Buffer 132, 133, 171-173, 177, 179, 184, 196, 251, 254
Buffer storage 69, 196ff
  cost 171ff, 184
Buffered terminals and multidrop lines 155-157, 158, 171-183, 184
Buffering:
  and concentrators 196-197, 202-203, 209, 251, 254
  and multiplexing 192
  and paper tape 157
  and response time 155-157, 158, 159, 171, 173, 177, 179, 181, 192, 196
Buffers:
  vs. concentrators 196
  shared terminal 133, 134, 172-173
Bursts of errors (*see* Error bursts)

C

CCITT 34, 35
Call directing code (CDC) 159, 160

283

284 INDEX

Central office (public exchange) 2, 138
Channel frequency 2, 4, 6, 9–12, 184, 185, 186–188, 191
Check bits for polynomial code 85, 86, 87–90, 91–93, 94, 95
Checking 79ff, 163, 164, 166, 176, 177, 199, 200, 206–207, 250–251
Checks 79ff
  parity 79–82
Circuitry:
  error check 77
  for polynomial coding 91–93, 95ff
Code (*see also* Parity, Redundancy, Loop check):
  coexistent dual directional transmission 111, 112, 113
  conversion 199
  criteria for choice of 76
  error-correcting 66–69
  error-detecting 25–37, 43, 54–57, 66, 67, 68, 69 (*see also* Error-correcting and -detecting codes)
  performance 8, 79, 80, 81
  polynomial 76, 77, 83–95
  transition vs. state 80, 81, 82
Compensating errors 80–81
Computer network design 218–232, 246–257
Concentrators 71, 77, 134, 170, 195–217, 220, 224, 248, 251, 257
  vs. computers 218ff, 225–227
  cost factors 195–196, 200, 207, 209, 210
  design factors 195–201, 202–205, 206–210, 211–217
  and multidrop 135, 207
  with multidrop lines 198, 200, 201–205, 207–209
  and polling 200
  and response time 206–207
  and storage 197, 199
Conditioning 12
Connections, types of 144, 145
Contention 158, 170
Control:
  on batches 71, 72
  bits 28, 30, 31

Control characters 28, 30, 31, 40, 72–75, 99
  for batches 72–75
  to establish synchronization 50, 51, 54–58, 99–101
  maintenance of synchronization 54–56, 57, 100, 101
  and multidrop 157, 158, 159, 160–161, 163, 167–168, 176–180, 181–183, 199, 203, 209–211
  for transmission errors 72–75
Control unit and systems design 170–175, 176–183
Conversational transmission 105–111, 112, 220–224, 226–229, 230–232, 254, 257
Core plane 113, 114, 115, 116
Cost:
  of communication networks 118ff, 129ff, 139, 140, 141–147, 169, 170ff, 184, 193–194, 195, 196
  of long vs. short lines 119ff, 141
  of multiplexing 184, 185, 186, 190–191, 193–194
  reduction 118ff, 129ff
  of short private line systems 119–120, 123, 124, 127, 129ff, 138, 145
Cyclical redundancy check 167–168

D

DC transmission 16, 124
Data mode 97
Data sets 141, 145 (*see also* Modem)
  and multiplexing 186, 187, 188, 191–192
Data transmission:
  asynchronous 43–50, 58, 60–61, 199, 201, 251
  synchronous 43, 50–54, 57–58, 60, 61, 99–103, 134, 167, 190, 199, 201, 251
Datel 11, 12, 23
Delete code 67, 73
Dial-up computers 141, 143, 144, 145, 219–220, 224, 225, 226, 227, 255
Dial-up terminals 140, 141, 143, 144, 145

Di-bits 80, 82
Dual-direction transmission, coexistent 111–113
Duplex 8, 13, 38, 141, 142, 157, 190, 201, 205
  echo or loop check 67
  message overlapping 111, 112
  and multiplexing 132
  point-to-point control 96, 99, 101, 105
  transmission and multidrop terminals 154, 157, 201–205
  two-way transmission on 39

E

Echo check 67
Echo suppressor 97
Editing logic 174–175, 252, 253
Error bursts 87–90, 91, 94
Error check circuitry cost 77
Error checking and multiplexing 73ff, 193
Error clustering 78–80
Error control 62–75, 250–251
  characters 72–75
  and radio circuits 67, 68
  system, example of 74–75
Error-correcting codes 66, 67, 72–74, 77ff
Error-correcting and -detecting codes 77, 78
Error detecting in batches 71–72
Error-detecting codes 55, 56, 65–68, 69–71, 72–75, 76ff, 104–106
  vs. correcting codes 76–79
  effectiveness factors 77, 79, 80, 81, 82
  vs. redundancy 77–82
Error detection:
  and polynomial codes 83, 84, 85–90, 91, 92–95
  probability of 86, 87, 88, 89–91, 94, 95
  retransmission 78
Error-free transmission, duplex line 109
Error miscorrection 78
Error rates 62, 63, 64, 68
  in radio links 68
  in subvoice- and voice-grade lines 79ff

Error retransmission 67, 68, 69–75, 78
Error signal, timing mechanism 105–107, 108
Errors 66, 69, 75, 104, 106
  compensating 80–81
  cumulative 64, 65
  double and single kit 79, 86, 87, 89, 91
  estimation of 65
  forward-error correction 78
  types of 62–65, 76–78
Escape characters 27, 28
Exchange vs. multidrop lines 130–131
Exchanges 2, 130, 139, 140

F

Federal Communications Commission (U.S.) 1
Ferrite core 113
Fire codes 83
Formatting 225, 229–232, 241, 252, 253, 254
Forward-error correction 66, 78
Four-out-of 8 codes 80–92
Frequency 4–6, 9–12ff
  restricted 5–6
  of voice channel 4, 5
Frequency-division multiplexing 40, 186–188, 189, 194
Full duplex (see Duplex)

G

Generating polynomials 84–95

H

Half-duplex and point-to-point control 96, 99, 101, 105
Half-duplex lines 8, 38, 40, 157, 190
Half-duplex transmission and multidrop terminals 155, 156, 157
Hamming codes 83
Handshaking 101–104

286 INDEX

Hertz, definition 4
High-speed pulse stream (see Pulse stream)
Hold and forward concentrator 71, 195, 207–209 (see also Concentrators)
"Home mode" operation 173–174
Home telephone line 123, 124
Horizontal parity check (see Parity checks, vertical)

I

Idling 101ff
Inhibit wires 116
In-plant wires and pulse-stream transmission 58, 59
In-plant wiring 119, 120, 121, 122, 123, 124, 127, 134
Interleaved codes 83
Interleaving 83, 128, 185, 186, 191

J

JOSS 221

K

Keyboard locking 155

L

Large-scale integration circuitry 117, 250
Leased lines 2, 3, 139, 140, 141, 145, 148
 vs. dial up, cost 127
 groups 148
 and point-to-point control 96
 vs. switched lines 3, 139, 140, 141, 145
Line adapter 186, 191
Line computer 135, 136
Line control 159–169, 176–183, 200–202, 206, 207, 209–211, 234, 247–249

Line mileage reduction 127–129, 130, 165, 170
Line switching 118ff, 233, 234
 vs. message switching 138
Line vs. terminal speed 170ff, 184
Links, transmission 138ff, 145
Logic editing and format 174–175
Logic syntax 174–175
Long-line network design factors 127, 128, 129–138, 165, 170
Long-line polling 165, 170
Loop:
 check 67
 paper tape 235–286
 wire 2, 122, 123, 124

M

M-out of-N codes 81–82, Fig. 2–7
Melas codes 83
Message address 157, 158, 159, 160–162, 163, 165–168, 169, 174, 175–183
Message files 238, 241, 243
Message overlapping 111
Message switching 129–136, 137, 138, 233–237, 238, 239–245
 cost factors 238, 239, 242, 243–245, 252
Microwave links 184, 185
Modem 5, 17, 20, 21, 22, 39, 124, 126, 127, 128, 140–145
 design and use 22, 127, 128, 140, 141–148
 for multidrop lines 157, 158, 169, 170ff
 for multiplexing 186, 187, 188, 191–192
 public use limited 23
 turnaround time, on full duplex 103, 108, 109
Modulation 16, 17, 20
 2 and 4 phase 82
 and multiplexing 17, 188
Multidrop lines 130, 131, 135, 137, 154–157, 233, 235, 250, 252

INDEX   287

Multidrop lines (*cont.*)
   and asynchronous transmission 155, 156, 167, 201
   and buffered terminals 155–157, 158, 171–183, 184
   and concentrators 135, 195–197, 198. 200, 201, 205, 207–209, 211–217
   vs. exchange 130, 131
   and message address 157, 158, 159, 160–169
   systems 154, 155–157, 158, 159, 160–169, 170–175, 176
   transmission system design 154–157, 158, 159–169, 170–175, 176
Multidrop and multiplexing 191
Multidrop systems:
   and batching 154, 157
   and transmission rates 154, 155–157
Multidrop telegraph line 159, 160–161, 170
Multiplexing 17, 131ff, 153, 154, 184–190, 191–194, 195
   channels 185–188
   and conditioned voice line 187
   cost 193–194
   error check 193ff
   frequency-division 186–188, 189, 194
   and multidrop 191
   pros and cons 193–194
   time-division 186, 188–190, 194
Multiplexors 131, 132, 134, 141, 185, 186–188, 190–192, 195, 200, 210
Multipoint (*see* Multidrop lines)
Multipoint line control 153ff, 159–169, 170, 176
Multitone transmission 40–43
Multiwire cables and terminals 120–122, 123, 134
   with application program 136
   with files 135

N

Network costs lowered 129–138, 139, 145, 146

O

Odd-even record count 113
Overlap (*see* Message overlapping)
Overloading, with multidrop lines 154, 155–157

P

PABX (*see* Private automatic branch exchange)
PAX (*see* Private automatic exchange)
PBX (*see* Private branch exchange)
PBX and dial-up computers 139ff–141, 142, 143, 144
Paper tape as a buffer 133, 134, 171, 174, 179
Paper tape and multidrop line transmission 155, 157, 158, 159, 174
Paper tape storage 234–238
Parallel transmission 40–42
   vs. serial transmission 40
   and synchronous transmission 58ff
Parallel wire transmission 40–43
Parity 28, 55, 56, 65, 104
Parity checks 79–82, 83, 89, 91, 94
   horizontal parity check 67ff, 80–82
   vs. polynomials 79–83, 89, 91, 94
   and redundancy rate 80, 81
   vertical parity check 67ff, 80–82
Parity and polynomial coding 91–93, 94
Point-to-point line control 96ff, 127, 157
Point-to-point line modems 157, 158
Point-to-point transmission 97ff, 127, 159
Polling 132, 135, 153ff, 158–162, 165–169, 170, 175, 197, 198, 200, 201–205, 206–209, 210, 211–217, 234, 252, 255
   and concentrators 200
   general vs. specific 175, 176, 179, 181
   hub 165, 166–169, 201–205, 206
   hub and response time 167
   list 158, 160–162, 163, 164–166, 167 169

Polling (cont.)
  message 158, 159, 160–162, 163, 164–165, 170
  roll call 162–164, 165
  and synchronous transmission 167
Polynomial check, on variable length message 94, 95
Polynomial codes 76, 77, 83–95
  check bits 85, 86, 87–90, 91–93, 94, 95
  and error bursts 87–90, 91, 94, 95
Polynomial coding circuitry 91–93, 95
Polynomial generating 84–95
Private automatic branch exchange (PABX) 140, 145
  use of 139–152
Private automatic exchange (PAX) 2
  use of 139–152
Private branch exchange (PBX) 2, 22, 140, 141, 145
  use of 139–152
Private exchange 129ff, 138ff, 200
  private in-plant lines 22, 40, 123, 124, 139, 141, 144
Private line systems 3, 6, 7, 139, 140ff
  cost of 119, 120, 121, 123, 124, 127, 129ff, 139, 140, 145
Private lines:
  conditioning of 7, 12
  cost 7, 8, 129, 139, 140
  switched 3, 129, 139–148
  vs. switched lines 3, 139ff
Public exchange 2 (see also Central office)
Public network 2, 139, 144, 145
Pulse stream 58, 122, 123

Q

Queuing 197ff, 207
  and message switching 235, 236, 240–241, 243
  and multidrop lines 156, 158–159

R

Rates 1
Receiver translator 112, 115, 117
Redundancy 65, 66, 77ff, 85, 94–95, 128
  check, vertical 28, 55
  vs. checking bits 77ff
  cost 128
  and parity check 80
  in polynomial codes 85, 94, 95
Response, canned 220–224, 226–232
  with concentrators 206–209
  conversational 153–157, 169, 175
  with multiplex 132, 133, 136, 184
  and polling 153, 155, 158, 161, 163, 165, 169, 171, 181
  in real time 154, 155, 156–157, 167, 169
Resynchronization characters 51

S

Satellite channels 148
Satellite transmission 169
Scanning and terminals 120, 121
Screen terminals 171, 173, 175–177, 179–183, 196
Security 252
Selective calling 157, 158, 159, 160–161, 175, 176
Serial vs. parallel transmission 40
Serial transmission 40–42
Series 8000 14
Short-line transmission 119–127, 134
Simplex lines, defined 8
Simultaneous transmission and concentrators 197, 199, 206
Start-stop transmission 134 (see also Asynchronous transmission)
State coding 80, 81, 82
Store and forward 134, 199, 207–209
Stored-program computers 218–226, 227–232
Subvoice-grade line 9, 10, 159–160
  error rate 63, 79

Subvoice-grade line (*cont.*)
  international terminology 11
  speeds of 9, 10
  transmission facilities 11, 140
Switched vs. leased lines (*see* Leased vs. switched lines)
Switched wideband data network 150
Switching 129, 131, 134, 137, 138, 139, 140, 145, 150, 233–238, 239–245 (*see also* Message switching)
  automatic 238, 239–245
  back up 238, 239, 245
  of broadband channels 3
  by computer 238, 239–245
  manual 141, 233, 234, 235–237
  private exchange 138, 140, 141, 234ff
  public network 138, 141
Synchronization:
  establishment of 50, 51, 54–58, 99–103
  maintenance of 5
  types of 57
Synchronization bits, multiplex 192
Synchronization characters 51, 56, 167
Synchronous transmission 43, 50–54, 57–58, 60, 61, 134, 199, 201, 251
  and buffers 50
  and multiplexing 190
  and polling 167

T

TWX 4, 15, 139
TWX types of access lines 15
Tariffs 148
  defined 1
Teleprinter (links) 137
Teletypewriter Exchange Service (*see* TWX)
Telex 4, 14, 15, 139, 234
  error rate 63, 69
TELPAK 12, 13, 14
Terminal address multidrop 157–164, 165–168

Terminal buffers 170–184
Terminal connections:
  high-speed 127, 130, 142–144
  low-speed 120–123, 127
Terminal control units 127–138, 195–199, 201, 203, 209, 220, 224, 248–250
  cost factors 122, 170–184 (*see also* Concentrator)
Terminal cost, for error checks 77
Terminal interchange 201ff
Terminal speeds 171
  vs. line speed 170ff, 184
Terminals 249ff
  cost factors 119–123, 127–137
  manual, use and cost 127, 128, 132, 135, 140, 141
  and multidrop lines 154–159, 162–163, 165, 169, 170ff
  and multidrop systems 153–159, 163
  for selective calling 157–160
  types of 142–146, 148
  use of 119–123, 127, 129–137, 141–146, 148
  wiring of 120–123, 130
Tie lines 2, 140–145
  broadband 3
  grouped 3
Tie trunk (*see* Tie line)
Time-division multiplexing 186, 188–190, 194
"Time out" 155–156, 163
Time-sharing 220, 255
Timing mechanism, for error signals 105–108
Torn-tape switching 234–237
Transition coding 80, 81, 82
Translator-receiver 112, 115, 117
Translator transmit 112, 116, 117
Transmission:
  coexistent dual direction 111–113
  conversational 105–111, 130
  error control characters 72–75
  high-speed 127, 128, 131, 133, 145
  links (*see* Links, transmission)
  low-speed 141
  without modems 124, 127
  point-to-point 97, 127

Transmission rate, to terminals 123, 124, 127, 129–137, 140–146
Transmission speed and multidrop lines 154–159, 171, 195–199
Transmit translator 112, 116, 117
Transparent code 37
Trunk 2
 interoffice 2
Trunk lines 234
Turnaround time:
 for modems 108–109
 in polling 165, 169

V

van Duren ARQ 68, 82
Vertical parity check (*see* Parity checks)
Voice channel frequencies 5, 6, 9–12, 13, 79–80, 184–185, 186–188, 191
Voice-grade lines 9, 10, 11, 63, 79, 139, 170, 181–185, 207, 210
 error rate 63, 79, 80
 international terminology 11
 speeds of 9, 10
 transmission facilities 11, 184–185

W

Wideband line 10, 11, 134, 145, 148, 150, 185 (*see also* Telpak)
 international terminology 11
 speeds of 10, 11, 145
Wideband network, switched 150